DISCARD

D0805222

Between Hope and Fear

Between Hope and Fear

A History of Vaccines and Human Immunity

Michael Kinch

PEGASUS BOOKS
NEW YORK LONDON

BETWEEN HOPE AND FEAR

Pegasus Books Ltd.
148 W. 37th Street, 13th Floor
New York, NY 10018

First Pegasus Books edition July 2018

Interior design by Maria Fernandez

Library of Congress Cataloging-in-Publication Data is available.

ISBN: 978-1-68177-751-1

10 9 8 7 6 5 4 3 2 1

Printed in the United States of America
Distributed by W. W. Norton & Company, Inc.
www.pegasusbooks.us

For My Parents, Sue and Frank Kinch,

for their inspiration and love and

for ensuring I was protected through immunization

Contents

Introduction

An underground conspiracy has been quietly targeting the scions of some of America's most wealthy, accomplished, and educated families. The headline of the December 1, 2016 edition of a key mouthpiece of the elite, *The Harvard Crimson,* captures one example: MUMPS OUTBREAK GROWS TO 5, CASES SUSPECTED AT YALE.[1] The article decried the growth of a disease that had already claimed dozens of victims at the vaunted Cambridge campus over the previous year. Otherwise healthy coeds were reporting to the health clinic in record numbers, complaining of a sore and swollen neck, malaise, fever, and difficulty swallowing. These were the textbook (though rarely encountered in the real world) symptoms of a disease known widely as the mumps (whose etymology reflects its Dutch ancestry for a word meaning 'grimace' or 'whine'). As the infection progresses, some patients require intubation, and about one in ten cases progresses to a swelling of the covering of the brain, a sometimes fatal condition known as meningitis.[2] As the number of cases at Harvard increased, reports speculated about the cancellation of the 2016 winter graduation ceremony and possible spread of the infectious disease to their rivals' campus in New Haven. This broadening of the outbreak was a consequence of the Elis' hosting of "The Game" (and an unintentional payback for the 21–14 loss handed to Harvard by their rivals just weeks before). The annual football matchup between Harvard and Yale recalls memories of raccoon coats and small bullhorns from a Jazz Age long since passed, but it might in the future be remembered as a turning point that foreshadowed the dramatic rise of a larger malady, which threatens to unleash not just one but a looming constellation of future pandemics.

Unlike most public health crises of the modern world, the affliction at the heart of this book is less likely to affect the blighted poor or dispossessed and more likely to plague the more genteel elements of society, which include individuals with the highest incomes, most prominent pedigrees, and advanced academic credentials. Nor are the symptoms being felt very far outside of the United States. While its roots are British in origin, the problem is manifestly American. The abilities to diagnose its symptoms were honed more than a century ago, but despite, and perhaps because of, the advent of the extraordinary scientific and public health breakthroughs of the late 19th and early 20th centuries, the newly minted physicians of the 21st century are less prepared to deal with the fundamental causes of this disease than their counterparts from the Jazz Age. These symptoms are indications long since written off and include measles, mumps, rubella, chicken pox, polio, bacterial encephalitis, diphtheria, and whooping cough. The disease itself is not these individual maladies, each of which is horrific in its own way, but manifests as a growing forgetfulness of the agonizing and terrifying ailments that have threatened man since time immemorial and are now returning en masse, often infecting an unwitting population that assumes they have already been protected.

The risk is not indeed any of these pathogens but instead reflects a knowing and intentional decision to avoid or delay vaccination against childhood and adult diseases. Many of the nation's most elite, wealthy, and progressive minds suffer from a false sense of believing they have special insight into truth. They have knowingly jeopardized themselves, their children, communities, and their country by exposing them to life-changing injury and death. Worse still, we now know individual decisions not to vaccinate unintentionally triggered a backlash that collectively threatens all children and young adults, regardless of whether their parents chose to have them vaccinated. The decision to oppose vaccines blurs political lines, as evidenced by a Republican president of the United States calling upon a senior member of the Kennedy family to lead a forthcoming campaign against vaccines.[3,4]

The rise of activism against vaccines is not a story of intended malevolence. Quite the opposite; the motives for avoiding vaccination are based on agonizing fears of a different spectrum of diseases, those associated with autism. Consequently, fears of a poorly understood disease rendered future generations susceptible to well-known and easily preventable disorders. This vulnerability has been greatly increased through the intentional fraud of one man, Andrew Wakefield, who has a dubious track record full of inaccurate research methodologies and error-prone data analysis. Although stripped of his credentials and discredited widely and repeatedly by the medical community, both for scientific and ethical lapses, Andrew Wakefield is but the latest symptom of a disease that has been manifest since a time even before the first vaccine was tested.

While the media has largely focused on a small subset of highly visible anti-vaccinators, the campaign against vaccination has rendered the entire American population, both those who have been vaccinated and those who have not, susceptible to a recurrence of common childhood diseases. Worse still, some preventable diseases that are merely annoying in young children can instead manifest themselves as life-disrupting and deadly diseases in older children and adults.

This book conveys a story of vaccine-preventable infectious diseases and the national and global implications of poor decision making, often by a relatively small number of highly educated and powerful elites. The title reflects a recurring theme in the book, which contrasts waves of discoveries of life-saving vaccines against a seemingly inevitable and irrational rejection by fringe elements in the public.

By blending an understanding of the scientific discovery with an evolving view of disease through the years, this book conveys the challenges posed by infectious disease and relates a story of unparalleled successes in vaccines that have raised both the quality and quantity of life for all people. The improvements have been so remarkable that many Americans have long forgotten or never experienced the dread that accompanied each seasonal cycle as different pandemic diseases accompanied changes in weather and interpersonal contact. For example, spring was associated with waves of chicken pox, summer with polio and whooping cough, autumn with mumps, and winter with measles and influenza, to name but a few. These diseases triggered not minor worries by parents about the need to take time off from work to care for a sick child but unspeakable horrors that could shorten or end their children's lives. Paralytic fear of polio frequently emptied the swimming pools and summer camps of Depression-era America, condemning children to remain indoors and away from friends all summer for fear of contracting a disease that might otherwise sentence a child to a lifetime (albeit a short one) in the dreaded "iron lung." Such anxiety persisted until the miracles of the 1950s, which witnessed the introduction of waves of seemingly miraculous vaccines.

To properly understand the impact of vaccines, it is necessary to recall the devastation wrought by infectious diseases before the invention of the vaccine. Going back as far as ancient Egypt, we rediscover the history of various infectious diseases, starting with smallpox. This disease shared the nefarious property of being both highly infectious and comparably deadly and likely killed more humans throughout history than any other cause. The early chapters convey the symptoms and transmission of smallpox as well as the personalities that both defined the disease and laid the groundwork for its eradication. The little-known events leading up to the discovery of the smallpox vaccine represent some of the more fascinating discoveries in the post-Renaissance era, sweeping away long-held superstitions. The characters involved in these discoveries are as interesting for their intrepidness, curiosity, and motivation to protect those they love as for their fundamental contributions to science or health. The 18th-century discovery of the vaccine laid the groundwork for a Herculean task two centuries later to utterly eradicate the disease from the planet, an achievement that arguably surpasses all other human achievements.

The early example of a smallpox vaccine provides a backbone to a larger story of a never-ending war against infectious diseases, a conflict as old as life itself. We evaluate the nature of our allies and enemies and the struggles against these opponents in a never-ending war. Our most important partner is a remarkably complex and adaptable immune system, a virtual organ comprised of trillions of individual proteins and cells that patrol the body in a manner nearly identical to the most sophisticated modern military organization, including the use of guided missiles and deadly forms of chemical warfare. This natural army within each of us evolved over the epochs of time to attack potential pathogens.

Counted as both friend and foe are massive numbers of microorganisms, including bacteria and viruses. As our understanding of these tiny pathogens has increased, so has our appreciation for the complexity of what it means to be human, as each of us is composed of more microbial cells and DNA than human cells and DNA. Most of the bacteria and even viruses we are in constant cohabitation with are benign and provide essential services. We are presently in the midst of an exciting renaissance, begun only within the past few years, of understanding ourselves afresh within the context of these interactions with our microbes. Accumulating knowledge reveals the essential roles microbes play in defining an ever-increasing array of "human" functions, from the most elementary roles in assisting the digestion of foods to regulating emotions and even personality. When these relationships go awry, as with any other relationship with an intimate cohabitant, a breakup or quick hookup with a new and different partner can spell disaster. Our knowledge of diseases caused by a misfiring microbiome is increasing, almost by the day, and includes obesity, heart disease, breast cancer, and susceptibility to various infectious diseases, the last of which remains historically the largest mass killer in human history.

The past decade has witnessed a dramatic uptick in the emergence of infectious diseases, not simply those caused by exotic pathogens such as the Zika or Ebola viruses but also those caused by familiar and preventable childhood pathogens. Rather than rising to meet this new challenge with existing and proven measures, the response, if any, has been lackluster. The airwaves are more likely to resonate with the concerns expressed by well-intended but misinformed big personalities, such as Jenny McCarthy or Donald Trump, both of whom embrace the "anti-vaccinator" movement. Likewise, a quick search of the local bookstore or web retailer is more likely to reveal long-debunked yet pervasive works glorifying disproven links between vaccines and autism or spectacular claims of conspiracy. The rising volume of vaccine denial, which tends to focus on the refuted link between vaccination and autism, has generated a self-propagating meme that has increased the tenor and virulence of conversation about the need for proper vaccination. The volume and advocacy of false facts by an obnoxious and loud minority has overwhelmed the fact-based attempts by credible sources to expound the extraordinary health benefits of vaccination.

Unfortunately, the scientific community has largely demurred from confronting these loud disagreements. Specifically, most scientists, physicians, and educators have been trained to objectively analyze data and reach principled and defensible conclusions and thus have correctly dismissed the call of anti-vaccinators as "crackpots." A recent conversation with a key vaccine specialist about the vaccine denial subject elicited a typical response that this has been scientifically refuted and thus is no longer an issue. Despite the technical accuracy of such responses, the messages conveyed by the anti-vaccinators continue to resonate with the key demographic, including educated and affluent parents. The consequences of the failure to follow through with keeping current with childhood vaccination recently passed a tipping point that could threaten the health of our nation if not reversed. In fact, I was in part motivated to write this book upon learning of an outbreak of mumps at nearby University of Missouri. Similar outbreaks have been documented at Harvard, Yale, the University of Washington, and many other institutions over the past year.

I serve as an associate vice chancellor and professor at Washington University in St. Louis, one of the top five medical schools in the nation. As such I have been witness to and have been actively engaged in the science, public health, and policy implications of vaccine denial and the real-world outcomes of failing to vaccinate. Since earning a doctorate from Duke University as an immunologist, I have split my career between the biopharmaceutical industry and academia, most recently leading drug development at Yale University before joining Washington University in 2014. My responsibilities have included leadership of two of the world's leading biomedical research centers, which share the responsibility of analyzing and supporting drug and vaccine research and development.

Prior to Yale, I lived in suburban Washington, D.C., working within its thriving biotechnology community. These experiences include leaving a tenured professorship at Purdue University to help guide a medium-sized biotechnology start-up by the name of MedImmune into a large biotechnology juggernaut of the industry. In 2006, I left MedImmune to take on the challenge of leading research and development for a start-up company focused on emerging infectious diseases. In this role my time was spent leading research and development of new medicines for diseases that were largely unknown to the public (at the time) but which we knew were percolating in the background, representing threats from nature and man (since a major focus of the company was bioterrorism, a subject addressed in this book). The primary focus of my work at the time was the Ebola virus, a pathogen that came to the public's attention due to a natural outbreak in West Africa. My charge at this company also included the development of medical countermeasures for Marburg virus, Rift Valley fever virus, chikungunya virus, Lassa fever virus, dengue fever virus, and pandemic influenza viruses, all of which have been implicated both as bioterrorist weapons and as naturally occurring events. To paraphrase one of my former CEOs, a past head of the secretive Defense Advanced Research Projects Agency (DARPA), "Mother Nature is a most dangerous and inventive terrorist."

During this same period, I was a witness to, and a reluctant participant in, a series of incidents that vaguely recall the situation with vaccines today. In the early autumn of 2002, while attending a company meeting, my administrative assistant quietly slipped me a note stating there had been multiple shootings in and around Silver Springs and Olney, two communities in northern Montgomery County, Maryland. These shootings were not far from my home outside Olney and the school where my daughter had just started kindergarten. Just after announcing the news to the team and dismissing the meeting, I found myself speeding towards my daughter's school while radio news constantly updated the ongoing story of the shootings, which tracked ever farther northwards and nearer the elementary school. Tracking on a parallel course just to the west, I arrived at the elementary school, ran to its front door, and was confronted with the terrifying reality that the school appropriately was subject to a lockdown. Standing alone in such a prominent and exposed place created an exceptional vulnerability. After a frantic spate of door knocking, I was eventually recognized as a parent (and not a sniper) and let in to fetch my daughter.

Over the next three weeks, the D.C. sniper attacks riveted the nation. An early lead in the investigation surfaced from a witness in Silver Springs, who noted a late-model white box truck near the first shooting.[5] Later witnesses of other shootings, both on that fateful first day and thereafter, confirmed the presence of a white box truck at the time of each shooting. Within hours, the entire Eastern Seaboard was actively seeking (or avoiding) white box trucks. Anyone living in Montgomery County during that time can recount stories of loading groceries or refilling gas tanks, when the appearance of a white box truck triggered a duck-and-cover instinct or the famous zig-zag walk, both of which became almost ritual practices meant to avoid being the next victim of "the sniper."

With hindsight, we now know the conveyance for the snipers (you may recall there were two, not one) was not a white box truck but a dark blue 1990 Chevrolet Caprice sedan (about as different a vehicle from the suspected white box truck as one can imagine).[6] As much angst as the sudden arrival of a white box truck had triggered in the Washington suburbs in that autumn of 2002, it must have been even more troubling for the many legitimate drivers of white box trucks, as the news and traffic reports of the day were replete with stories of countless vehicles pulled over (snarling the already atrocious traffic of the Beltway) and drivers interrogated by nervous officers at gunpoint.

From the first moments of the attack, the idea that a white box truck was responsible for the sniper attacks was burned into the minds of well-intended witnesses all throughout D.C., the Eastern Seaboard, and nationwide. The initial reports were corroborated by honest witnesses seeking to assist in the investigation. However, if anything, the "chatter" generated by such spurious leads only served to delay the identification of the correct vehicle. Indeed, the linkage of the snipers with the blue sedan provided the essential breakthrough that triggered the suspects' arrest and abrupt conclusion of the crisis.

A similar phenomenon has occurred with the vaccine-autism link. In recalling the first time their child demonstrated overt symptoms of autism, many parents linked the disease with a recent vaccination, particularly when prompted to do so. However, pediatric vaccination schedules are about as ubiquitous as white box trucks in metropolitan Washington and about as relevant to the investigation of the causes of autism. Nonetheless, the belief in the vaccine connection is often staunchly defended and perhaps represents a tangible source for blame by frustrated and devastated parents. Such is the genesis of a challenge that has furthered the magnitude of the overall pain of autism by promoting an avoidance of vaccines in a misguided attempt to protect children. Instead, these choices have endangered not only their own children but also their friends, siblings, neighbors, and, if allowed to fester, quite possibly the entire nation's health.

My feeling of vulnerability while standing in front of the elementary school and believing the snipers were nearby pales in comparison with the vulnerabilities being faced every day by innocent victims who have no realization of the health dangers they face. It may be shocking for many to learn that most of the collegiate victims of the recent and ongoing resurgence in measles, mumps, rubella, and other outbreaks were in fact immunized against these diseases as children. Like virtually all types of medicines or any other product one can envision, vaccines wear out. As we grow older and gain experience with an expanding number of foreign microbial intruders, the immune system can focus less and less on any individual pathogen (or vaccine), particularly those last boosted years or decades before. Eventually, it becomes a "use it or lose it" proposition, and absent using it, the immune system tends eventually to lose its ability to respond.

The slow decay in the ability of the immune system to recall recognition of pathogens is compounded by a phenomenon discussed among vaccinologists and referred to as "the herd effect" (or social immunity). Sparing the reader the agonizing details and mathematical modeling that distinguishes the field of epidemiology, the herd effect can be visualized as a protective shield that arises when a large fraction of a population is rendered insensitive to a particular infection. If enough of "the herd" (or any community of individuals) is adequately protected, then even the unprotected will find a safe harbor from that pathogen. However, when the herd is thinned, sometimes even by just a small number, the consequences for the entire population can be disastrous, instigating a dangerous domino effect. This explanation is demonstrated by the current situation with measles, mumps, rubella, and a cadre of additional infectious agents that threaten not just our children but all Americans and indeed the entire planet.

Part of the concern with the rise of childhood diseases reflects the fact that while some pathogens, such as the virus that causes chicken pox (varicella zoster virus), cause a relatively minor rash that tends to resolve after a few days, the disease is far more aggressive in teens or adults.[7] Not only is the severity of the skin rash increased but common responses include a potentially fatal swelling of the brain (known as encephalitis) and inflammation of the joints. Much worse is the situation with a mumps virus infection of teens and adults, which can trigger bouts of inflammation in or near the testes that render their victims sterile.

My goal for this book is to convey the stories of the remarkable history of science, technology, and disease that helped eradicate many of the deadliest plagues known to man. I also intend to convey the reality that the victory against vaccine-preventable diseases is not durable and they could reemerge like B-movie antagonists to kill or maim more victims. The ground covered will also highlight current and future challenges being confronted by the vaccine community, including old threats and new, including Ebola, Zika, antibiotic-resistant infections, and other deadly emerging and reemerging pathogens.

Not only do I seek to present the history of vaccines alongside the history of deadly pathogens and the role they've played in human history (toppling empires as well as causing intense heartbreak and loss on individual levels) but I also seek to shine a light upon the long history of vaccine hostility. Many readers might be surprised to learn that anti-vaxxers have always been around, even before the first vaccine was introduced in the 1790s. It might seem like a modern phenomenon, but in fact the history of vaccines is impossible to tell without discussing how each breakthrough has been hindered by a vocal pushback. The fears underlying this resistance have too often counterweighed the hope a new vaccine might bring. Such pushback was prominent even at times when diseases like smallpox or polio could devastate entire families or doom innocent children to short lives spent in a black lung. Too often, we dismiss the fringe elements of the anti-vaccine movement because they have not directly experienced the devastation wreaked by maladies such as polio or measles. However, the damage has clearly been done, as evidenced by the rising incidence of these deadly scourges. Although efforts have been expended to convey the benefits of vaccines, the hopes referred to in the title of this book have been trumped by a fear of an invisible menace perceived as worse than infectious microbes—namely, autism. These more negative sentiments have been winning the day and now present very real dangers to our societies and our families. So perhaps we need to look back at history—specifically to the history of vaccine naysayers and fear mongers—to help us as members of our community, in education, medicine, or in public health. In doing so, perhaps we can help develop a better approach to convey and appreciate the extraordinary benefits and hope that vaccines have imparted upon modern society.

1

Pox Romana

Most historians concur that the middle of the 2nd century of the current era (C.E.) was the apex for the most dominant realm the world had yet known (and would not witness again for millennia). The Roman Empire had emerged from the crisis of the Roman Republic and a period of intense civil wars which were finally concluded with the victory of Gaius Octavius (soon to be known as Augustus Caesar) over Marc Antony at the Battle of Actium in 31 B.C.E. Over the following century, the empire continued a campaign of merciless expansion, geographically, militarily, and in terms of what we today refer to as soft power, including cultural, architectural, and artistic contributions. A two-century period of relative calm demarked the Pax Romana (from the conclusion of the Battle of Actium through 250 C.E.). Unbeknownst to its citizenry, the end was nigh.

A strong central government (it was a dictator-run empire, after all) had committed substantial investments in vital infrastructure, including roads paved with innovative forms of a breakthrough composite material known as concrete (*opus caementicum*) that allowed for the building of large and durable buildings and roads, many of which remain fully functional two millennia later.[1] Larger buildings increased the density of modern cities, and Rome itself is estimated to have housed as many as a million people in the 2nd century,[2] a feat that would not be reproduced in Europe until the latter days of the Industrial Revolution. However, the combination of unprecedented mobility and population density would ultimately conspire against the empire.

The revolutionary new road system permitted Roman citizens to travel and emigrate peaceably from as far north as the city of Eboracum (the present-day city of York, England) to Hieraskaminos on the Upper Nile (near contemporary Aswan, Egypt). Throughout this four-thousand-mile trip, a Roman citizen could interact with merchants using a common tongue and utilize the same currency throughout her travels and remain confident of her personal safety under the protection of a Roman militia, whose garrisons or relay stations were interspersed at fixed points and protected the traveler from the privations of highway robbers. This system in turn facilitated trade, both within the empire's provinces as well as with distant lands such as the Indian subcontinent (by land or sea) and China (via the Silk Road). This same transportation system also greatly hastened the speed by which these travelers could spread disease throughout the Western world.

A period of such remarkable unity brought forth by these technologies was thus fated to implode, largely under its own weight. While the greatest causes of the Roman Empire's decline have been the subject of considerable erudition, from Edward Gibbon in 1776 onwards,[3] the smallest causes were quite literally microscopic and tied to a part of the empire that rarely if ever entered the history books or thoughts of even the most erudite Roman statesmen.

The ancient city of Seleucia is located on the west bank of the Tigris River, deep within the heart of the ancient Fertile Crescent. Lest the modern reader mistakenly assume that local turmoil in this region is a feature unique to our own time, this region, twenty miles southeast of downtown Baghdad, has remained a hotbed of political and military instability for millennia.

Three centuries before the usurpation of the Roman Republic by the empire, Seleucia was a hinterland of an empire aggregated by Alexander the Great.[4] Having secured his Hellenic possessions within a greater Macedonia,

Alexander crossed the Hellespont in 334 B.C.E. with the goal of challenging the power of the Archaemenid Empire of Persia, which was ruled by Artashata, also known as Darius III. The Persian ruler and his vast holdings had been the target of Alexander's father, Phillip II of Macedon, who used the Persian desecration of the Athenian temples a century earlier as an excuse for conquest. After Phillip's assassination, an act for which Alexander is occasionally and probably unfairly implicated, Alexander began to realize the opportunities arising from a Persian conquest.

Leading a group of brilliant generals (later known collectively as the "Diadochi," from the Greek word for 'successors'), Alexander bested army after army, first at the Battle of the Granicus, near the site of ancient Troy, and a year later at the Battle of Issus in southern Anatolia.[5] The Battle of Gaugamela in present-day Iraqi Kurdistan sent Darius into retreat, this time for good as the disgraced commander was murdered by his cousin, the Satrap Bessus. Rather than being relieved of a burden of his most dire enemy, Alexander was angered by Bessus's rash actions, in part because greater prestige could have accompanied Darius's becoming his prisoner and because the Macedonian leader had gained great respect for Darius. Consequently, Alexander had Bessus tortured and executed for his crime.[6]

After destroying the primary Archaemenid force and subjugating Bessus's army, Alexander's forces entrenched at a minor village on the western bank of the Tigris River in preparation for an invasion of the Indian subcontinent.[7] During this period, Alexander's expansionistic urges were slowed by the homesickness of his generals. The Diadochi were overcome not just with a longing for their homeland but by concerns that Alexander had embraced the habits of the civilizations he'd conquered perhaps a bit too much and, as a result, had "gone native." As the armies prepared for one long, last push into India, the army was also fighting a malady altogether different from homesickness.

Alexander's troops were encountering a disease endemic to the region. Specifically, the occupation of the Tigris River valley region was accompanied by a regional infection, characterized by a contemporary as "a scab that attacked the bodies of the soldiers and spread by contagion."[8] This is generally presumed to be an early written description of smallpox, a disease that slowly marched through Alexander's army and would continue to play a prominent role throughout much of history.[9] The much-anticipated Indian campaign itself would grind to a halt in 327 B.C.E. in large part because of the toll taken by smallpox upon Alexander's troops. The commander himself may not have been exempt from the suffering (the records are insufficiently precise to verify such a diagnosis two and a half millennia later).

Though Alexander and his army survived the 327 B.C.E. smallpox epidemic, both were greatly weakened. The Indian campaign was abandoned, and Alexander focused his efforts closer to home (though not Macedonia, much to the chagrin of the Diadochi, but rather the region known as the Levant). Still not fully recovered from the strain imposed by smallpox four years earlier, the 32-year-old commander began complaining of fever and exhaustion in the early days of June 323 B.C.E. These symptoms progressed rapidly, and the young general was dead within a week. The premature passing has prompted all types of explanations, ranging from natural causes (malaria, typhoid fever, West Nile fever) to man-made (poisoning and alcoholic liver disease). While two and a half millennia precludes a definitive diagnosis, it seems likely the physical toll paid by the previous encounter with smallpox contributed to his later susceptibility and early demise.[10]

The sudden death of the world's most charismatic and successful young dictator was unsurprisingly followed by a period of extended turmoil. Within days, the Diadochi turned upon one another in an attempt to sway the succession.[11] Waves of intrigue, assassinations, and internecine fighting failed to resolve the vacuum left by Alexander's absence, and the once-great empire fractured into a series of successor states that dotted the region throughout the remaining Hellenistic period. Despite periodic attempts, these never again coalesced into anything resembling a unified domain.

One Diadochi commander of the cavalry, Seleucus, was appointed Satrap of Babylon and quickly began to consolidate and strengthen his grip upon the central regions of the former Persian Empire. Seleucus renamed the site of Alexander's former resting spot on the west bank of the Tigris after himself. Seleucus and a string of his successors progressively extended the domains of the Seleucid Empire for the next two and a half centuries. At its peak, the empire encompassed most of modern-day Iraq, Iran, Pakistan, Afghanistan, Israel, Lebanon, Syria, and Turkey, as well as parts of India. While impressive when viewed on a map, the political and military power of the Seleucid Empire was largely illusory and constantly overstretched.

The fragile grip of the Seleucids became apparent in the late 3rd century B.C.E. when delusions of grandeur drove its shortsighted leaders to emulate the deep power of Alexander's realm and establish a partnership with a new generation of Macedonians. The now-minor Hellenistic kingdom of Macedon, led by a descendent of another of the Diadochi, struck an alliance not just with the Seleucids but also with the Carthaginian general, Hannibal.[12] Hannibal, one of a handful of personages in history whose strategic and tactical prowess could accurately be compared with Alexander's, was the bane of the Roman Republic. The Carthage-Macedon partnership came at a high-water point during the Second Punic War, as Carthaginian troops occupied much of the Italian peninsula. It seemed merely a matter of time until the upstart city-state of Rome would succumb to Hannibal's offense.

Yet looks could be deceiving. Despite the appearance of a winning position, Hannibal was in the third year of an arduous attempt to engage the Roman consul Quintua Fabius Maximus Verrucosus in a decisive battle.[13] The Fabian strategy avoided a pitched battle at all costs and instead sought to wear down Hannibal through attrition, a tactic that has been successfully replicated many times, including by the American generals George Washington and Robert E. Lee. The tactic remains quite effective in modern times as evidenced by experiences in Afghanistan. Nonetheless, the avoidance of a decisive battle was highly controversial, and the Roman Senate, hinting at cowardice, sacked Fabius in favor of another commander who would espouse a more direct approach to dealing with Hannibal. Such folly led to the appointment of the more aggressive Gaius Terentius Varro as consul. The sought-after battle was soon gained and resulted in a decisive defeat of Varro in 216 B.C.E. at the Battle of Cannae (a name synonymous with a resounding victory still to this day). Returning to a Fabian strategy, Rome survived Cannae and outlasted the Carthaginian invaders, whose troops were far from home with overextended supply lines and surrounded by hostile locals. By 201 B.C.E., the Romans had brought the war across the Mediterranean Sea to Carthage, to the defeat of Hannibal. These events allowed Rome to become the unchallenged superpower of the central and western Mediterranean.

All the while the Roman and Carthaginian forces were wrangling over the Iberian Peninsula and North Africa, the eastern Mediterranean remained a cauldron of geopolitical instability. In the months following the extirpation of Carthaginian power in the Second Punic War, Roman concern turned eastwards. The Romans had scores to settle with the old Carthaginian allies, the Macedonians. In the years the Romans were forced to focus their efforts upon containing Hannibal, the Macedonian threat had been countered through an alliance with the Aetolian League, a loose confederation of Greek city-states in central Greece and a longtime rival to the Macedonians (employing the approach of *the enemy of my enemy is my friend*). Victory against Hannibal meant that the Romans could now concentrate upon the Macedonians, and they were swiftly and soundly defeated in 197 B.C.E.

The Aetolians might have been rid of their long-standing Macedonian rivals, but they now had to contend with the powerful and ambitious Roman victors. The Aetolians had always viewed the Roman alliance as one of convenience based on a shared enemy, and the presence of Iberian soldiers on Greek shores soon dissolved whatever friendship had existed. As the alliance between the two realms rapidly deteriorated, the southern Balkan peninsula again dissolved into chaotic political and military clashes.

The resulting perception of a power vacuum in Greece might have repelled a pragmatic leader of a paper tiger such as Seleucia but instead triggered visions of grandeur by the Seleucid king, Antiochus III. The delusional leader clung to the belief that expansion into Europe in general, and Macedonia in particular, would create an opportunity to equal the glories of Alexander the Great. However, Antiochus's views of his own greatness were soon ended by decisive defeats at the hands of the Romans at the Battle of Thermopylae in 191 B.C.E. (this battle is not to be confused with the more famous Spartan-Persian battle near the same site a few centuries before, nor the six other battles of the same name fought since then). In the Romano-Seleucid version of the Battle of Thermopylae, the domination of a small Roman army against a much larger Seleucid force foreshadowed the even more strategically important Battle of Magnesia a year later. The resulting peace treaties with the Roman Republic stripped the Seleucids of much of their conquered lands, including not only the loss of their European possessions but the loss of most holdings on the Anatolian peninsula as well.

The Seleucids were not only humbled by the Romans, but the revelation of their inherent impotence rendered them subject to intermittent cycles of discontent, civil war, and then insurrection. Repeatedly, new pretenders promised to restore the glory of earlier days, only to be later disabused of the notion by the superior arms of neighboring Roman forces.

Although not a direct vassal of Rome, the reduced Seleucid "empire" was well within the Roman sphere of influence, and Seleucid leaders remained constantly on tenterhooks in recognition that restiveness in the Levant could cause their larger neighbor to snuff out what remained of their fledgling yet marginally independent realm. Unfortunately for the Seleucids, the chaos that continues to characterize the Tigris and Euphrates river valleys even today has ancient roots, and the Seleucids were no better at quelling unrest than any other occupying power, before or since. By the beginning of the 1st century B.C.E., Seleucia had devolved into a failed state manipulated by larger neighbors to the south (the Ptolemaic Egyptians, a Hellenistic colony) and north (Pontus and Armenia), who sought ways to distract and bleed forces from their mutually shared Roman enemy.

Eventually, these distractions drove Pompey, the great military general and member of the ruling triumvirate (along with Julius Caesar and Marcus Licinius Crassus), to put an end to the Seleucid nuisance once and for all. The utter eradication of the Seleucids and the subsequent carving up of their remaining lands at last quieted some of the disorder as the Romans ceded the lands of Seleucia to other vassal states and minor regional powers.

The Parthians were one example of a minor client state that benefited from the fall of Seleucia. These people arose from a group of nomadic tribes in the north and eastern regions of Persia, and they had been a long-standing rival of the Seleucids. The weakening and eventual elimination of the Seleucids by the Romans provided an opportunity for the Parthians to expand their holdings. The lands encompassing Seleucia transferred into Parthian hands. As it was with the Seleucids, however, the term *empire* is an overstatement, as the Parthian holdings were arguably more a confederation of satraps and lacked a strong central power.

Consequently, the Romans were again faced with instability on their strategic southeastern flank, this time from the Parthians. This required intermittent reintroduction of forces to police the regions of the Fertile Crescent over the next two centuries. An otherwise humdrum round of local interdiction in the mid-2nd century C.E. would unexpectedly undermine the mighty Roman Empire and catalyze its downfall.

Gaius Avidius Cassius was the scion of a powerful family with a strong and prestigious bloodline on both sides.[14] His father was a Roman politician, whose maternal ancestors included Herod the Great, Gaius Cassius Longinus (the same Cassius as profiled by Shakespeare, whose intrigues culminated in the assassination of Julius Caesar), and Gaius Octavius (father of Caesar Augustus). Gaius Avidius Cassius embodied a genetic lineage that encompassed both the major protagonist and antagonist in the Roman transition from republic to empire. In a world not known for outbreeding, his maternal lineage included both Cassius and Octavius, as well as Marcus Vipsanius Agrippa, an architect of notable fame and the general who plotted the defeat of Mark in the Battle of Actium.

Given his high birth and marshal familial history, it is unsurprising that Gaius Avidius Cassius, who was born in Syria, rose quickly through the military ranks. In 161, he was commanded to lead Roman legions to quell an uprising by the Parthians and to sack their capital in Ctesiphon. During this campaign, he marched his troops down the Euphrates River valley and attacked the Parthians at the town of Seleucia (just across the river from Ctesiphon) in the year 165. A few days after a pivotal battle in which his troops decisively defeated the Parthians, a small number of soldiers began to fall ill with symptoms that included fever, diarrhea, and eruptions of the skin. Looking at the geography (the same location where Alexander's army had encountered the same disease), symptoms, and rapid spread of the disease, modern epidemiologists have largely concluded that smallpox was again responsible for the outbreak.[15,16] Within weeks, this so-called Antonine Plague (later named because it began during the reign of Emperor Marcus Aurelius Antoninus) would begin to burn up and through the empire.

The plague spread quickly by exploiting the same innovations that facilitated the growth of the empire during the Pax Romana. A combination of improved transportation, urbanization, and emigration propagated the expansion of smallpox from a localized epidemic to an empire-wide pandemic. Evidence these improvements accelerated the disease is supported by the fact that within a year following the occupation of Seleucia, the Greco-Roman physician Galen described the symptoms of the disease (facilitating the attribution to smallpox by epidemiologists two thousand years later) as it struck the city of Rome in 166.[17] Undoubtedly, the higher population density that had been facilitated by improved building technologies and urbanization unintentionally expedited transmission of the pathogen among the citizenry. The exact extent and impact of the Antonine Plague remains a subject of academic disagreement, but virtually all concur as to its devastation. Low estimates of the plague's mortality rate cite 7–10 percent of those infected, which would translate into approximately three to five million people killed across the empire (though census figures are unreliable, given the large number and mobility of citizens, slaves, and emigrants within the expansive empire).[18] At the other extreme, the 19th-century German historian Otto Seeck claims the Antonine Plague killed more than half the population of the Roman Empire within the fifteen-year period spanning 165–180.[19] Putting this into perspective, a comparable plague in the modern United States would cause more than 150 million deaths, which would be equal to the deaths of every person in forty-three states (all save Ohio, Pennsylvania, Illinois, Florida, New York, Texas, and California).

Without question, the plague altered day-to-day life for virtually all living in the empire. A leading 19th-century historian of Roman law and economics concluded, "The ancient world never recovered from the blow inflicted on it by the plague which visited it in the reign of Marcus Aurelius."[20] Among other things, the living and dying were preyed upon by charlatans. Quack remedies for the plague proliferated and took advantage of a terrified populace. Such behavior is alluded to by the Emperor Marcus Aurelius, who, in his *Meditations*, conveyed that the disease was less deadly than the lies and malicious intent of those lacking understanding of the pestilence.[21] Sadly, the emperor himself would ultimately be counted among the victims of the plague, succumbing to the disease in the year 180. His death brought an end to a period characterized as the rule of the Five Good Emperors: Nerva, Trajan, Hadrian, Antoninus Pius, and Marcus Aurelius.[22] Thereafter, a suffering nation was governed by a series of selfish, imperious, and ineffectual leaders, dooming the Western world's only superpower.

The Spotted History of Smallpox

As historically devastating as the Antonine Plague was, it was hardly the first time that humanity, or even the Mediterranean basin, had experienced the desolation caused by smallpox. Modern genetic-based modeling suggests the disease jumped from a rodent to humans between 16,000 and 48,000 years ago, somewhere in or near the Gold Coast of Africa.[23] From there, the disease began a relentless northeastern progression through Egypt and the Middle East, into the Caucasus and China. Despite and perhaps because of the familiarity of our species with smallpox, which might have seemed commonplace to many, descriptions of the disease are largely undocumented by most of the oldest extant written sources, such as the Old Testament or Egyptian papyri. However, lesions consistent with smallpox have been detected on Egyptian mummies, including that of Ramses V, who died more than a millennium before the Antonine Plague.[24] The high density of people in the Nile River valley likely facilitated the propagation of the disease, and trading with other civilizations perpetuated its spread throughout the ancient world. Another river valley, that of the Indus in modern-day India, also hosted endemic smallpox, and trade with the ancient Egyptians or Chinese might have introduced the pathogen to the subcontinent.[25]

Some modern historians speculate that the Plague of Athens might have arisen from a smallpox pandemic that had its beginnings in Egypt.[26,27] This particular plague stands out in history because it arose in the midst of, and influenced the outcome of, the Peloponnesian War. As an overly brief summary of that conflict, the two antagonists, Sparta and Athens, utilized very different strategies centered upon their individual strengths. Sparta was a land power, and their soldiery was feared throughout the region. In contrast, the Athenians were a seafaring power that flourished largely because of trade with the many other city-states and civilizations found throughout the Mediterranean world. The Athenian leadership recognized the mutual mismatch: Athens could not compete with Sparta on the land, and Sparta was no match for the Athenian fleet. In response, the celebrated Athenian leader Pericles adopted an approach of building great defensive embattlements around Athens and succoring the city with seaborne supplies. The surrounding farmlands were to be harvested and the farmers brought inside the massive fortress with the intention of waiting out the inevitable Spartan sieges. Pericles was confident in this approach, as Athens could be provisioned by trade with other coastal city-states, while the Spartans would be compelled to squander precious manpower and resources on maintaining the siege.

Pericles could not have realized that his innovative strategy would be undermined by unwanted microbial guests. As documented by the father of modern history, Thucydides, the plague was rumored to have started in Ethiopia before entering the Hellenized lands of Egypt, which was a major trading partner for Athens. Thus, a Periclean strategy based on trade with Egypt and other Greek city-states hastened and magnified the spread of disease. Compounding the issue, a high population, which was exacerbated by the refuge granted to the peoples outside the Athenian defensive walls, increased the efficiency of transmission once it gained a toehold within the Athenian redoubt. The spread of the disease was so rapid and pronounced that the besieging Spartans became disconcerted by the constant burning of funeral pyres within Athens and lifted the cordon. Their fears further caused them to temporarily suspend all military and nonmilitary interactions with the Athenians in an attempt to prevent the infection from spreading to their own camps.

Based on written reports and mass graves excavated in the area, the victims of the Plague of Athens might have included more than half the Athenian population. Prominent among these were the great leader Pericles himself, who was joined in death by his entire family. Despite the lifting of the siege, the disease combined with the sudden loss of Pericles's leadership to undercut the influence of the Athenian resistance to Sparta thereafter. Indeed, Thucydides writes that the social fabric of Athens disintegrated during the epidemic as citizens stopped respecting authority and obeying the law and social conventions, such as caring for afflicted family and neighbors and following religious authority. The societal unrest effectively neutered the once-mighty Athenians for at least a generation. The Plague of Athens was unquestionably a disaster for the Western world. While smallpox seemed to have played a role, additional or alternative microbial pathogens, such as typhoid fever and typhus, likely also contributed to the epidemic.[28] Indeed, the lowering of individual and public health by one disease, especially during times of war, often encourages the rapid expansion, and synergy with, other infectious diseases. Nonetheless, these results suggest smallpox might have played a role in destroying both of the great pillars of ancient Western civilization.

A New World

Through a combination of luck and circumstance, new microbial pathogens are constantly challenging hosts new and old. A virus that lives in one species, say a bird, might change (mutate) ever so slightly such that it can now infect a human. Alternatively, a virus-infected member of one species might introduce the disease to another, for example by serving as its prey. Many such examples have and will continue to occur, as evidenced by an ongoing avian influenza (H5N1) epidemic throughout much of Asia, as well as the recently established monkey-to-human transmission of HIV (most likely the result of a hunter being bitten by bush meat) in the early 20th century. We will return to this subject in greater detail in subsequent chapters, but the dynamism of infectious disease has been and will remain a constant feature afflicting human civilization.

A long-standing theory of infectious disease epidemiology is that the introduction of a new pathogen into a population or species conveys a more pathogenic or deadly form of disease. One tenet of this idea holds that if a particular microbe became too aggressive, in its frenzy to feed it would likely kill or incapacitate its host in a short enough time that its food would be exhausted before the pathogen could be propagated to its next victim (and so the first meal would be its last). Such a high-morbidity and -mortality virus would "burn itself out." Such outbreaks would therefore remain local and easily maintained, much as we had seen with the Ebola virus until recent years.

Following this line of logic, a less lethal form of a microbe may ironically cause a more dangerous disease to the larger population. Viewed a different way, more danger can arise when a pathogen becomes slightly less obnoxious so that it does not kill the host until after it can spread to others. If the disease is easily spread and death or sterility (i.e., the failure of the host to reproduce) is the outcome, then the microbe could again destroy the entire host species, its food source, in a relatively short period of time. Thus, a well-mannered and long-term-minded pathogen will seek to farm its food and harvest only when needed.

As a consequence of these dynamics, a microbe is best served if its virulence or ability to infect the host is moderated (known as attenuation). Clearly, microbes are not sentient beings that make conscious decisions to slow their growth or diet. Under the assumption we are all at least vaguely aware of DNA and the ideas of mutation, then this process could offer an explanation. To readers familiar with the concept of genetic mutation, attenuation explains why. Worse still, a well-mannered microbe still faces competition for this same food source from its more obnoxious cousins, who may still be in a frenzied state of feeding. Thus, the microbe encounters a situation that is identical to the economic problem known as the tragedy of the commons, first described by William Forster Lloyd in early Victorian England and popularized by Garret Hardin in the 1960s.[29]

Fortunately, a parallel form of evolution in the host species can help resolve the microbial version of the tragedy of the commons. It is widely appreciated that genetic diversity is a good thing. Anyone who has visited a major zoo has likely heard the story of the cheetah, a species where a past population crisis (i.e., where many animals die off) about ten thousand years ago caused a genetic bottleneck—an event that limits genetic variation in a population. This fundamentally threatened the future viability of the entire species.[30] Although the causes of the cheetah crisis are not clear, the bottleneck event might have consisted of a sudden environmental change and/or contact with an obnoxious new microbe. Regardless, the genetic diversity of the survivors quickly dropped by at least 75 percent. As a second example, we are well aware of the problems associated with consanguineous mating when siblings or close cousins produce an offspring. Indeed, most modern countries have laws limiting the ability of an individual to marry a close relative, since the progeny of such matings tend to display recessive traits that rarely improve the stock.

This raises the question of why genetic diversity is so important. Let's consider an example inspired by mixing and matching a few theological viewpoints: If God created an individual (or two) specifically to live within a privileged environment (the Garden of Eden), then why bother to continue the process of DNA mutation and evolution? The best answer, to which any Buddhist can attest, is that the world is subject to constant change. The future will be peopled by those who can adapt. Another pretty good response to the question derives from the reminder that all individuals (even bacteria, as we will soon see) are constantly forced to fight off pathogens.

Though we will return to the subject of genetic bottlenecks many times throughout this book, it suffices to say that genetic diversity is one way to increase the likelihood that a species will not be wiped out by a new microbial pathogen. If a population of organisms (let's say people) is sufficiently diverse from the standpoint of genetics, then the odds are (and it is all a statistical gamble) that some individuals will be more prone to survive an epidemic than others. When a large enough population has sufficient genetic diversity, then a Darwinian-like selection should allow the more robust population bearing a trait (also known as a phenotype) to survive, and perhaps even thrive, in the face of an obnoxious microbe. The survival of both the host and pathogen thus depends on the ability of the host to maintain enough genetic diversity to provide enough time for the microbe to learn how not to destroy its host.

A comparable form of parallel evolution of both the pathogen and the host has occurred with smallpox. As the virus continued to burn through the Middle East and Europe throughout the Middle Ages, the death toll slowly waned, and those people who were less prone to die (even slightly so) had more of an opportunity to pass along this decreased susceptibility to their progeny. These dynamics progressed towards an equilibrium in which the virus could be stably maintained in the human population without threatening the fundamental survival of the human species. Individuals might succumb to the disease, but mankind itself would survive. As such, the populations of Europe, Africa, and Asia experienced a rapid (known as punctuated) form of evolution over the past few millennia, in which the survivors were somewhat more adapted to survive smallpox than individuals who lived ten thousand years ago would have been. Something very different happened when smallpox was turned loose in a population in which the most susceptible individuals had not been subjected to multiple generations of selection. Such a disaster was experienced just over a half millennium ago all throughout the Western Hemisphere, with tragic consequences.

Coming to America

From its beginnings, America has always been a nation of immigrants. Even the "native Americans" and other indigenous peoples of the Western Hemisphere are relatively recent arrivals. Thus, while many Americans refer to the "Old World" in terms of European migration, the reality is that North America was subject to rounds of immigration from the "Old-Old World" in the form of migration from Asia. Until quite recently, it was believed that ancient humans began migrating to the New World no more than fourteen thousand years ago, largely as a consequence of the last great ice age.[31] The widely held idea, conveyed by countless textbooks and documentaries, was based upon the hypothesis that the colder climate caused much of the planet's northernmost oceans to become encased in ice, which served to both lower the ocean levels and create an ice bridge, known as the Bering land bridge or Beringia, between eastern Siberia and Alaska. As the ice bridge melted, but before it disconnected the two continents, a warmer climate provided just enough vegetation and animal life to nourish the travelers during their expeditions to the New World. This idea recalls depictions of a great migration by thousands of people trudging relentlessly over the ice sheets through gales of blowing snow, intent upon the great opportunities waiting for them in the Western Hemisphere. Until recently, there had been general agreement within the scientific community that some travelers embarked upon an odyssey that progressively led them south through North, Central, and finally South America. More recent findings cast doubt as to whether this image accurately conveys the primary mode of conveyance.

The Clovis peoples (named for an archaeological site near Clovis, New Mexico), used tools that readily identified them as immigrants to the New World. By analyzing the tools and other unique aspects of the Clovis culture, archaeologists and anthropologists could track their spread not just from north to south but also from east to west as they populated the large landmasses of the Western Hemisphere. Given the vast distances involved and the time (measured in scores of generations) needed to cross these on foot, it was largely assumed that the regions farthest from Alaska, namely the steppes of Patagonia, could not have been populated more recently than 12,500 years ago.

A convergence of information from many fields has identified inconsistencies in the popular theory associated with the Clovis peoples and their crossing of the great ice land bridge. First, carbon dating of archaeological evidence reveals bones, feces, and other detritus of humans in Patagonia more than 14,600 years ago (before the presumed opening of the Bering land bridge). From the results of recent meteorological studies, the timing of the Bering land bridge was inconsistent with the environmental conditions that the travelers would have required to access the necessary vegetation and animal sources needed to provide food, clothing, and shelter. Limitations in the availability of food, clothing and shelter would also have limited the number of people who could be provisioned to cover the considerable distance from habitable portions of Siberia to comparable climes in North America. Perhaps the most basic inconsistency is that artifacts linked with the Clovis culture have never been found in Alaska or the Canadian Yukon even though these have been found in Oregon, the American Southeast (as far as Florida), and in Patagonia.[32]

Although a frozen land bridge cannot be excluded and might indeed have contributed to the population of the Western Hemisphere by some early humans, many scientists now favor an idea that the Clovis population, or perhaps earlier settlers, might have entered the Americas along the coasts in boats in multiple waves of migration rather than within the limited timeframe needed for a land bridge crossing.[33] Indeed, a 2017 bombshell study of an archaeological site in Southern California by scientists from the San Diego History Museum suggested that humans might have arrived in the New World 130,000 years ago, presumably by sea.[34]

Analyses of genetic ancestry reveal that many native peoples of North and Central America share genetic material with north Asians (especially the native people of modern-day Siberia and Taiwan). In contrast, some Amazonian tribes are more closely related to Australasians.[35] A sea-based migration, at least by modern standards, would likely have involved small boats, which is consistent with the limited genetic diversity of the immigrants to the New World, whose founding numbers might have been measured in the hundreds or thousands.

The mode of transportation is not a minor consideration for our story, because understanding the means of movement impacts our understanding of genetic diversity. Rather than swarms of people crossing a land bridge, much of the Western Hemisphere might have been peopled by handfuls of migrants arriving from north or south Asia on rickety boats. This is combined with the unfortunate reality that these Asians had split from their Eurasian relatives before the latter were infected with smallpox. Both were ingredients that put a disaster in place. In addition, while the number of immigrants able to make the journey was limited by the size of the boats, a further reduction of genetic diversity arose as some individuals did not survive the arduous journey before passing along their genes to a new generation. Make no mistake, these were a hearty people, who proved able to conquer two entirely new and hostile continents. Indeed, Charles C. Mann's opus, *1491*, makes a strong case that far from a racist European view of Amerindians as "noble savages," these first peoples were capable of extraordinary technical and engineering feats.[36] However, the selection pressures on the surviving population that first populated the Western Hemisphere had not included resistance to diseases like smallpox.

Within a relatively short time (in a geological and anthropological sense), the population of the Western Hemisphere grew rapidly as people spread from west to east and north to south (and south to north if some travelers arrived by boat and landed in South America). Estimates of the number of pre-Columbian native Americans vary widely. On the low end, the population of the New World might have peaked at a low level of 1,000,000.[37] At the other extreme, the 1968 book *The Population of Miexteca Alta, 1520–1960* states that more than 25,000,000 people lived in the Mexican plain alone (not including the rest of North America or South America).[38]

The new tenants of the New World were likely not the primitive noble savages romanticized by many European conquerors but a wide variety of sophisticated cultures. For example, early Amerindians shaped and cultivated the land extensively. Consistent with this idea, archaeological evidence reveals that the system of mounds located just across the river from St. Louis, known as Cahokia, was a sophisticated metropolitan and religious center for the Mississippian culture, which extended from the northernmost lands of modern Minnesota to the Gulf of Mexico and from the Atlantic Ocean to the Missouri River valley.[39] The ruins at Cahokia reveal an advanced culture with sophisticated astronomical landmarks that aided agricultural planning (much as Stonehenge did for the Bronze Age Britons) and copper metalworking that supported the manufacture of intricate religious items.

Despite these achievements, the relatively low genetic diversity of these thriving civilizations (as compared with their Eurasian counterparts) made them susceptible to outside microbial challenge. The landing of Columbus and other Europeans triggered a comprehensive and rapid collapse in the population and infrastructure that had supported human life on the two continents of the Western Hemisphere.[40] Much as we saw with the Romans, the existence of a more modern transportation and urbanization infrastructure provided a superhighway to facilitate the spread of diseases brought along by the newly arrived Europeans. Within the half century separating 1492 from the time in which Hernando de Soto became the first European to gaze upon the almost uninhabited regions of the lower Mississippi River, much of the population had already been decimated by waves of disease. These diseases were spread by the vital interactions among tribes that ranged up and down the Atlantic and Pacific coasts and into the heartlands. The considerable contacts among the different native peoples, much like the Roman highways, facilitated the spread of diseases, decimated entire civilizations, and rendered those remaining susceptible to subjugation. Chief among these pathogens was smallpox.

Smallpox was endemic throughout the classical world by the time of the ancient Egyptians, Greeks, and Romans. While bouts of disease continued to claim many lives, the population had been culled such that the threat to human civilization was problematic, to be certain, but no longer existential. From the cold, analytical standpoint of a population scientist, smallpox in the Old World had been rendered into something greater than a Eurasian annoyance, albeit a fatal one, but less than a society-ending apocalypse. For the natives of both Americas, who lacked prior exposure to the virus, and who were subject to a relative genetic bottleneck, the virus was to prove much more problematic.

Smallpox was introduced into the New World by a Spanish sailor most likely in or around 1507, though rather dubious accounts of the time tend to point the finger at African slaves as the culprits.[41] By 1520, smallpox had been transported beyond the Caribbean islands and entered the continental Americas. Just over a quarter century after Columbus first set sail to the New World, a pandemic was raging throughout the Americas. As we have seen with the Romans, the organization and technologies that facilitated the transportation and urban sophistication of the natives rendered them particularly susceptible to the spread of disease. Evidence for this vulnerability can be seen by the fact that disease efficiently decimated virtually all native cultures on two continents, ranging from the extreme northeastern provinces of Canada to the tip of Patagonia, all within a few generations. By its conclusion, a lethal combination of smallpox and other Eurasian diseases likely claimed the lives of as many as 90–95 percent of all New World natives.[42] The physiological collapse of individual smallpox victims mirrored governmental and societal collapses, much as was reported by Thucydides during the Plague of Athens. This rendered the few survivors susceptible to conquest by the likes of Hernando Cortes and Francisco Pizarro. Rapid depopulation might in part have explained the open spaces witnessed by de Soto and other early European explorers and settlers; they were unknowingly witnessing infrastructures that had been created and maintained for centuries but which had quickly fallen into disrepair after their stewards were killed by a stew of infectious microbes introduced in the early years of the 16th century.

The irony is that while the devastation of smallpox was wreaking havoc in "the New World," the Old World was embarking on the first of a series of scientific revolutions that would eventually eradicate the disease altogether. Sadly, these achievements were not to be put into widespread practice until after the microbe-based genocide had taken its toll on the first peoples of the Americas.

Variolation

The first intentional and successful intervention in the long war against smallpox was recorded more than a thousand years ago. A scholar of East Asian history, Joseph Needham of Cambridge University, attributes the first attempt to prevent smallpox to a basic tenet of Taoist medical philosophy.[43] According to Chinese tradition, the medical community had, since at least the year 1000 CE, adopted a practice of "nasal insufflation" to prevent smallpox. The idea behind this procedure was to isolate scabs from individuals who had suffered relatively mild cases of smallpox. This material was dried and refined into powder that was blown into the nose of healthy children. Over time, this practice became a ritual to mark a milestone of a child's life (probably marking five years after birth). These children might display some or all of the symptoms of a mild form of the disease, but the ancient Chinese recognized they would be spared the severity of extreme scarring and death that might accompany an infection later in life.

The geographic proximity of the abutting Turkish civilization, assisted by the ease of transport afforded by the Silk Road, eventually allowed them to learn of the practice of nasal insufflation.[44] It appears that the details of the procedure had been carefully safeguarded for as long as a half millennium by passing along the knowledge in an oral, but not written, form. In a tragic case of poor timing, nasal insufflation was first introduced into the Eastern parts of Europe at a time roughly coincident with the beginning of the Columbian voyages to the New World, which originated at the other end of the continent. Consequently, the practice arrived too late to prevent the tragedy conveyed by Spanish soldiers and sailors that would ravage the native population of the Western Hemisphere. Over the next two hundred years, the Ottomans increasingly experimented and refined the practice, preferring a subcutaneous introduction of the infectious material (jabbed just under the skin) rather than up the nose. This practice came to be known as variolation.

Despite the heavy toll smallpox continued to impart upon western and central Europe, leading minds were not particularly inclined to embrace what seemed to be a highly unhygienic practice. Indeed, the venerated Royal Society of London was inundated with reports from multiple sources of the Chinese and Ottoman practices by 1700 but chose not to act upon this information. Nonetheless, individual acts of bravery and foresight allowed a handful of quite remarkable personalities to convey the life-saving procedures that would save the lives of thousands.

Prominent among these early advocates was the remarkable Lady Mary Wortley Montagu, wife of the British ambassador to the Ottoman Empire and a talented figure, who excelled as a writer and poet.[45] Lady Montagu had lost a brother to smallpox and herself had suffered severe scarring from an infection with the pox in 1715. During her travels throughout the Ottoman Empire, she learned of variolation, a technique preferred by the Ottomans in which smallpox material was introduced into a scratch in the skin. While variolation would cause an infection, the symptoms were generally less severe and conferred immunity thereafter. As a demonstration of her belief in the practice, Lady Montagu in 1718 volunteered her four-year-old son, Edward, for inoculation by an experienced and elderly Greek woman.[46] Apparently, she strong-armed the embassy surgeon, Dr. Charles Maitland, to witness and document the procedure. The doctor reluctantly agreed and watched with considerable discomfort as the old woman introduced the dried scabs into the child's arm with a rusty and dull needle. Dr. Maitland then utilized a more pristine lancet to do the same to Edward's other arm. Over the following few days, Dr. Maitland remained discomforted, as he had been sworn to secrecy. As it happens, Lady Montagu had elected not to inform her husband, Ambassador Edward Wortley-Montagu, of the risky procedure that had been conducted upon his only male heir until at least a week had passed and the fear of danger to the child had expired.

Upon returning home, Lady Montagu broadly advocated for variolation and, as a person of considerable prominence, gained the attention of her friend Caroline of Anspach, the Princess of Wales and future queen to George II. Amidst a particularly obnoxious London epidemic in 1721, Lady Montagu demanded that Maitland inoculate her daughter, Mary, who was four years old at the time. Back home in Britain, Maitland initially resisted this request, since variolation was regarded to be an "eastern" or "Asian" practice, which could sully his reputation.[47] Ever subservient to Lady Montagu, Maitland eventually agreed to do so but only if the procedure was witnessed by prominent members of the Royal College of Physicians. At least one of the witnesses, Dr. James Keith, was so impressed that he had Dr. Maitland variolate his only remaining son (all others had died from smallpox).[48]

Within weeks, the news of variolation spread through the London medical community and among the gentry. Soon thereafter, the Princess of Wales (who possessed an intellect for science and a strong propensity for advocacy every bit as strong as Lady Montagu's) demanded that an experiment be conducted on prisoners held at London's Newgate prison. Three men and three women prisoners were subjected to inoculation and observation. To verify the protective effect, one of the women, a nineteen-year-old by the name of Elizabeth Harrison, was compelled to care for patients in the town of Hertford, where a particularly aggressive outbreak of smallpox was burning through the region. Elizabeth was in close contact with at least two patients, including sleeping every night for six weeks in the bed of a ten-year-old infected boy. Elizabeth remained healthy, and she was later released from bondage for her service. The following weeks bustled with experimental activity, including a somewhat unsuccessful and widely publicized attempt at the Chinese practice of nasal insufflation. In the end, Maitland's approach (actually, the elderly Greek woman's approach) of subcutaneous delivery of smallpox residue became accepted practice.

The use of smallpox variolation was initially adopted by the wealthy and educated population of London, largely based on the advocacy and prominence of Lady Montagu and the Princess of Wales. As we will soon see, this outcome sits in stark and ironic contrast to a modern wrinkle in which wealthy and educated individuals tend to resist vaccination. As with any new medical procedure, there were many bumps along the road, as evidenced by high-profile deaths and improper technique. The conventional (for early-19th-century Europe) but inaccurate understanding of the immune system suggested that deep punctures would confer a more lasting immunity. However, this deeper form simply increased the degree of discomfort and lowered the efficacy of variolation, and the elderly Greek woman's technique (credited to Dr. Maitland) ultimately regained favor.

Meanwhile, on the other side of the Atlantic, variolation was advocated by a personality and intellect comparable to that of Lady Montagu. Cotton Mather is best remembered today as a 17th-century paragon of intolerance. One of his earliest publications, the 1689 treatise *Memorable Providences Relating to Witchcrafts and Possessions*, detailed the possession by evil spirits of every child—save one—born to Bostonian mason John Goodwin.[49] The literal evildoer was identified as the family's neighbor and housekeeper, an Irish Roman Catholic indentured servant by the name of Ann Glover. The housekeeper was accused of possessing Mr. Goodwin's eldest child after the child accused her of stealing the linens. The evidence included the unfortunate statement that the servant's husband had claimed she was a witch just prior to his death. Mather continued that the accused old woman (described in Mathers's writings thereafter as "the Hag") responded to the child's accusation of stealing her laundry with an outburst of vulgar language (though the record suggests Ann did not speak English and the child did not speak Irish). Nonetheless, the trauma triggered a series of convulsive episodes in the girl. More troublingly and in a seeming chain reaction caused by the supposed hex, some of Mr. Goodwin's other children began to act out in the coming days. Despite the clergy's attempts to exorcise the demons by reciting biblical passages, the possessions seemed a form of particularly dark witchcraft, as it caused the children not to hear the sermons or other parental requests. Upon questioning, Mrs. Glover apparently was tricked into admitting she was an atheist (i.e., a Roman Catholic) and that she prayed to a set of figurines (idols of Catholic saints). As pious Puritans, Mather and the prosecutors recognized the figurines as powerful conveyors of witchcraft and acted swiftly to contain the danger to the community. During her trial, one pious neighbor recounted that Mrs. Glover routinely came down their chimney and put a hex on his wife. Laying to rest the worries of the God-fearing families of early Boston, the witch was hanged on November 16, 1688, and her story was memorialized by Cotton Mather as a means to identify future witches. Ann's story became the prototype for a series of events that would occur four years later and just down the road from Boston in what would become known as the Salem witch trials. Ultimately, history would recognize Ann as the first Catholic martyr in New England, and her memory is preserved by a plaque at the site of her Boston home.

What this story fails to convey is that Cotton Mather was actually one of the most progressive and iconoclastic thinkers of his time. Evidence for this arose from a conversation regarding variolation that Mather had with his slave, Onesimus, who was gifted to Mather by his congregation.[50] The reality that Mather owned a slave doesn't seem progressive to modern readers (in his diary, he records he did not seek out this gift but later refers to his good fortune to be given a slave as a "mighty Smile of Heaven"). Mather, despite being a slaveholder, at least had some regard for Onesimus, which distinguished him as relatively progressive for this time and place. As it happened with many of the unfortunate victims of slavery, it is difficult to trace the lineage of Onesimus, who was named so by Mathers based on a slave mentioned in the Bible.[51] A common attribution cited by some academic sources suggests that the man was a "Guaramante," which has been referenced by some to indicate a person from the Akan or Twi people of the Gold Coast (now Ghana) in West Africa.[52] Such an origin for Onesimus would be consistent with the fact that most slaves sent to North America were native to a swath of coastal lands ranging from west Africa or west central Africa. However, the practice of variolation at the time was largely restricted to the extreme eastern and southern regions of the African continent, as well as the lands north of the Sahara that were occupied by the Ottoman Turks (who, as we have seen, were relatively early adopters of variolation).[53] Other sources suggest that the Guaramante people represent a tribe in what is now southern Libya.[54] This alternative origin would be geographically closer to Ottoman lands but raises the question of how Onesimus had had the misfortune to become enslaved and shipped from the Gold Coast.

Regardless of these geographical considerations, it is clear that Onesimus conveyed the advantages afforded by variolation. In 1714, Mather wrote a letter to the Philosophical Transactions of the Royal Society (of London), to which he was the first American colonist to be elected:

> *I had from a servant of my own an account of its being practised in Africa. Enquiring of my Negro man, Onesimus, who is a pretty*

> *intelligent fellow, whether he had ever had the smallpox, he answered, both yes and no; and then told me that he had undergone an operation, which had given him something of the smallpox and would forever preserve him from it; adding that it was often used among the Guramantese and whoever had the courage to use it was forever free of the fear of contagion. He described the operation to me, and showed me in his arm the scar which it had left upon him.*[55]

Mather further maintained that Onesimus instructed him as to the best means of conducting the procedure, which again demonstrates the exceptional intellect of the slave. Mather sought out and spoke with other African slaves, who similarly advocated the advantages of inoculation. In the early spring of 1721, Mather inspired a period of experimentation with variolation that was triggered by an obnoxious outbreak of smallpox in Boston. Mather invited the town's physicians to attempt the new technique, but, despite his renown in old and New England, Mather was soundly rebuffed. He continued advocating for variolation to Boston's leading families and haranguing local physicians as a group and individually. Eventually one Bostonian doctor, Zabdiel Boylston, relented and inoculated his six-year-old son and two slaves.[56] By that time, Boylston had already gained a reputation as a bit of a maverick. In 1710, he was the first American-trained physician to perform surgery (removing a gallbllader stone). In 1718, Boylston became the first surgeon to successfully remove a breast tumor. Word spread of the audacious experiment with inoculation and became the talk and consternation of Bostonian society. Within a few days, Boylston felt pressured enough to announce the success of the procedure in a June 1721 edition of the *Boston Gazette.*[57]

Almost immediately, letters and cries of denunciation were directed at both Boylston and Mather. The primary accusation was that the procedure would propagate smallpox. Quite different arguments questioned the morality of intervening in the providence of God.[58] Unbeknownst to Boylston, Mather had already obtained reports from his Royal Society colleagues about the successes being achieved in London with variolation (recall the efforts by Lady Montagu). In an attempt to assuage nerves, Boylston also announced the English findings in a later edition of the *Boston Gazette.* By then, the local furor over variolation grew to the point where a grenade was thrown though Mather's bedroom window. The bomb failed to explode but contained a note with the words:[59]

> *COTTON MATHER, you Dog, Dam You: I'll inoculate you with this, with a Pox to you.*

In an act of defiance, Mather and Boylston conducted a large study of at least 280 persons (presumably volunteers) in which only six ended up dying from smallpox. This survival rate was a considerable improvement upon the non-variolated population. A preponderance of objective data eventually quieted the riotous Boston medical community and society.[60] Because of the increasingly obvious advantages conveyed by variolation, the practice was adopted, again among the most learned members of the community, and spread thereafter throughout New England and the remaining colonies. Likewise, variolation gained acceptance throughout the rest of the 18th century in much of Europe. However, this practice was doomed to obsolescence, as the closing years of the century would witness a breakthrough with unprecedented abilities to eradicate not just smallpox but many of the deadliest plagues that had ever afflicted mankind.

2

Vaccination & Eradication

Just over a century after Columbus initiated his voyages to the New World and unwittingly triggered genocidal waves of disease, an emerging playwright by the name of William Shakespeare penned a tragedy about young love in the far-off land of Verona, Italy. In the first act of scene three, Mercutio utters his dying words as a curse upon the families of Romeo and Juliet, whom he correctly blames for his death. "*A plague o' both houses!*" Since its composition in 1596 and for reasons not entirely clear, this line has been rendered by many into "*A pox on both your houses.*" These inaccurate reproductions may reflect the stigma associated with smallpox. A POX UPON BOTH PARTIES was the headline for an editorial lamenting the Republican and Democratic Parties for nominating unworthy candidates in the 2016 election.[1]

Exactly two centuries after Shakespeare composed these lines, and fifty miles as the crow flies from his beloved Stratford-upon-Avon birthplace, an Englishman by the name of Edward Jenner was unknowingly writing the first lines in a drama that would ultimately end a real pox and for all houses. Perhaps because our society prefers simple solutions and triumphant heroes (and because a founding myth often needs to be neat and tidy), much acclaim has been placed upon the individual of Edward Jenner and his work to discover a vaccine for smallpox. The heavily propagated story goes something like this:

Edward Jenner was a bird lover, whose first major contribution to science was *Observations on the Natural History of the Cuckoo*, a letter he composed while practicing as a physician in the quaint parish community of Berkeley in Gloucestershire, England, midway between Bristol and Gloucester on the Little Avon River.[2] As a fourteen-year-old student, Jenner started his training in medicine in 1763 under the mentorship of Dr. Daniel Ludlow. Among his other studies, Jenner was instructed in the art of smallpox variolation.[3] Seven years later, Jenner apprenticed in surgery and anatomy at St. George's Hospital in London and then returned to Gloucestershire. The key moment in our tale arises during a conversation between Jenner and a local milkmaid, who informed the physician that ladies of that profession were rarely, if ever, afflicted with smallpox. This revelation provided an epiphany moment for Jenner, who deduced that the milkmaids were prone to cowpox, a skin infection that caused minor lesions somewhat akin to smallpox. Jenner then connected the dots (or the spots, if you will) and presumed that cowpox would protect the general public from smallpox.

As a practitioner of conventional variolation, Jenner postulated that the transfer of material from a cow's or milkmaid's cowpox lesions might confer protection against smallpox. This idea was successfully tested on May 14, 1796 when Edward Jenner inoculated (a word chosen to distinguish the transfer procedure of cowpox rather than variolation with smallpox) eight-year-old James Phipps, the son of a poor laborer who tended Jenner's garden.[4] The source of the infection was a pustule on the hand of a milkmaid by the name of Sarah Nelms, who in turn had been infected by a cow named Blossom. James Phipps developed a low-grade fever but was otherwise healthy. A few days later, Jenner intentionally infected James Phipps with smallpox, one presumes *via* variolation rather than by a means that could have conveyed a more lethal outcome. The normal inflammatory signs of variolation (swelling and fever) were absent, and the boy remained healthy without any signs of localized inflammation or infection—a sign that the original inoculation had protected the child from smallpox.

Jenner then performed similar inoculations and infection schema with a total of two dozen people and published his research in an 1801 report to the Royal Society of London.[5] The Royal Society remained cautious at first, likely based on their remembrance of the opposition targeted at Lady Montagu and Cotton Mather. Soon, however, they embraced Jenner's approach. To honor his landmark achievement, the medical community adopted the term *vaccine*, which is based upon the Latin term for cow, to honor Jenner (and Blossom) for their contributions to the discovery of immunity in general and the smallpox vaccine in particular. Sadly, the end of the story for Blossom was not as favorable, as evidenced by the fact that her hide is prominently displayed at St. George's Hospital in London. (A pelt was donated by Jenner's family more than fifty years after the seminal event, but much speculation suggests the hide is not Blossom but faux fur.)

As tends to be the norm in science, Jenner certainly deserves credit for advancing and widening the impact of a pivotal discovery. Nonetheless, serious doubts surround the question of whether Jenner truly was subject to an Archimedes-like epiphany moment in 1796 or if he ever had the oft-described conversation with the milkmaid. As we will see, it seems more plausible that Edward Jenner's work built upon millennia of experience with prior observation and was highly influenced, if not entirely motivated, by a chance meeting he had at the beginning of his medical training. At the extreme, Jenner might not have innovated the smallpox vaccine at all but simply took credit for the work of others. Before returning to Jenner, we will need to take a minor detour to gain some understanding of why the over-attribution to Jenner is so common.

Conceptions

On September 16, 2011, President Barack Obama signed into law the America Invents Act (AIA), which represented the largest overhaul of the American patent system since the middle of the previous century.[6] The major change of this legislation was a pivot from a system known as first-to-invent to one known as first-to-file. Up to this time, the claim for exclusivity for an invention was granted to an inventor so long as they could convince the patent examiner they were the first to conceive of the idea. This restriction led to many interesting examples, such as the notorious "bar napkin," which did not refer to the legal bar but rather to a case in which the inventors came up with an idea while sipping cocktails. Despite having quaffed a few rounds, the inventors in this case retained the presence of mind to draw up their invention on the back of a bar napkin, and each then signed and dated the napkin. The thin sheets of paper provided incontrovertible evidence in a later trial to demonstrate the inventors had come up with their idea before their adversaries, whose ideas were documented with a date that was later than the napkin. In a single swoop, the bill signed by President Obama overrode such events and eliminated an entire field of patent law known as opposition. The first-to-file provision meant that the first entity to file a patent, regardless of who first conceived the idea, would thereafter be considered the inventor (even if another conceived the idea first).

How does this relate to Jenner and the discovery of the smallpox vaccine? The attribution of the smallpox vaccine to Edward Jenner is consistent with the implementation of the AIA. Although he might not have been the first, even in southwest England, to conceive of the smallpox vaccine, Jenner was the first to put his invention into widespread use. Almost at the same moment that President Obama was signing AIA into American law, an article penned by Robert Jesty of Hampshire, England, and Gareth Williams of the University of Bristol, appeared in the scientific literature.[7] This report amplified an argument that Edward Jenner was not the inventor of vaccines. Going back to our patent analogy, their argument was that the bar napkin had been written and signed by others. Jenner, by analogy, was the first to file and has mostly retained credit for discovering the smallpox vaccine. This article continued a centuries-long debate within the scientific community as to the paternity of vaccines. We will thus attempt to piece together the parts in an objective chronological overview.

The earliest mention of what we now know to be smallpox immunity was deduced and recorded not by Jenner but at least as far back as ancient Greece (two hundred years before Archimedes), by the father of scientific history himself.[8] Thucydides was a military commander, philosopher, and historian, who recounted his experiences during the Plague of Athens in the 5th century B.C.E. In his descriptions of the plague (which might indeed have been or included smallpox), Thucydides conveys the first recorded notation that individuals who had survived one round of infection could care for others without danger of becoming re-infected. The physiological basis for this protection was not known, and little was done to either gain understanding of the phenomenon or utilize the idea to protect the population.

A few short years after Lady Montagu successfully advocated for the widespread use of variolation among the gentry of England, an entrepreneurial family of Suffolk physicians capitalized upon the opportunities in conveying the technology to the wealthy. The family patriarch, Robert Sutton, optimized the technique for variolation, whose quality and reproducibility had in the past been rather spotty.[9] The Suttonian technique entailed a process involving specialized pre-treatment and post-treatment activities, which included rigid control over diet, exercise, and, to be polite, purgatory events.[10] As part of the process, it was essential to verify that the treatment had taken hold by assessing the degree of overall fever and inflammation (redness, swelling, and pus) at the site of local infection (including large pox near the site of variolation). This refinement became a family business overseen by Robert and his six sons.

Not only were these improvements fundamental to increasing the volume of patients that could be variolated, but they also gave rise to an enterprising business opportunity. You may recall that a newly variolated person, harboring an infectious form of smallpox, necessarily must be quarantined from the rest of the community to avoid spreading the disease. Thus, the Sutton family opened a series of "variolation houses," which provided not only the medical service of variolation and subsequent care but also a location for quarantine and all the medical care, food, and libations needed for a month-long stay. Soon these package deals were popularized as a sort of spa and vacation opportunity. Indeed, the practice grew so lucrative that they began to franchise both the "Suttonian technique" and "variolation houses" to other physicians throughout the country. One entrepreneurial franchisee was John Fewster, who was based in Thornbury, a village in Gloucestershire.[11] Fewster apparently oversaw a lucrative business utilizing a Suttonian variolation house located midway between the towns of Thornbury and Berkeley.

The year 1763 was a landmark for smallpox. In addition to a natural outbreak in the mother country, English soldiers in the far-off North American colonies commenced a horrific new era of biological warfare by attempting the eradication of local Shawnee and Delaware Indians. These tribes were engaged in Pontiac's war against Britain and had barricaded an English force within Fort Pitt (modern-day Pittsburgh). The British captain, Simeon Ecuyer, had received orders from his superiors, Colonel Henry Bouquet and General Jeffrey Amherst (the latter memorialized by a city and college in Massachusetts), to break the impasse by spreading disease to the besiegers. In a 1763 letter, Amherst wrote to Bouquet, "Could it not be contrived to send smallpox among these disaffected tribes of Indians? We must on this occasion use every strategem in our power to reduce them."[12] During a lull contrived to facilitate negotiations, Delaware tribe emissaries were given a gift of blankets and a handkerchief as a sign of English intent. This was indeed a quite accurate sign of British motive, as the cloths had been purposefully infected with smallpox and were intended to exterminate the surrounding tribes. These actions did ultimately break the siege. Amherst was recalled to London to be rebuked, not for the barbarous use of initiating biological warfare but rather because his lax oversight had allowed the tribes to rise against the British forces in the first place.

Back in Gloucestershire, 1763 witnessed a perplexing mystery that would redeem England to some degree and counteract just a tad of the negative karma arising from Amherst's fatal judgement. The enterprising John Fewster was attempting to variolate two brothers but was perplexed when they showed no signs of fever or infection. Thinking he had not performed the procedure correctly, Fewster subjected them to two additional variolation procedures, but none of the variolation procedures triggered redness, swelling or fever.[13, 14] Upon questioning, the brothers responded that while neither had been previously infected with smallpox (which would be one explanation for why variolation had failed to trigger a response), both brothers had been infected earlier in life with cowpox.

Still perplexed, Fewster reportedly returned home that evening and prepared for an evening of food, drinks, and discussion with colleagues at the Ship Inn near Alveston. The engagement was part of a monthly meeting of a group of physicians known as the Convivo-Medical Society. The party consisted of Fewster, Joseph Wallis, Daniel Ludlow, and Ludlow's fourteen-year-old apprentice, a young man by the name of Edward Jenner.[15] The Convivo-Medical Society meetings were partly professional, and, still perplexed, Fewster used the occasion to discuss the experience from earlier in the day with the two brothers, who'd failed to respond to variolation. The observation linking a failed attempt at variolation with cowpox apparently intrigued young Jenner, and through a remarkable coincidence, both would recollect the puzzling outcome at virtually the same time and place more than three decades later.

Meanwhile eleven years passed and waves of smallpox came and went. In the summer of 1774, a key breakthrough came from an utterly unlikely source: a dairy farmer in Yetminter, in the southern England county of Dorset, approximately sixty-five miles south of Fewster's hometown of Thornbury.[16] The farmer, Mr. Benjamin Jesty, became alarmed at reports of smallpox in the region. Since time immemorial, dairy maids had been known for their complexion, as evidenced by sayings like "as smooth as a milkmaid's skin." To modern ears, milky skin is synonymous with an unblemished countenance, and advertisers have used this analogy to peddle many high-priced, milk-based cosmetics. Rather than referring to the properties of dairy products, a historical derivation of milky skin refers to the unblemished skin of milkmaids. Epitomes of beauty based on smooth skin were advanced by a variety of 16th-century Dutch artists, most famously Johannes Vermeer (such as in his 1658 oil on canvas portrait *The Milkmaid*). The concept of milkmaid as object of desire was also famously personified by Pilt Carin Ersdotter, an early-19th-century Swedish dairymaid. Pilt was so widely desired that she drew crowds of men to her stall in the central square market of Stortorget in central Stockholm.[17] Her beauty was such that the commoner was propositioned by the Crown Prince and was later employed simply to sit in the parlors of the rich and famous. The maids featured in popular Dutch art and Swedish parlors were widely known for flawless skin that was unblemished, which meant that the skin was not covered in smallpox scars. A legend developed and spread implying that milkmaids were somehow exempt from infection with the pox and its disfiguring outcomes.

Only they weren't. While milkmaids did not usually suffer disfigurement or perish from smallpox (known today as variola major or minor), they were frequently infected with a relatively minor and more localized skin infection of the hand or lower arms, caused by a related pathogen known as cowpox (or vaccinia). Although unsightly during peak infection, cowpox spreads less and leaves little scarring, compared to smallpox.[18] It was understood by dairy workers that cowpox readily spread from the udders of an infected cow to the maid's hands and arms during the daily milking process. This usually occurred within months after a milkmaid had initiated her vocation, and the disease did not return thereafter.

Benjamin Jesty, who was undoubtedly aware of this long-standing reputation, became intrigued by the fact that one of his dairymaids, named Anne Notley, seemed resistant to smallpox.[19, 20] Anne was the caregiver for a family stricken with smallpox and had repeatedly been in contact with the sick but did not ever become ill herself. In talking with Anne, Jesty further realized that another of his employees, a milkmaid by the name of Mary Reade, had likewise nursed sick family members but did not show any signs of smallpox.

With the return of smallpox in 1774, Jesty made the conscious decision to ignore variolation, which still held the prospect of extended isolation and severe, if not fatal, side effects. Instead, the farmer asked his friends and relatives if any cows in the region were currently displaying symptoms of cowpox.[21, 22] None of Jesty's own cows were infected at that time, so he walked from neighbor to neighbor, eventually locating a herd grazing miles away in Chetnole. This herd was experiencing an active infection with cowpox, and Jesty resolved to walk with his wife, Elizabeth, and two young sons, Robert (age two) and Benjamin (age three), to the herd and carry out an audacious act that would forever change humanity (an infant daughter, Elizabeth, was presumably deemed too young for such a daring experiment). Referring to the almost five-mile distance that Benjamin Jesty and his family had to travel, Patrick Pead wrote for the *Lancet*, "What happened in this meadow in 1774 was not the result of a simple farmer's fleeting daydream—the act required inspiration, a firm resolve, and physical effort."[23]

What did happen was that Jesty, who had no prior medical training, borrowed one of his wife's stocking needles and scraped some material from the udder of an infected cow.[24] He intentionally infected his wife, just below the elbow, using a procedure remarkably similar to variolation. He then repeated the procedure with his two sons. Although there is no record as such, we can probably assume Jesty himself had been variolated and that he recalled enough detail from that procedure to replicate it on his family.

Despite an ongoing smallpox epidemic in the area, none of the treated family members became infected. In the days after the procedure, Mrs. Jesty did suffer from inflammation, presumably arising from a procedure performed with dirty knitting needles surrounded by the muck of a Dorset cowshed. We know this because Jesty, who was not particularly interested in recognition for his bold experiment, was eventually required to take Elizabeth to the local doctor. The doctor chastised Jesty for his irresponsible actions, and Elizabeth recovered.

In a series of events reminiscent of the initial public reaction to Cotton Mather's experiments with variolation, word leaked out about Jesty's experiment. The local community became outraged, though perhaps not for the reasons one might think. The crowd "feared their metamorphosis into horned beasts."[25] Jesty and his family were excoriated by his neighbors, who threw rocks (and various types of organic matter) at them when the Jesty family dared to show themselves outdoors. Perhaps for fear of the beasts lurking within, the family was forced to move far away, eventually isolating themselves on the Isle of Purbeck. The choice of this location was symbolically consistent with the fact that Purbeck, which is technically a peninsula and not an island, had housed the English king Edward the Martyr after his overthrow as king of the English centuries before.[26]

Rather than becoming martyrs or transforming into minotaur-like creatures, Elizabeth and the children lived long and healthy lives, avoiding a later smallpox outbreak that devastated the region. We know this because Benjamin Jesty intentionally exposed his sons to smallpox a few years later. In 1789, fifteen years after being infected with cowpox, Jesty had his sons variolated.[27,28] Unlike a naïve patient, who would react with a localized and less virulent form of smallpox, neither of the two inoculated boys reacted to variolation—no local inflammation, no fever, and no signs of disease. Lacking a scientific background, Jesty might not have realized that the lack of infection by smallpox variolation meant that his earlier vaccination with cowpox had proven effective and was still working.

While Jesty was pioneering cowpox vaccination in Britain, a parallel story was unfolding in the northern German state of Schleswig-Holstein, just outside Kiel. In this case, our protagonist was a peripatetic school tutor by the name of Peter Plett.[29] While teaching the children of a Mr. Weisse in the parish of Schonweide in mid-1790, Plett had what must have been quite an engaging conversation with Mr. Weisse's mother-in-law. A former milkmaid herself, she relayed to Plett that she had been infected as a child with cowpox but was never bothered by smallpox. Like the story told to

Jesty by his milkmaids, Mrs. Weisse related that she had remained healthy despite frequently serving as a caregiver to many others in the community suffering from the disease. Plett was intrigued and discussed this fact with other dairy maids, who confirmed similar experiences.

Ever the itinerant teacher, he found himself a year later near Holstein (famous for their eponymous cattle). In the employ of a Mr. Martini in Hasleberg, Plett was tutoring two of Martini's daughters, whose names were Hedwig and Margaret. By coincidence, a smallpox outbreak was raging in the area, and Martini's daughters were gripped with terror that the disease would seek them out and disfigure their faces. The two girls learned of the connection between cowpox and smallpox (presumably based on a conversation with Plett). On their own initiative, the sisters went to a local dairy barn, sought out cows with active pox lesions, and rubbed the material over themselves. Despite the rather nauseating practice of covering their faces and hands in the revolting goop, the girls were disappointed to experience no eruption of cowpox pustules. Recalling his own experience with variolation, Plett eventually gave in to their pleading and performed a variolation-type technique on the girls with material oozing from cowpox sores. Specifically, the tutor introduced the infectious material just under the skin using his penknife to infect the hands of the two anxious sisters (as well as their younger brother, Charles). This action was taken without the foreknowledge of their parents. There does not seem to be any record of how the parents reacted to what certainly must have been a conspicuous infection between the thumb and index finger. What is clear is that Plett did not remain employed by Mr. Martini much longer. In 1794, the two coincidentally met in Kiel, where Martini conveyed his joy that while a particularly bad strain of smallpox had indeed infected the village, "the children he [Plett] had vaccinated had escaped all infection." Working now in Kiel, Plett tried to convey his experiences to the faculty of the local university. However, the learned scholars brushed off any suggestions from a mere tutor, and variolation with smallpox remained the dominant countermeasure to manage the epidemic.

Two years after the Kiel encounter between Plett and the grateful Martini, John Fewster, still practicing variolation, was dispatched to see the eldest son of a local and well-connected villager, John Player, who was suffering from early signs of smallpox. Here the details become rather fuzzy, primarily because they were related decades after the event. One version relates that a local polymath, John Player, had learned that Fewster successfully immunized multiple children in the village of Tockington with cowpox exudate and sought Fewster to vaccinate his son as a means to treat an active smallpox infection. Another version of the story suggests that Fewster had an epiphany during a conversation with Player while treating his son that night, which motivated Fewster to experiment with cowpox in place of variolation.[30, 31] This second version states that Fewster then immunized three children. The motivations, machinations, and recollection of what occurred that night remain foggy but suggest Fewster took some time away from his "day-job" variolating to experiment during that spring of 1796. Fewster himself remains mostly silent on the subject, especially since the controversy was not aired until more than a decade after his death in 1824.

In the same year Fewster was revisiting the concept of cowpox immunization, we return to the central figure of our story, Edward Jenner. Jenner's contribution to vaccines began in 1796, forty-one years after Fewster's puzzlement with the smallpox-resistant brothers (who had previously had cowpox), five years after Jesty, and coincident with Fewster's renewed experimentation with cowpox. Despite not being the first to conceive of the idea, Jenner is rightly recognized as a key contributor to the adoption of vaccination by the medical community. His work is recognized largely because he was the first to invoke proper scientific method to evaluate multiple subjects and analyze the results prior to disseminating his work to the medical and scientific communities. In particular, Jenner's work was cited and reviewed by the world's most prestigious medical association of the time, the Royal Society of London.[32]

As word of the discovery spread, Jenner was placed in positions of escalating responsibility to expand the adoption of smallpox vaccination. Given the increasing burden on his time, Jenner petitioned Parliament for a grant to pay for his services. While discussing how best to reward Jenner for services to his country, Parliament sought assistance from George Pearson, a prominent physician and advocate for vaccination.[33] Pearson relayed rumors arising from southeast England that Jenner had not discovered the vaccine and that this was indeed the act of a previously unknown farmer from Dorset. These stories were largely set aside and Jenner received the bulk of credit from Parliament, along with grants of more than thirty thousand pounds sterling over the next ten years.

Back in Dorset, Benjamin Jesty came to learn of the accolades being showered upon Jenner. Remembering the abuse that had triggered his family to flee their home in Yetminter, he was understandably loathe about bragging of his own experiments years before. Nonetheless, Jesty was encouraged to do so by his local rector, Andrew Bell, with whom he had privately shared his earlier experiment.[34,35] Jesty refused, so Bell began advocating for Jesty, approaching members of Parliament on his behalf. Ultimately, Bell and Pearson connected, and the two convened an 1805 meeting of the Vaccine Pock Institute, a government organization established to advance vaccination.

Jesty was asked to appear before the Vaccine Pock Institute to describe his actions from thirty-one years before. Approaching seventy years of age, the farmer related his experiences and brought along his eldest son, Robert, as evidence.[36] Robert volunteered to be variolated, which demonstrated continued immunity (though in practice, the fact that Robert had been variolated in 1789 overrode a conclusive demonstration that the 1774 vaccination had been efficacious). Based on this testimony, the committee agreed that Jesty had contributed to the discovery and should share in the recognition. While there is no evidence that Jesty was ever compensated financially (short of fifteen guineas to cover his expenses for his trip to London), the Jennerian Society recognized Jesty with a strongly worded statement affirming Jesty's seminal contributions. They commissioned a portrait of the great man, who, ever the country farmer, posed in decades-old clothing and was loathe to sit still for the portraitist.[37] Nonetheless, the recognition of Jesty's contributions might have influenced the fact that when Edward Jenner was later elected to the prestigious Royal Society of London, it cited his groundbreaking work: the penning of *The Natural History of the Cuckoo*.

Within a few years, the practice of vaccination with vaccinia (also known as cowpox virus) overtook variolation, due to its improved safety. By 1840, variolation was banned first in the United Kingdom and eventually across the globe. With the establishment of a safe and efficacious method for preventing smallpox, people developed an ambitious goal in the early years of the 19th century to fully eliminate the disease—an odyssey that would ultimately prevail after more than a century and a half of arduous work.

Spreading Like a Virus

A few clicks on your favorite search engine reveal that the term *spreading like a virus* is replacing the age-old saying *spreading like wildfire*. The author's bias as a biologist supposes that the term *virus* usurped conflagrations as the object of choice based on recognition that respiratory diseases like the cold or influenza can disseminate quite rapidly (within days or weeks) throughout a community (a subject to which we will soon return). However, it must be conceded that millennials and other web citizens are undoubtedly more aware of, and actively concerned about, the digital form of viruses, whose propagation at the speed of light does admittedly outpace the velocity of a cough or sneeze.

In the case of the first vaccination, this new vernacular is both literally and figuratively accurate. As we have seen, Jenner's discovery of using the cowpox virus for vaccination aroused considerable attention within the United Kingdom in the months following its revelation to the scientific and medical communities. The Jennerian approach with cowpox, unlike the experience of Benjamin Jesty years before, was fervently embraced as it provided an alternative to variolation. Recall that variolation utilized a highly pathogenic and readily transmissible infectious agent that was potentially life-threatening and highly inconvenient, as it required a monthlong quarantine and considerable support from caregivers. In contrast, vaccination in the truest sense of the word—with vaccinia virus—was generally safe. In the unlikely event that the infection was transmitted to another, this would simply serve to increase the coverage of individuals protected from smallpox. Within weeks, a government commission called the British Sick and Wounded Board mandated the vaccination of the Royal Navy and Army.[38] Likewise, the civilian population of the United Kingdom largely embraced vaccination, and variolation was rendered obsolete, then barred within a few decades.

Jenner himself devoted the rest of his life to spreading the word and teaching the practice of smallpox vaccination. To quote the man himself, "It now becomes too manifest to admit of controversy, that the annihilation of the Small Pox, the most dreadful scourge of the human species, must be the final result of this practice."[39] Jenner was handsomely rewarded by Parliament for his devotion, though the pillars of British government might not have been as enthusiastic had they realized some of the strategic implications of Jenner's humanitarian views. Specifically, Jenner believed that the world should benefit from vaccination, and he readily communicated the details and shipped samples of cowpox (vaccinia) around the world.

An early adopter of vaccination was a paragon of the Age of Enlightenment, Napoleon Bonaparte.[40] Ever a consummate strategic thinker, Bonaparte realized that expansion beyond France's historical borders necessarily involved intimate contact with an ever-increasing number of people and that war tended to increase the breadth and severity of infectious diseases. Thus, the emperor ordered that his troops be vaccinated for smallpox in order to provide a strategic advantage over adversaries who were less progressive-minded.

Napoleon retained an extraordinary fondness for Edward Jenner, who was allowed to travel at will in Europe. This amity flew in the face of the Continental System, which otherwise forbade all British commerce or travel. Likewise, Napoleon allowed Jenner to repatriate prominent British citizens who were isolated on the Continent in 1803 after a resumption in hostilities that concluded a yearlong peace as negotiated in the Treaty of Amiens. Once hostilities resumed, many prominent British citizens were trapped in Napoleonic Europe, including the Earl of Yarmouth, members of Parliament, and prominent academics.[41,42] In requesting their liberation, Jenner wrote to compatriots at the National Institute of France: "Gentlemen, Pardon my obtruding myself on you at this juncture. The Sciences are never at War. Permit me, then as a public body with whom I am connected to solicit the exertion of your interest in the liberation of Lord Yarmouth."[43] Comparable communications succeeded in gaining the freedom of others, as evidenced by Napoleon's statement, "*Ah, Jenner, je ne puis rien refuser a Jenner*" (Ah, Jenner, I can refuse him nothing).[44] Ironically, and perhaps as a result of active repulsion against Napoleon and his dictatorship, the French army actively rejected smallpox vaccination in the years following Waterloo, a decision that would prove deadly to many thousands of French troops and contribute to their 1870 loss in the Franco-Prussian War. (The decision was made worse by the fact that Bismarck had compelled his troops to be vaccinated during the same campaign.)[45]

On the other side of the Atlantic, a once and future English adversary, the United States, was also quick to embrace vaccination, driven by another

Enlightenment thinker. Thomas Jefferson advocated vaccination. While serving as president in 1806, he wrote to Jenner, "Medicine has never before produced any single improvement of such utility . . . You have erased from the calendar of human afflictions one of its greatest . . . Mankind can never forget that you have lived."[46] The manner by which Jefferson learned of Jenner's achievement is worth relating, both to demonstrate the brilliance of the man and, consistent with other events that have created considerable ambiguity, actions that raise serious questions about his moral choices.

Benjamin Waterhouse was a key thought leader in the early United States. The native Rhode Islander trained at the Universities of Edinburgh and Leiden (where he received his medical degree in the spring of 1780). Waterhouse was a degreed physician, rather than a physician who obtained his qualifications by shadowing a practicing doctor.[47] As such, his credentials were a relative rarity upon his return to the new United States in 1782 (which, despite the defeat of Lord Cornwallis at Yorktown the year before, was technically still at war). Unsurprisingly, Waterhouse was highly sought after and accepted a position as a founding faculty member of the new medical training program at Harvard University in September 1782. This made Harvard only the third medical school in the fledgling United States, after the founding of comparable institutions at Benjamin Franklin's University of Pennsylvania (1765) and Columbia University (1767).

By the time of Waterhouse's return, the United States had already gained a reputation as a progressive nation, at least in terms of smallpox prevention. In the first year of the Revolution, a disastrous attempt by the thirteen colonies to invade Quebec failed, largely as a result of a rampant smallpox outbreak that decimated General Richard Montgomery's troops, causing a loss of life that was equaled by a crushing blow to morale on the eve of battle.[48] Most of the British defenders, in contrast, had been variolated, lengthening the mismatch that inevitably capped Gates's failure to capture Montreal. Learning from this outcome, Washington ordered all troops in the Continental Army to be variolated. This was a bold step, given the high death rate at the time; approximately 12 percent would succumb as a result of the procedure, even under ideal conditions, and the conditions in Washington's camp were hardly optimal.

The country to which Waterhouse had returned was just taking shape, and the control of the federal government was still largely in limbo. Disarray reigned supreme as Waterhouse helped establish what would eventually evolve into the Harvard School of Medicine. Waterhouse, a native of Newport, Rhode Island, had close ties to Europe, having attended college at the University of Edinburgh and then training as a physician at Leiden University. Waterhouse proved a most capable physician and scientist and his career vaulted forward, eventually reaching the high honor of election to the American Academy of Arts and Sciences. His election to this nascent but extraordinary society was preceded by a handful of important Americans such as John Adams, his cousin Samuel, and John Hancock.[49] To put this in perspective, the new academy elected a handful of new members per year, and Waterhouse's class of 1795 came a year after the election of James Madison.

Through the years, Waterhouse kept abreast of the latest medical breakthroughs. Immediately upon learning of Jenner's success, Waterhouse penned an article to the *Columbian Sentinel*, a Boston periodical, on March 12, 1799, extolling the virtues of this new procedure.[50] In parallel, he reached out to John Haygarth, a prominent British physician specializing in smallpox, to request a sample inoculum to be tested in the United States.[51] By July of the following year, the requested samples arrived, and Whitehouse immediately vaccinated his wife and children.[52] Having a monopoly on the procedure (with the only samples and experience in conducting immunization in the Western Hemisphere), Waterhouse capitalized upon this fact. Drawing upon his prominence and many powerful connections throughout the young nation, Waterhouse functioned as the sole distributor of the new vaccine, requiring that the clinicians administering the vaccine share a fraction of their proceeds with Waterhouse.[53]

This entrepreneurial approach elicited a firestorm of media protest, which was enflamed further when Waterhouse, in the long summer months of 1800, petitioned his old housemate from his medical school days. That man happened to be John Adams, the sitting president of the United States.[54, 55] While Adams had been ambassador to the Dutch Republic, Waterhouse had been conducting his medical training in Leiden and residing in Adams's ambassadorial home. The two had established an enduring correspondence, which now allowed Waterhouse to urge Adams to advocate for Jenner's new vaccination strategy, presumably for the public good but which would enrich Waterhouse personally. Adams, perhaps aware of the controversy, acknowledged receipt of Waterhouse's request and forwarded it to the aforementioned American Academy of Arts and Sciences, which sat on it for the remainder of the second president's term.[56]

Regardless of his reasons for tabling the smallpox vaccine, Adams was deeply engrossed in his own problems, which centered upon a contentious reelection campaign against his former best friend, newly sworn enemy, and sitting vice president, Thomas Jefferson. This would serve to be the first truly contested reelection in American history, embroiling the two in the most caustic and derisive election campaign in American history (even measured by today's low standards), with accusations that Jefferson fathered a child with one of his slaves (accurate) and that Jefferson and Adams were pawns in the employ of the French and British governments, respectively (both inaccurate). As the mudslinging degenerated into a virtual civil war among Federalists in the Electoral College, the Democrat Jefferson prevailed after an equally contentious vote count that rang with recriminations of "defective electoral ballots" from the Georgia delegation (surprisingly reminiscent of the "hanging chads" controversy two hundred years later).

Hedging his bets, Waterhouse bided his time. Amidst the turmoil, he reached out to Jefferson at his Monticello home with a letter dated December 1, 1800, in which he appealed to the vice president's patriotism to advocate for broad use of the smallpox vaccine in the United States.[57] Waterhouse included in this letter a pamphlet (which may be viewed cynically as an advertisement) for Jefferson to ponder. As Jefferson read the pamphlet on December 24, he found himself more inclined towards action and replied enthusiastically on Christmas Day.

Jefferson connected Waterhouse to a network of physicians throughout Virginia and initiated experimentation at Monticello, as well as within the confines of the White House itself. In an act that would blanch even the most strident modern spin doctors, Jefferson oversaw the intentional infection of slaves with the most deadly disease known to man. The first "volunteer" was a slave by the name of Ursula Granger, the daughter of two of Jefferson's Monticello slaves and a member of Jefferson's kitchen staff at the Executive Mansion.[58] Unfortunately, the May 29, 1801 experiment failed to elicit an immune response, but Jefferson was undeterred. Using fresh material, Jefferson successfully oversaw the immunization of two other slaves while at Monticello that summer, including that of his butler

(Burwell Colbert) and blacksmith (Joseph Fossett).[59] Unlike the experience at the White House months earlier, both men displayed the characteristic swelling and redness, suggesting the vaccine was effective. The exudates from Colbert and Fossett were harvested for use in additional experiments.[60] Over the remainder of the summer, almost fifty Monticello slaves served as guinea pigs. As an even more audacious act, many of these slaves were subsequently challenged with native smallpox to ensure that the vaccine had indeed conferred protection (which, thankfully, it had). Based on these positive findings, Jefferson then proceeded to vaccinate two dozen of his own family members. By the early autumn of 1801, he began distributing smallpox vaccine throughout his native state of Virginia. Later that fall, the president personally advocated the expanded use of smallpox vaccine to other states, soliciting the support of other prominent physicians such as Dr. John Redman Coxe of Philadelphia and Dr. Edward Gantt of Washington. As a sign of friendship, Gantt vaccinated the chief of the Miami Indians, Little Turtle (or Mihšihkinaahkwa), during his 1802 visit to Washington.[61]

Within the United States, the popularity of smallpox vaccination expanded rapidly. By 1809, vaccination was mandatory in Massachusetts. School systems began to require children to be vaccinated for smallpox prior to enrollment; by the end of the century, the disease was largely restricted to the urban poor. Despite these encouraging outcomes, periodic outbreaks continued, with no fewer than eight epidemics recorded in Philadelphia between 1800 and 1850, with six in Boston and three in Baltimore. The situation was far worse for the native Amerindians, who (unlike Little Turtle) remained largely unvaccinated. Terrifying outbreaks in 1801–02 and 1836–40 exterminated entire tribes and depopulated much of the United States west of the Mississippi River. An outbreak in Brooklyn, New York, among the city's homeless in 1893–94 sparked particular public outrage. Individuals trained to give vaccinations flourished as these "vaccinators" were rewarded with thirty cents for each person vaccinated (though this same lure also caused less scrupulous vaccinators to treat the same patients multiple times).[62, 63]

Despite high-powered backing for vaccines and massive outbreaks in 1865, 1871, and 1881, much of the American population still remained unvaccinated. While the army that faced the British in the War of 1812 was compelled to be vaccinated against smallpox, its counterparts on both sides of the Civil War a half century later were not.[64] In part, individual and collective decisions were influenced by a vigorous anti-vaccine campaign that blossomed in the 19th century and nearly impeded the use of smallpox vaccine. A fringe element emerged from the same Luddite-inspired movement responsible for the rock that had crashed into Cotton Mather's house a century and a half before. Similar to Jesty's experience, an 1802 cartoon by the British caricaturist James Gillray conveyed an ironic depiction of commoners vaccinated with cowpox taking on bovine features.[65] The irony was lost on many, who believed this an accurate depiction.

Financed by a wealthy and outspoken English businessman and demagogue named William Tebb, an international campaign preached the evils of vaccination against smallpox on both sides of the Atlantic.[66] Tebb was a self-admitted British radical who faithfully followed the teachings of characters such as John Bright (best known for battling the British Corn Laws) and the utopian Robert Owen. Today Tebb might be labeled a liberal libertarian. He embraced many causes such as abolition of slavery, vivisection, and the premature burial (for fear many still-living beings were interred). However, Tebb invested the bulk of his passions, as well as a fortune made in the chemical industry, into opposing the use of vaccines. His actions were based on the practice of mystical and occultist philosophies known as Theosophy, and he advocated the rejection of vaccines as a right of the individual over government. A high-profile figure with a penchant for making unjustified claims, Tebb claimed that smallpox vaccinations were responsible for more than 48,000 deaths in England and Wales alone (utterly fallacious figures that are still cited today).[67] Such outrageous and unfounded claims incited a scared and confused public, spurring more than 80,000 protestors to march in Leicester in 1885, burning effigies of Edward Jenner and carrying child-sized coffins.[68]

In the United States, the wealthy firebrand and industrialist John Pitcairn Jr., founder of PPG Industries, took over the mantle from Tebb and fought against vaccination. Pitcairn was an avid follower of the Swedish mystic Emanuel Swedenborg (who reported frequent conversations with angels and demons as evidence that he was the embodiment of the biblical resurrection). Among the beliefs advocated by Pitcairn was homeopathy, and the wealthy industrialist deployed his wealth and power to become a trusted voice of opposition to vaccination. The motivation for his venomous opposition to vaccination was based on a minor blood infection experienced by his son, Raymond, in 1885.[69] The blood infection coincided with, and might have resulted from, poor technique during vaccination. However, the Swedenborgian view of the body maintained that contamination (infection) left a scar on the soul and was therefore morally reprehensible. Indeed, the only concoctions felt appropriate to be used therapeutically would be those in which a perceived toxin had been minimized or eliminated because of dilution. For such reasons, homeopaths such as Pitcairn were early converts to the Swedenborgian church.

It is worth expanding a bit on the story behind Pitcairn and the Swedenborgian church. While PPG was originally founded in Pittsburgh, Pitcairn spent much of his life across the state in Philadelphia. The industrialist donated most of the funds used to erect the stunning Bryn Athyn Cathedral just outside Philadelphia. The cathedral's emphasis on individualism and good deeds gave rise to the legendary Helen Keller and John Chapman (better known as Johnny Appleseed). The church itself was a mainstay of the Philadelphia spiritual community.[70]

A Little Lie

Another pious, turn-of-the-century anti-vaccinator was Lora Cornelia Little. Born in a log cabin in what is now Waterville, Minnesota, Lora married an engineer who specialized in bridge design. They moved to the East Coast and began a family, giving birth to a son, Kenneth Marion Little, in January 1889.[71] Six years later, Kenneth was about to start his primary education in Yonkers, New York, and was legally required to be immunized against smallpox. The cycle of shots was completed in September 1895. In the months thereafter, Kenneth was exposed to many other children, even more so when the family moved from Yonkers to Philadelphia.[72] During that tumultuous time, Kenneth was beset by multiple insults and injuries, including persistent ear and throat infections; later he developed measles and diphtheria (a disease to which we will return in a later chapter). According to the certificate of death issued by the Children's Hospital of Philadelphia, a weakened Kenneth finally succumbed to diphtheria in April 1896. His grieving mother sought for an explanation for this tragedy and concluded that "the artificial pollution of his blood had weakened his constitution and left him at the mercy of the subsequent infections." Reflecting the Swedenborgian views of the time and place (Philadelphia in the time of John Pitcairn, Jr.), she concluded that the artificial pollution that claimed the life of her son must have been the result of the smallpox inoculation. Thus began a lifelong and tireless crusade to preach the evils of the smallpox vaccine. As this cause largely involved children, the most vulnerable population, Little's campaign appealed to the fears of a public that was particularly sensitive to perceived overreaching of the state during the early years of the 20th century.

Among the vehicles used by Little to spread the word about the anti-vaccinator campaign was the *Truth Teller*, a subscription-based periodical printed in Battle Creek, Michigan.[73] The paper was the re-branding of the *Peril*, a weekly paper touting the advantages of homeopathy, managed by a rather sketchy group known as the American Medical Liberty League (formerly the National League for Medical Freedom). This group had previously advocated against pure food and drug laws that had been put in place following multiple revelations of corrupted foods and snake oil drugs as depicted in Upton Sinclair's best seller, *The Jungle*.[74] The *Peril* primarily served as an advertising opportunity for peddlers of cures for unreal maladies such as "disappointment in love," "floating kidney," and "locomotor ataxia."[75] The publication and its products, however, were soon put out of business by the federal passage of the Pure Food and Drug Act, which was signed into law by President Theodore Roosevelt in 1906.[76]

Having lost the campaign to fight against pure food and drugs, the group redirected itself against vaccines. The *Truth Teller* peddled outrageous stories of the evils of vaccination as described by an expose in the July 29, 1922 *Journal of the American Medical Association*. In an investigation conducted by the American Medical Association's Bureau of Investigation, the prestigious medical organization refuted the supposed "authentic facts" conveyed in *The Truth Teller*.[77] This expose also revealed advertisements featured in the *Truth Teller* were strikingly similar to those from the *Peril* except that they focused on alternatives to vaccines that were not subject to regulation by the 1906 Federal Pure Food and Drug Act. Sadly, the very real suffering and animus felt by the followers of Lora Little were exploited by the *Truth Teller* and Lora Little herself, who sold a series of additional publications on the perils of vaccines to a gullible public.

There was some validity to the argument that vaccination could be harmful. As we will see in a future chapter, the state of the art in the manufacturing of vaccines was wholly unregulated in turn-of-the-century America. Like any industry, vaccine distributors covered a wide range, from large factories to smaller companies run out of a kitchen or garage. Many were not much more evolved than the situation at the beginning of the 19th century, when Jefferson collected the pus from his vaccinated slaves to propagate vaccination in subsequent waves of "volunteers." An unregulated practice meant that some vaccines were indeed unsafe, and the specific sources supplying the vaccines used on Kenneth Little and Raymond Pitcairn have long since been lost to the ages. Such concerns about the manufacturing and quality of vaccines would remain for years to come, until they were finally resolved by a calamity that would claim the lives of children in St. Louis, Missouri, as we will sson see.

Despite the fact that neither Pitcairn nor Lora Little had advanced educations or a scientific or medical background, Pitcairn's resources, combined with Little's sincere passion and a tiny grain of scientific support, gave rise to the anti-vaccinator cause. These resources incubated the growth of an opposition that grew vigorously throughout the mid-1800s. The evidence that this opposition was quite effective is demonstrated by rising rates of smallpox infection in the United States in the second half of the 19th century (the disease had almost been eradicated but made a deadly resurgence).[78] Taking the position that the common man should decide whether to be vaccinated, the movement was underwritten by a handful of wealthy demagogues until it was finally stopped in its tracks by the 1905 Supreme Court decision *Jacobson v. Massachusetts*, which ruled that compulsory vaccination was in the best interest of the state.[79]

Although largely eliminated in the United States by the early years of the 20th century, smallpox made a frightening return eighteen months after the end of the Second World War. On February 24, 1947, Eugene Le Bar and his wife boarded a bus in Mexico City. They were returning to Maine after a successful Central American vacation. During the bus ride, Eugene started feeling poorly, but when they arrived in New York, he felt good enough to check into a Midtown hotel and go shopping on Fifth Avenue. By May 5, Eugene had developed a rash and was admitted to Bellevue Hospital in Manhattan with a diagnosis of chicken pox.[80] Mr. Le Bar was then transferred to another hospital, Willard Parker, but within days, the patient was dead. A few days after his passing, a worker at Willard Parker was diagnosed with a similar rash, which was recognized as smallpox. Hoping to avoid panic, the officials conducting Le Bar's autopsy had listed the cause of death as hemorrhagic bronchitis, but a series of people who had encountered Eugene, either directly or otherwise, were diagnosed with smallpox. Officials scrambled to identify people within the city of New York with whom he had had direct or indirect contact (at the hotel, department store, or hospitals), as well as fellow bus riders from Mexico City. In an attempt to vaccinate individuals potentially exposed to smallpox, newspapers and radio stations communicated the message to an anxious public. Unsurprisingly, anxiety grew, and the health department was unprepared to procure or distribute millions of vaccine doses. In what is still regarded as a masterful demonstration of a high-functioning public health system, the city was able to deploy its limited supply of vaccines in a manner that protected the most vulnerable population while calming the public and avoiding panic.[81] This admirable performance during the 1947 outbreak would also prove to be the final chapter in America's long history with smallpox.

Utter Eradication

Over time, the use of the smallpox vaccine would gain acceptance worldwide. Given the geography and ties to Jenner's England, many European countries were the quickest to embrace the practice of vaccination, both for their homeland populations and beyond. The Spanish king Charles IV was an early adopter. His motivation was based in part on the experiences of his daughter, Maria Teresa, the Infanta of Spain, who died from smallpox on November 2, 1794 at the age of five. Upon learning of Jenner's discovery, Charles directed the Spanish government to initiate a successful domestic immunization campaign. Given its many ties to the New World, Spain then launched the three-year Balmis Expedition, so named for its leader, physician Dr. Francisco Javier de Balmis, who was dispatched in 1803 to vaccinate millions in Spain's American colonies in South and Central America, as well as Spanish possessions in the Philippines and China.[82]

Most of the developed world likewise utilized a gauntlet of immunization and monitoring to eliminate or limit what had been periodic and devastating epidemics of smallpox. However, the disease lingered in many of the less affluent and developed countries. A century and a half after the Balmis Expedition, the 19th World Health Assembly of the World Health Organization (WHO) launched a measure intended to eradicate smallpox forever. The rationale behind this daunting challenge was formulated in the first meeting of the organization in 1948 (coincidentally just after that the Le Bars were planning their Mexico trip that would later cause the last case of smallpox in the United States). The goal of the program was unprecedented: To intentionally and forever eliminate an organism from the face of the Earth.

The ability to eradicate the disease altogether was a fortuitous opportunity based upon a peculiar aspect of smallpox biology. While we will discuss the biology of viruses in a subsequent chapter, a key feature that rendered smallpox susceptible to eradication is the fact that smallpox can only be propagated in humans. This feature likely represents a dramatic development within the past fifty thousand years or so. Initially a rodent virus, the smallpox virus gained the ability to grow in humans but lost the ability to do so in rodents.[83]

The selectively of smallpox for humans meant that there was no disease reservoir, other than people, where the virus could hide. Stated another way, stop smallpox in humans and the virus goes away—forever. In the midst of the Cold War, the United Nations initiated a campaign that would unite humanity with the goal of eliminating the deadly scourge of smallpox. However, it didn't necessarily start that way.

Both the United States and the Soviet Union actively sought to eliminate smallpox in their own countries and then around the world. As part of each rival's outreach to court nonaligned nations, they launched their own programs in part to inspire goodwill from potential strategic partners. The United States launched smallpox eradication efforts in South America and western Africa, both of which were paying considerable dividends in terms of public health successes, goodwill, and propaganda value.[84] Not to be outdone, the Soviet Union vocally lobbied the United Nations to direct its public health entity, the World Health Organization, to eliminate smallpox across the globe. The WHO, ever a highly political and rather insecure organization, was averse to an endeavor whose risk level was comparable to its high-profile nature. Indeed, the WHO had previously failed in their attempts to similarly eliminate regional outbreaks of yellow fever and malaria, and the prospect of a global program caused eye and stomach rolling in many in the WHO leadership.

Nonetheless, the pressure to support the program intensified as the United States joined its Soviet rivals in advocating for a global program to eliminate smallpox.[85] As American pressure increased, the Brazilian director-general and infectious disease specialist Marcelino Candau called upon an American to lead the program. In an open message to the United States surgeon general, William H. Stewart, in early 1966, Candau demanded, "I want an American to run the program because when it goes down, when it fails, I want it to be seen that there is an American there and that the U.S. is really responsible for this dreadful thing that you have launched the World Health Organization into, and the person I want is Henderson."[86]

The Henderson that Candau insisted upon was Donald A. Henderson, known to his friends as "D.A." Henderson was a physician, who admitted that his first "real job" began after he completed his medical residency in the mid-1950s and started working for the Epidemic Intelligence Service of the Communicable Disease Center (later known as the Centers for Disease Control and Prevention, or CDC).[87, 88] While at the CDC, D.A. became enamored with the field of epidemiology, which is tasked with public health in general and sleuthing out the cause and spread of disease in particular. He earned a master of public health from Johns Hopkins in 1960 before returning from Baltimore to CDC's headquarters in Atlanta, Georgia. There he worked with another Johns Hopkins alumnus and expert epidemiologist, Alexander D. Langmuir. Together, Langmuir and Henderson developed a plan that built upon ongoing experiences in South America. Specifically, the CDC had been working in partnership with the Pan American Health Organization (PAHO) since 1950 to eliminate smallpox in Latin American nations. By 1960, the last reported cases of smallpox were recorded in Bolivia, Chile, French Guiana, Guyana, Paraguay, Peru, Suriname, Uruguay, and Venezuela (within six years, Colombia, Ecuador, and Argentina would join the list of smallpox-free nations). By 1965, Henderson and Langmuir also set their sights on eliminating smallpox in west and central Africa, and they gained support from the U.S. Agency for International Development (USAID) to do so. Within the year, Candau demanded Henderson be given responsibility (and blame) for the impending WHO program.

Building upon the demonstrated successes of the Balmis Expedition, CDC, and PAHO programs, Henderson and the WHO targeted South America, which was a slightly ironic choice given the devastation wrought upon the natives of that continent in the post-Columbian era. The WHO initiated the Intensified Programme in 1967, focusing largely on Brazil, which documented the vast majority of smallpox cases in the Western Hemisphere.[89] By 1971, the number of reported cases of smallpox in South America had shrunk from a peak of ten thousand cases per year in 1962 (even with PAHO participation) to nineteen in 1971, all of which were in Brazil. No natural infection with smallpox has been observed since that landmark year.

In parallel with the success in South America, the smallpox eradication program spread like wildfire (or a virus) through Africa and Asia. Indonesia was the first Asian country targeted, starting in June 1968, and the last case was recorded on January 23, 1972.[90] South Asia followed, as did west, south, and east Africa, more or less in that order. This geographic trend also reflected the fact that under Henderson, an effort had already begun in Africa under the auspices of USAID. The eradication of smallpox by USAID was not only a laudable humanitarian goal but it also helped to serve the need to enhance friendship and cooperation, with the goal of wooing a restive region away from pro-Soviet regimes in Congo or Angola. Henderson's leadership in eradicating smallpox while at the helm of the WHO program was made possible by a grand coalition of public health, scientific, and political muscle, which also facilitated the training of armies of vaccinators. Rather than immunizing every person on the planet, vaccinators were efficiently dispatched to hot spots of infection. These individuals practiced the strategy of "ring vaccination" by vaccinating and evaluating individuals near an infected patient.

Immunizing the peoples of east Africa proved to be the most problematic, due to endemic warfare among the Ethiopians, Somalis, and various tribal antipathies, and because a nomadic culture, which runs contrary to the idea of ring immunization, has predominated for millennia in this corner of the world. As a result of a Herculean effort involving dedicated volunteers and unprecedented domestic and international cooperation (every nation of the world), the last case of naturally acquired smallpox was documented in Somalia on October 26, 1977, almost ten years to the day after Henderson was appointed the head of an enterprise that was, according to the Director-General of WHO, doomed to fail.

The final victim, Ali Maow Maalin, ironically worked as a cook and laborer in a Somali hospital in the southern city of Merca, southwest of the Somali capital of Mogadishu.[91] The cook was presumed to have been vaccinated (there were conflicting reports at the time) but later admitted, in a 2006 interview with the *Boston Globe*, that he avoided being immunized because "it looked like the shot hurt."[92] Maalin had been exposed to the disease because of a series of unfortunate events. Two months prior, a handful of children in a nomadic tribe began to display the classic symptoms of smallpox: fever, malaise, headache, back pain, and the later eruption of flat, red spots on the face, hands, and arms, later followed by the trunk and legs. The wanderers made their way to an encampment at Kurtunawarey, where the symptoms were reported to a WHO team about sixty miles away in Merca. On October 12 the family was cautiously escorted by immunized WHO officials from Kurtunawarey to an isolation camp that was carefully controlled to avoid spreading the disease. In addition, WHO officials isolated the encampment and anyone who had come into contact with the nomads. Maalin, supposedly protected, was dispatched in a Land Cruiser to gather the sickened from the isolation camp for transfer to the Merca hospital, a trip that lasted as little as five minutes.[93] Sadly, one of the children, a six-year-old girl by the name of Habiba Nur Ali, perished from the disease and was the last to die from naturally occurring smallpox. The WHO officials carefully isolated and monitored all susceptible people who had come into contact with the infected children. However, these same officials remained unaware that Maalin had not been immunized.

The conclusion of the disease unfolded like a Hollywood thriller. On October 22, Ali Maalin displayed a fever and headache and was treated for malaria, which is endemic in the region. The symptoms persisted, and the doctors treated Ali for chicken pox before releasing him from the hospital. Over the next few days, it became obvious that smallpox, not chicken pox, was the culprit responsible for his symptoms. Maalin, fearing isolation, fled. Officials began a manhunt that ranged more than seven hundred square miles, both to isolate Ali and immunize all individuals who came into contact with him. More than ninety people had direct contact with Ali during his flight from the hospital, but through masterful detective work, each was tracked down and quarantined. The runaway himself was sought using, among other things, a reward of two hundred Somali shillings (about forty US dollars), an amount of remuneration sufficient to entice one of his coworkers to turn him in. For the next six weeks, officials isolated 161 people who might have come into contact with Maalin. Collectively, they held their breath. More than 54,000 people in the immediate locale were vaccinated in a final push to ensure that the disease would spread no further. Ali Maalin survived with minimal complications. Over the following four months, WHO officials and volunteers initiated a door-to-door campaign to seek any other cases of smallpox, an action that was gradually expanded to include most of Somalia. On April 17, 1978, regional officials in Nairobi, Kenya, sent a telegram to the Geneva headquarters of WHO saying, "Search complete. No cases discovered. Ali Maow Maalin is the world's last known smallpox case."[94]

Although Maalin had been a source of considerable consternation for the WHO staff in Somalia, his later good works redeemed him in the latter months of 1977. Again reminiscent of a Hollywood film, he remained in Merca and volunteered for a later WHO immunization campaign to eradicate poliomyelitis, avowing that, "I'm the last smallpox case in the world. I want to help ensure my country will not be last in stopping polio." Empathizing with the local villagers' fear of immunization and relating his own past fears with smallpox vaccination, Ali Maow Maalin could draw upon his experience and educate his countrymen as to the benefits of polio immunization. His audiences included some of the most nefarious, best-armed and skeptical people in the world, including an ever-changing variety of warlords, terrorists, and militias. Yet because of the efforts of Ali and other WHO officials and volunteers, Somalia was declared free of polio in 2007. In the spring of 2013, polio returned to Somalia (from an unknown source), and Ali immediately rose in response. In a region with seemingly endless conflict, most recently propagated by the Al Shabaab militant Islamist organization, the public health system was close to shattering. Nonetheless, the last smallpox survivor returned to his personal war against polio. However, the regional killers, large and small, were not limited to terrorists, smallpox, or even polio. Rather, on a hot summer day in mid-July 2013, Ali Maow Maalin began again to display signs of fatigue, headaches and fever. Dedicated to his polio vaccination campaign, he delayed seeking medical health but ended up in the hospital, where he was again, and this time accurately, diagnosed with malaria. Unfortunately, the 58-year-old husband and father of three was unable to overcome a different and still elusive microbial adversary inside him, and he died of malaria on July 22, 2013.[95]

A taut Hollywood plotline might have ended with Ali Maow Maalin, but the story took an unexpected turn in 1978. On August 11 of that same year, four months after the triumphant telegram declaring Ali Maow Maalin to be the last case of smallpox, an Englishwoman by the name of Janet Parker complained of a migraine headache and intense spasms of muscle pain.[96, 97, 98] Janet was a forty-year-old professional photographer, who had formerly served as a forensic photographer for the West Midlands police and was now employed by the University of Birmingham. Brushing off the symptoms as the early signs of a simple cold, she continued working until noting a rash, which led her to seek medical attention. Nine days after her first symptoms, Janet was admitted on the late morning of August 20 to the East Birmingham Hospital (now known as Heartlands Hospital) with complaints of an unremitting headache. Her ending up in this particular hospital was fortuitous, as Dr. Thomas Henry Flewett, consultant virologist to the West Midlands NHS Trust hospital, was a leading authority on smallpox diagnosis using electron microscopy. This relatively new technology provided ultrahigh resolutions to detect some of the world's smallest pathogens, such as viruses.[99] Within a few hours, Flewett and his colleague Alasdair Geedes, himself an expert in bacterial infections, were dumbfounded to diagnose Janet Parker with a rampant case of smallpox.[100] Later that evening and in recognition of the lack of adequate containment facilities at the East Birmingham medical facility, Janet was transported by secure ambulance to the Catherine-de-Barnes Isolation Hospital in Solihull, which had been recently abandoned but was quickly revitalized to accept an unexpected entourage.

Janet Parker's admission to the isolation ward occurred amidst a scrum of activity.[101] For one thing, it was essential to identify and isolate all individuals who had encountered the photographer. With her health quickly deteriorating, her ability to communicate was ever more limited, but officials could identify and isolate more than five hundred people, who were quarantined at home or sequestered within the Solihull isolation facility. In parallel, health officials identified and sterilized all of Janet's belongings, both at home and at work. At the same time, public health and law enforcement officials were challenged by the fact that by 11:00 P.M. on August 20, members of the press had been alerted that smallpox could be running rampant in one of England's largest cities. The problem with containment—not just of the virus but of widespread panic as well—was compounded by the fact that a photographer with no obvious exposure to smallpox or other high-risk behaviors had suddenly been diagnosed with one of the most deadly and infectious agents ever known to man, thought to have been eliminated in the United Kingdom years before.

The detective work began in earnest and revealed that while Janet had not knowingly engaged in any behavior, either in her personal or professional life, that might have exposed her to smallpox, by coincidence, the location of her darkroom was immediately above a laboratory belonging to Professor Henry Bedson. Bedson had a history of working with smallpox but was equally known for not meeting the high safety standards needed to contain the pathogen. For example, the WHO informed the investigator that Bedson's application to become a collaborating center had been rejected based on safety concerns that his facilities did not meet their minimal standards.[102] Indeed, the Dangerous Pathogens Advisory Group, an expert committee of the Department of Health, had inspected and rejected Dr. Bedson's laboratory on two separate occasions. Although Bedson had assured the WHO and local authorities that he was winding down his research on smallpox, a subsequent government inquiry (known as the Shooter Report, so named after its lead author) exposed the truth. Not only was the Bedson laboratory using unprecedented amounts of virus for their studies but the laboratory facilities necessary to contain the deadly pathogen lacked airlocks, adequate shower facilities, changing facilities, or specialized clothing. Likewise, the virus was handled and stored in open-air facilities and outside standard biological safety cabinets, which is a cardinal sin for any biological research, and even more so for a legendary killer such as smallpox.[103] Consequently, the widespread use of the deadly pathogens without sufficient precautions or facilities had allowed the virus to enter the ventilation system and infect Janet Parker, who had the misfortune to work immediately above the laboratory.

As authorities pieced together the string of events, a series of tragic episodes ensued. On September 5 (just over two weeks after Janet had been admitted), her 71-year-old father, Fredrick Whitcomb, died while in quarantine at the isolation hospital.[104] The cause of death was presumed to be cardiac arrest. This was never confirmed, however, as an autopsy could not be performed, given the potential that the body might have contained smallpox. The next day, Professor Henry Bedson was found in his garden shed, bleeding to death from a self-induced wound while in quarantine at his home. Despite the potential for disease spread, he was transported to another Birmingham hospital but eventually passed away from his wound. Finally, Janet Parker lost her struggle against smallpox and died on September 11, 1978. A later investigation revealed that while Parker had been vaccinated for the disease years before, time had decreased the level of protection to a point where the amount of smallpox exposure she experienced proved fatal.[105]

The 1978 smallpox outbreak in Birmingham led to an outcry to eliminate research on the virus and extirpate it once and for all. Specifically, all strains of smallpox, including those used for research or the production of vaccines, were destroyed in a systematic campaign. A mere two vials were allowed to survive, both were intended to serve as the seed stock for a future treatment or vaccine if needed. One vial was kept in strict isolation and under armed guard at the Centers for Disease Control and Prevention (CDC) in Atlanta, Georgia.[106] The other vial was in Soviet Russia. As we will see towards the end of our story, this effort to limit smallpox to two vials was largely, though not completely, successful. While one of the greatest killers of humankind has been contained in isolation, like an evil Hollywood villain, the potential for a sequel is hauntingly real.

Now that the extraordinary benefits of vaccination to eradicate diseases such as smallpox and polio have been elaborated, we will turn our attention to a brief discussion of how vaccines work and, following that, the array of antagonists and adversaries to vaccines.

3
Becoming Defensive

In the interest of full disclosure, it is important to warn the reader that much of this chapter is devoted to perhaps one of the least attractive concepts and words in the English language: pus. For as long as people have suffered wounds and infections—in other words, from time immemorial—pus has been a subject of considerable speculation, though understanding of its function, composition, and scientific beauty is comparatively recent.

According to the venerable Oxford English Dictionary (OED), the first written description of pus is cited in the seminal manuscript *Chirurgia Magna*, first published in 1363.[1,2] This opus detailed contemporary knowledge of medical and surgical techniques. It was broadly used throughout Europe from the late 14th through the 17th centuries, having been translated from its original Latin to local languages. In part, the popularity of *Chirurgia Magna* reflected the fact that its author, Guy de Chauliac, was a bit of a celebrity physician, famous among the intelligentsia and courts of most western European monarchies.[3] By the publication of the work, the low-born physician from the Lozere region of south central France who answered to the name Guido rose well beyond his expected station. His own personal story is at least as interesting as the subject of pus.

Guy de Chauliac was born in 1290 to a family of peasants in a small town near Lyon. An only child, Guy demonstrated considerable interest and proficiency in the study of medicine. This fact became known to the local nobility, the Duke of Mercoeur, who underwrote Chauliac's medical training in Toulouse.[4] The promising Chauliac again impressed his mentors and began advanced training in medicine in Montpellier. Later he obtained what would today be considered a fellowship in Bologna under another luminary physician of his day, Bertuccio Lombardo (also known as Nicolo Bertucci).[5] Bertucci's fame as an educator derived in part from his technique of performing dissections on cadavers. The practice of adulterating sacred human flesh had been widely shunned as a Christian blasphemy, but the enlightened thinking of the early Renaissance ushered in many changes. Consistent with the iconoclastic nature of Renaissance thinking, Bertuccio's own tutor at Bologna, Mondino de Liuzzi, had reintroduced dissection of the human body in January 1315, the first time the questionable practice had been openly performed since the adoption of Catholicism by the Roman Empire. Such practices launched a new age in which medicine would evolve from a medieval superstition into an empirical science. At the epicenter of this transition was the University of Bologna, whose reputation was established by faculty such as Liuzzi and Bertucci. Having thrived in the Bologna environment, Guy de Chauliac honed his training further with a stint in Paris before returning to south central France and beginning his practice in Lyon. The promising young mind was utterly unaware that established and brewing international crises would soon propel the humble physician into greatness.

Two centuries before King Henry VIII of England broke with the Vatican, the 14th-century king of France, Philip IV, had his own quarrel with the Holy See.[6] Philip was a notorious spendthrift, supporting the Crusades until the fall of Acre in 1291 and then declaring overt and covert war upon England, ostensibly over lands on the Continent that the French king felt rightfully belonged to him. To finance these excursions, Philip became deeply indebted to many creditors, including the powerful Knights Templar and Jewish financiers. Philip largely dealt with these financial burdens using force.[7] First the French king expelled all Jews (including his creditors) from France in 1306. Later he staged the assassination and outlawing of the Knights Templar on the infamous Friday the thirteenth in October 1307 (which gave rise to the contemporary superstition still associated with that date).

As one means of procuring income, Philip imposed a tax on the Catholic clergy. This action precipitated a violent reaction from Pope Boniface VIII, who was already concerned about supporting a war between two Catholic countries at a time when the Crusades in the Holy Land were critically failing. In response, Boniface issued a papal bull, *Clericis laicos*, which forbade the transfer of papal property to any state (thereby preventing Philip from utilizing church holdings) in addition to threatening to excommunicate those who ignored the pronouncement.[8] After a failed attempt at a diplomatic solution, the newly excommunicated Philip ordered the detention of Boniface and sent an army to Rome in September 1303 with the order to capture the pontiff and force his resignation. The French forces overwhelmed the weak standing army of the Vatican states. Although denied at the time, modern scholarship suggests Boniface was tortured in a failed effort to force his abdication.[9] Perhaps realizing the brutality of their tactics and not wishing to be directly associated with the death of a pope, Philip released the 73-year-old pontiff from French custody. Boniface died within days.

The vacancy was quickly filled by the Italian pope Benedict XI, who, unsurprisingly, was favorably inclined towards the French, whose troops remained bivouacked in Rome. The new pope rescinded the *Clericis laicos*. However, the installation of Benedict conveyed only a short-term solution, as he too would be dead within a year (though of more natural causes). Continuing the Roman occupation, the French king pressured the College of Cardinals to elect a French pontiff, and the Gascon-born Clement V was named pope in 1305. Rather than continue an expensive occupation of Rome to hold sway over the Vatican, the French king simply relocated the Vatican, or at least the head of that institution, to Avignon in southern France. This era became known as the Babylonian Captivity, a period that lasted from 1309 to 1377 and witnessed a succession of seven consecutive French popes.

In 1342, the French pope Clement VI invited the preeminent local surgeon, Guy de Chauliac, to become his personal physician, an appointment that would include three more popes and continue until Guido's own death in 1368.[10] Arguably, the most defining event in the physician's life was not the elevation to this prestigious position but the witnessing and recording of the greatest plague the world has ever experienced. The Black Death devastated Avignon in 1348.[11] No one was immune, including Guido's benefactor, Pope Clement VI, who fell victim in 1352. Indeed, even Guy himself became infected. Though Chauliac survived, he suffered from a chronic "axillary bubo," a painful condition in which the infection is localized to a site, where a simmering infection continues to brew and destroy local tissues while being targeted by the body's immune defenses.[12] In Guy's case, the infection was localized to what we now call a lymph node just below the shoulder near the armpit. Lymph nodes, as we will see, strain the body's fluids to identify and eliminate infectious or cancerous cells. They are normally restricted to a few millimeters (a fraction of an inch), but an infectious agent localized to a lymph node can elicit a symphony of chemical and cell-based defense measures that cause dramatic and painful enlargement of that and other nearby lymph nodes. Indeed, these engorged lymph nodes were known as *bubos* and gave rise to the modern name of the plague.

Ever an observant scientist, Chauliac recorded his observations and distinguished between a bubonic form of the plague (characterized by the swelling and occasional bursting of lymph nodes) and an even more dangerous pneumonic form, which devastated the lungs and was almost invariably fatal. Beyond discriminating between the two forms of the disease, Guido was intrigued with the creamy white pus that exuded (or suppurated) from the burst lymph nodes in plague patients.[13] Chauliac contrasted clear or white secretions against the characteristic scarlet hue of blood or colored secretions (ranging from red to yellow and green). In recording his observations, Guy described the fluid emerging from burst lymph nodes as "pus."

Although Chauliac was the first to utilize the world *pus* in the European medical literature, these secretions were well known to the Athenian medical writer Hippocrates. Hippocrates' fame for many millennia (beyond his oft-repeated oath) rested largely on a bold assertion that most diseases were caused by natural processes rather than as a punishment from the gods or a curse from past or present enemies.[14] These ancient notions had held sway for most of history, and Hippocrates' radical but more correct view is rightfully still celebrated. Nonetheless, many of Hippocrates' underlying beliefs in how the body and disease worked were less accurate. In the 5th century B.C.E., Hippocrates described an exudate arising from battlefield wounds suffered by soldiers during the Peloponnesian Wars.[15] Like his 14th-century counterpart, Hippocrates noted that white pus (one of the four bodily fluids, according to contemporary Greek thinking) related to a positive outcome, whereas a smelly, turbid, and colorful fluid, which he labeled as an "ichor" (leading to the modern term "icky"), generally boded poorly for a wounded soldier.[16]

These views of pus and infection date from the 5th century B.C.E. in Athens. They remained almost unchanged for more than two millennia. According to Queen Victoria's personal physician, Frederick Treves, "Practically all major wounds suppurated. Pus was the most common subject of converse, because it was the most prominent features in the surgeon's work. It was classified according to degrees of vileness."[17]

Evolution & Immunity

By the time Hippocrates and Thucydides were reporting their experiences with battlefield wounds and the Plague of Athens, respectively, the biological system known today as immunity, which includes pus, had been evolving for billions of years. As long as life has existed on planet Earth, competition among organisms has reigned supreme. This rivalry extends well beyond conventional passive interactions, such as two individuals competing for the same food source, and often degrades to active combat among mortal enemies.

Recent evidence reveals that life began as single-celled organisms emerging as far back as four billion years ago (almost immediately, in geological terms, after the planet solidified from its molten beginnings—a suggestion that life is a default, rather than an exception). The emergence of the first free-floating organisms initiated a life-versus-death competition entailing ever-evolving means of attack-and-parry tactics. This interspecies arms race continues to this day. As evidence of the importance of this combative tendency, as many as 10 percent of the genes found in even the oldest known organisms were designated for defense against invaders.[18] The targets of these defense mechanisms included both bacterial and viral attackers. One way to think of this is that the fraction of genes in a cell is roughly proportional to the energy usage of that organism. Consequently, at least a tenth of the resources utilized by our earliest ancestors were devoted to fighting off predators. This is quite an investment of energy, given the many other things that life must accomplish (not the least of which are simple survival and reproduction).

The study of bacterial genetics is extraordinarily complex, and we will restrict our discussion to bacterial structures that will play a role in our story in this chapter and beyond. The primary genetic element in a bacterium is a piece of circular deoxyribonucleic acid (DNA) known as a chromosome. Chromosomes in bacteria remained steadfastly present in virtually every cell in the body, and this trait continued over the eons. In humans, twenty-three pairs of linear (rather than circular) chromosomes are found in virtually every cell in the body and provide the blueprints needed to produce all our molecules and cells.

Over time, an additional type of DNA element evolved. Unlike the large mass of chromosomal DNA, which encodes for thousands of different genes, miniature circles containing only a few genes, a structure known as a "plasmid," began to be produced and passed among different microorganisms through intimate physical interactions.

The genes encoded by plasmids often conveyed the means to deal with a changing environment. Indeed, the discovery of plasmids in the late 1940s arose from efforts to determine why some bacteria could adapt to antibiotics—a new set of wonder drugs developed during the Second World War. These studies revealed that some plasmids encoded for genes that allowed the bacteria to escape the toxic effects of antibiotics such as penicillin. Indeed, penicillin is yet another example of the interspecies arms race; it is produced by certain types of mold in order to combat bacterial interlopers. Humans merely borrowed this trick, which was useful until plasmid-based conjugation events passed along genes that rendered penicillin largely ineffective.

Another interesting and quite old example of a defense mechanism is the toxin-antitoxin system. Briefly, a gene on a plasmid might encode for a protein that is highly toxic unless neutralized by an antidote, known as an antitoxin.[19] Instructions for producing an antitoxin in turn may be embedded on the bacterial chromosome. Thus, when conjugating with a "friend" (which can produce the antitoxin), the toxin will be countermanded by the production of the antitoxin. However, a close embrace with an "enemy" could be lethal if one lacks the instructions to produce the antitoxin. Such tricks are often deployed by bacteria, who, like modern-day street gangs, seek to dominate the neighborhood and eliminate the competition.

An even more elaborate set of defensive measures has evolved to protect bacteria from viruses. As we will soon see, viruses are at least as obnoxious and problematic to bacteria as they are to people—perhaps even more so.[20] One feature that renders some viruses particularly problematic is their ability to insert their genetic material into the host chromosome. Like a monster from science fiction, the instructions to construct viruses can be snuck into a host chromosome and lurk unseen until some signals trigger the virus to reemerge and resume its rampage. To combat this menace, bacteria have evolved extraordinary mechanisms to identify and eliminate viral genetic materials. One set of evolutionary defenses, known as restriction enzymes, can slice the viral DNA into small pieces. As described in my book *A Prescription for Change*, recognition and reproduction of these gene-modifying events gave rise to the modern field of biotechnology, which followed upon the discovery of these enzymes and their potential use to cut up and modify human genes.[21]

Another defense mechanism that has gained considerable notoriety within the past few years is known as *clustered regularly interspaced short palindromic repeats* (or CRISPR), which also evolved to eliminate embedded viruses.[22] As a brief interlude, some bacteria have evolved a system whereby DNA sequences that are unique to viruses are captured and organized into repetitions of DNA that are stored by the bacterium as a means to recall viruses that they (or their ancestors) have encountered. These sequences provide an early-warning mechanism the bacteria can use to identify invading viruses. By utilizing these unique sequences as a guidance strategy for enzymes, known as a DNases (the most famous of which is known as Cas9), the CRISPR system can seek out and obliterate the unique signatures of viral DNA (thereby killing the virus as well). Over generations, the CRISPR-based immunity allows bacteria to recall previous encounters with viruses and to develop a defensive strategy to minimize harm from future attacks. In the hands of contemporary molecular biologists, CRISPR holds promise likewise to identify the unique signature of diseased DNA to correct these defects, thereby leading to opportunities to prevent or reverse established diseases in humans.[23] As such, the technology has recently gained great acclaim (and substantial investment dollars) for start-up biotechnology companies seeking to modify everything from inborn genetic diseases to cancer and/or eliminate certain infectious diseases.[24] A more detailed overview of CRISPR-Cas9 is beyond the more limited scope of this book but it nonetheless suffices to state that this system holds the potential for a future revolution in medicine.

From Eating to Fighting

For a billion years after life arose, the single-cell organisms—bacteria, archaea, and the viruses that infest them—continued their relentless warfare. Excluding viruses, all single-cell organisms derived their nutrition (and energy) from either eating bits of proteins, lipids, and carbohydrate (e.g., detritus from organic life) or synthesizing their own materials using solar (photosynthetic) or chemical (chemosynthetic) power. All the while, life was evolving sundry ways to improve its offensive and defensive capabilities. One example was a modification of an engulfment technique that single-celled organisms used to ingest foodstuffs. Analogous to how the largest forms of life (blue whales) ingest and filter some of the smallest (plankton), single-cell organisms evolved to filter their immediate environment to capture nutrients. Given the efficiency gains from targeting larger prey, this system was improved to capture and kill entire microorganisms (rather than individual molecules). The prey in turn developed ways to defend themselves using specialized techniques such as the aforementioned toxin-antitoxin approach.

Constant warfare among single-cell organisms reigned supreme until 2.7 billion years ago, when a nano-sized but groundbreaking peace treaty was achieved by only two of the countless cells on the planet. At the time, the event went largely unnoticed, but the implications were profound. Like the events of any other day, this one began when one single-cell organism ingested another. Rather than the victor feasting upon the various proteins, fats, and carbohydrates of its victim, the two organisms arrived at a mutually beneficial arrangement. The ingesting organism, which we will call the host cell, gained a specialized function that was provided by keeping the prey alive (rather than killing and digesting it). The cohabiting organism, which we will call a hitchhiker, in return provided ongoing benefit to its host and gained from the encounter not only by not being killed but also by being protected by the host from attack by other pathogens—namely, protection from the harsh outside world, where the next host cell might not be so benign. Indeed, the hitchhiker might have gained the better part of the deal because it could itself proliferate within this new protective cocoon provided by the host.

As this new companionship continued, it was inevitable that there would be a split. Specifically, the host cell eventually needed to divide. For its progeny to survive, the hitchhikers needed to be divided up as well. Thus, both the host and hitchhiker evolved a means such that when cell division occurred, the hitchhikers would be equally dispersed between the two daughter cells. Variations on a theme continued as changes in the local availability of nutrients required the number or function of both the hosts and/or hitchhikers to change. Over time, a system evolved to assure a proper balance of hitchhikers.

The process described above is an example of endosymbiosis, a theory substantiated by the American scientist Lynn Margulis.[25] The evidence for at least one endosymbiotic event resides in each cell in our body and powers all aspects of human life. Indeed, all animals and plants are the direct progeny of that first peace treaty. The combined structure is known as a eukaryotic cell, a name derived from a Greek term meaning 'true nut.' Eukaryotic cells have a nucleus, a specialized structure that evolved as a centralized storehouse for DNA storage. In contrast, the DNA in bacteria and archaea float around the cell interior. These cells are known as prokaryotes (meaning 'before nuclei').

The hitchhikers in our story are more commonly known as mitochondria. Mitochondria are a component of virtually every cell in our body, and the specialty they provided to their host cells was the production of energy. Like their independent ancestors, mitochondria encode for their own DNA and replicate themselves independent of, but in coordination with, the host cell. Mitochondria are thus in constant communication with the larger cell. Sometimes they will increase in number or become degraded to recycle their key building blocks, as fits the continued survival and propagation of the overall cell, in response to changes such as lean times or old age. While all known species of eukaryotes contain mitochondria derived from that initial pairing, a subset of eukaryotic cells underwent additional endosymbiosis events. The most prominent example arose when a eukaryotic cell engulfed a bacterium that had the capacity to generate its own food (or more specifically sugar) in response to sunlight. The resultant cell gained a photosynthetic element known as a chloroplast and was the progenitor of all plants.

These two small accommodations went almost unnoticed within the internecine fray that persisted all around the planet as microorganisms fought to secure a future for themselves and their progeny. As we will see in a future chapter, some cooperative events involved bacteria and eukaryotes. Other benefits were derived by partnering with one's own progeny. For most of evolution, cell division was followed by the two daughter cells going their own way. Eventually, some daughter cells decided to remain attached and work together. As these collections of eukaryotic cells organized and evolved into what we now refer to as multicellular organisms, specialization became possible. Like the well-known transition in human civilization that facilitated specialization of work when some individuals remained in place to farm the land (rather than constantly chase wild game), the idea of remaining attached to the collective could allow some cells to hone certain specialties. Amidst the chaotic need for defense, some of these cells were specialized to defend against outside aggression.

In many cases, the primary defense utilized by our early ancestors simply entailed the creation of a physical barrier such as skin, which prevented unwanted invaders from gaining access to the organism's interior. Over time, other adaptations in the multicellular organism included further specialization resembling the use of something akin to modern-day firearms, which could project protection at a distance to repel or discourage potential dangers. One example can be seen with the hydra, a relatively simple multicellular organism.[26] Complementing a protective outer barrier (i.e., its skin), the hydra evolved a means to produce and secrete a surprisingly complex array of antimicrobial peptides (very small proteins). These defenses comprised some of the first examples of antibiotics, a term that literally translates from its original Greek into 'anti-life' and accurately reflects the goal of killing other organisms. This adaptation was crucial, given the fact that the hydra's environment was swarming with bacterial pathogens. One interesting detail is that the hydra's defense was selective, targeting some bacteria while encouraging interactions with others.[27] This approach is perhaps the most ancient form of a microbiome, the collection of microscopic organisms within the local environment intimately interacts with a macroscopic organism such as you or me. The symbiotic relationship people have with microbes is a relatively new discovery despite increasing evidence the microbiome crucially controls virtually all aspects of life and death and as we will see, will weave intimately throughout the rest of our story.

Chemical Warfare

Further evolution gave rise to more sophisticated defense systems. As multicellular animals became ever more complex, specialization became even more important. Given the hostile nature of the environment, technologies to increase defense continued to improve. A subset of cells called phagocytes, named for a Greek term meaning 'eating cells,' appeared. Their primary responsibility was to ingest potential enemies. Whereas in their ancestors such mechanisms had primarily served to gather nutrients, the new adaptations repurposed the activity to kill trespassers.

Further adaptation amidst an ever-changing environment of comparably adapting enemies compelled more sophisticated needs for attack and defense, and thus even greater specialization. Among the phagocytes, some cells were tasked with more efficient ways to kill intruders, while others were tasked with manufacturing and releasing specialized chemicals that could sense or immobilize potential invaders. Rather than remaining as static cells that simply awaited a chance encounter with potential invaders, many phagocytes gained the ability to actively move and thus patrol the entire host in constant search of potential threats. This newfound mobility also allowed multiple defenders to swarm upon the enemy. These adaptations were greatly assisted by the development of a complex set of chemicals collectively known as cytokines.

The name *cytokine* (a Greek term meaning 'cell movement') accurately captures their function. Cytokines can be thought of as early-warning messengers produced by cells threatened with attack from a foreign pathogen. Much as Paul Revere helped mobilize Minutemen to slow the British march on Lexington and Concord, cytokines alert the host defense network to mobilize and localize the forces (including phagocytes) and direct the attack (producing more cytokines as the assault continues or tapering down as the threat diminishes). As the number and types of threats increased (such as different bacteria, fungi, or viruses), the cytokines themselves began to specialize. The discovery of new cytokines or variants of existing molecules continues today. One example of this further specialization is a subset of cytokines known collectively as interferons.

In the mid-1950s, two British scientists, Alick Isaacs and Jean Lindenmann, were studying how influenza virus infects chicken egg embryos (many flu vaccines are grown in eggs). They noted a peculiar finding.[28, 29] If the scientists killed the virus with heat and used this material to infect the embryos, the eggs resisted subsequent infection with live virus. This was attributable to a host defense molecule produced by the infected cells. This molecule with the ability to interfere with virus infection gave rise to the name "interferon." As often happens in science, other laboratories were reporting similar findings at roughly the same time. For example, independent investigation at the University of Tokyo meant to improve smallpox vaccine development reported the same finding, as did a team led by the American polio vaccine researcher John Enders (whom we will encounter again later in our story).[30, 31] These seminal observations opened the door to investigation, which revealed that there was indeed not only one interferon but a range of different interferons, many of which shared the same ability to interfere with viruses but each of which had their own ways of doing so. Over time, investigators also realized that the immune-activating functions of interferons were not limited to viral infections but likewise alerted the immune system to the presence of cancer and other diseases.[32, 33] Indeed, interferon-based therapeutics are now routinely deployed in the fight against an array of diseases, ranging from infections and cancer to various autoimmune disorders.

About a half billion years ago and deep in the world's oceans, which were and are brimming with microorganisms, an invertebrate, jawless fish developed an even more complex and efficient means to eliminate potential pathogens.[34] By today's standard, jawless fish are quite primitive. One example is the lamprey eel. Yet by the standards of the day these fish were true trendsetters. Five hundred million years ago, these jawless fish developed the ability to not only eliminate unwanted pathogens but also to remember which organisms were friends and which were foes, and to deploy selective ways to rid themselves of the latter. In short, this new system increased the number and efficiency of how their defenses could recognize and eliminate potential disease-bearing microorganisms. Furthermore, the advance meant that these fish could efficiently keep themselves free of infectious disease, while the new system could remember past infections and use this information to mount a faster and more robust defense against future invasions should the pathogen be encountered again.

These chemical-wielding invertebrates evolved into aquatic and, later, terrestrial organisms with a backbone (known as vertebrates). Even later they evolved into mammals (fur-covered vertebrates that feed babies with milk produced by their mother). As they did so, their defense mechanisms became ever more sophisticated, powerful, and complex. In parallel with improvements in their defensive chemical warfare, comparable changes were improving the efficiency and lethality of the cellular component of the evolving immune system, a subject to which we now turn.

Revolutionary Discoveries

Francois Joseph Victor Broussais was born in 1772 to a physician in the Brittany city of Saint-Malo in northwest France.[35] The young Francois learned the basics of medical practice from his father but was drawn to revolution with the storming of the Bastille on July 14, 1789. Inspired by the new revolutionary tricolor flag, Broussais joined the infantry of the French Revolutionary Army, but poor health forced him to temporarily abandon military service. He used this hiatus to finish his medical training in Paris (though he later returned to the army as a practicing surgeon). Ironically, this sojourn by the low-bred Brittan led to a life-changing interaction with a bourgeoisie scion of the establishment.

Marie-Francois Xavier Bichat was also the son of a physician and the mayor of Poncin (a town in east central France near the Swiss border).[36] Though only a year older than Broussais, Bichat had already amassed an impressive academic record. His father had sent him to the top schools in Nantua and Lyons, underwriting both his education and his seemingly constant travel to other parts of the country, in part to stay one step ahead of the revolutionary chaos. Ultimately, Marie-Francois was subsumed by the revolutionary wave and indeed went with it, also serving as a surgeon in the army of the republic near his home in the Alps.

After the Reign of Terror, Bichat moved to Paris as a trainee under the famous surgeon Pierre-Joseph Desault in July 1794. Bichat adored Desault and his family. Under his mentor's tutelage, he quickly rose through the ranks. However, the sudden and unexpected death of Desault in 1795 not only emotionally devastated Bichat but added an additional burden, as Bichat was tasked with even greater responsibilities at work and in his private life, where he helped fill the gaping void in Desault's family left by the surgeon's death. The accumulated stresses took a terrible toll, and Bichat later died at the tender age of thirty, being buried in the same crypt as his friend and mentor, Desault.

Despite the prematurity of his demise, Bichat made extraordinary contributions to medicine and our story. For example, he is generally accepted as "the father of histology," a field he pioneered upon recognizing that each body is comprised of a complex number of different tissues, each of which is tasked with a different function. Together, these disparate pieces collectively work together to provide overall health and well-being and if any one system falters, the overall organism can be in jeopardy.[37] Furthermore, Bichat and his work made an indelible impression upon one of his last students, Francois Joseph Victor Broussais, who trained under Bichat from 1799 until his untimely death in 1802.

Building upon Bichat's idea that the body was composed of different tissues with different functions, Broussais added his own ideas, including the radical concept (at least for early-19th-century Europe) that diseases arose when normal tissue function went awry or failed altogether.[38] Although this idea has stood the test of time, Broussais went a bit off course in advocating that "irritations" caused by emotional distress or other stimuli could trigger the digestive system to initiate a sort of domino effect that eventually compromised organs throughout the body. Specifically, Broussais advocated for a "sympathetic" response that was propagated by the blood and tainted blood would cause the organs to fail.[39] He further advocated that the best means of countering these "sympathetic" responses was to remove blood via the liberal use of sucking leeches.[40]

Though the idea of removing "sympathetic" responses by leeching blood thankfully did not survive, Broussais' intensive discussions, arguments, and occasional harassment of a junior colleague at the University of Paris proved to be one of his greatest contributions. Gabriel Andral was born a generation after Bichat and Broussais. Like these two older figures, Andral's father had served as a military physician in the Revolutionary Army, being the personal physician to Marshall Joachim-Napoleon Murat (Bonaparte's brother-in-law).[41] Unlike these solider-physicians of an earlier generation, Andral was armed with a far more powerful weapon than anything wielded by Bichat or Broussais: a microscope. As we will see in the next chapter, the microscope had existed for many generations before but had rarely been deployed, due to the secrecy of its inventor. Once the technological challenges were solved and the instrument became widely used, the field of microscopy radically revolutionized our understanding of health and disease.

For many decades, it was known that blood contained troves of small red blood cells (erythrocytes). Unreported until Andral was evidence that much more was present in blood than could possibly have been anticipated. For example, Andral noted that a small minority of blood cells lacked the erythrocytes' red hue, which is due to iron in the hemoglobin of red blood cells.[42] Red blood cells shared an almost universal size, color, and shape (the rare exceptions being cells from patients with sickle cell anemia), but Andral noted the presence of leukocytes (a Greek term meaning 'white cells'), which lacked a fixed shape or color. Unlike erythrocytes, they also had a prominent nucleus (a structure that we now know contains the cells' DNA; erythrocytes uniquely lack a nucleus or genomic DNA). Furthermore, Andral noted a variety of different types of white blood cells, as well as the presence in the blood of what he referred to as "humors," which regulated the composition and actions of the many different types of cells in the blood.[43] This idea of bodily humors was ridiculed by many contemporaries, most notably Broussais. Ultimately, Andral prevailed, both because he replaced Broussais at the University of Paris and because his description of the different cells in the blood and the factors affecting them gave rise to a new medical field focused upon diseases of the blood: hematology.

Since the time of Andral, our understanding of hematology and the immune system has continued to grow exponentially. Typical academic texts on even the most basic aspects of the subject routinely exceed many hundreds of pages. Here we will convey an overly brief and admittedly superficial summary of a host defense mechanisms and how they function to keep the body safe from dangers, both internal and from a hostile world.

The human immune system comprises a series of layered defenses structured in a manner surprisingly like modern warfare. The front line in the war against invaders is an ancient physical barrier comprising human skin and the lining of our internal tissues. Although skin is the focus of billions of dollars in cosmetic, pharmaceutical, and surgical enhancement procedures, it is greatly underappreciated in terms of its importance as the first line of defense in preventing disease. At the risk of offending those enamored with the exquisite and undeniable beauty and complexity of the human body, consider that each of us is simply a tube within a tube. The outer tube is the body. This tube has a large surface area (e.g., skin) and occasional holes (e.g., eyes, nose, and genital orifices), which must be patrolled to halt myriad pathogens in an environment that even the most maniacal hygienist cannot (and should not) keep clean. A second, inner tube comprises the digestive tract, which is a large hole that penetrates all the way through the outer tube. Interspersed over this extensive framework, on both the inner and outer surfaces, are arrays of cells tasked with detecting and eliminating harmful microorganisms within a wet, warm environment ideally suited to most germs. Thus, the body has developed an array of mechanisms to help detect and eliminate harmful invaders.

Like the guard posts that line the defensive perimeter of a battlement, clusters of immune tissues are interspersed all along each tube in the body and its entry points. Prominent among these are structures known as lymph nodes, small kidney-shaped structures normally smaller than a half inch in diameter. As we have already seen, the detection of a foreign invader can trigger the production of cytokines, which mobilize the cells of the immune system to swarm into a nearby lymph node in an early attempt to limit the infection. As the defense mounts, these outposts enlarge. Examination (palpitation) of the lymph nodes has become a routine part of a clinical examination. Returning to our story, the breaching of these fortifications (*i.e.,* lymph nodes) was responsible for the exudates of pus that Chauliac witnessed as they erupted from the bubos (lesions) of plague victims.

The lymph nodes are well-organized fortresses designed to sample the fluids of the body to detect, isolate, and destroy potential invaders. This process entails a complex association of many different white blood cells. The catch-all term of *white cells* distinguishes them from the red blood cells witnessed by Anton van Leeuwenhoek. While we focused on discovery of leukocytes by Andrai in France, the practice of discovery often arises in multiple places at once, and comparable contributions were made by Gottlieb Gluge in Belgium, William Vogel in Germany, and William Addison in Britain.[44] Many of the leukocytes identified by these investigators include adaptations and improvements of ancient phagocytic cells as detailed above. Not only do these cells engulf and destroy pathogens such as bacteria but they also produce cytokines to alert the body that it is under attack and drive host defense cells to gather at these sites of infection. To understand how we gained understanding of the many cells that contribute to host defense, we now turn to one of the most important figures in the history of medicine, Paul Ehrlich.

Enter Ehrlich

In 1854, Paul Ehrlich was born in a small town in Lower Silesia (now Poland) to a family of distillers and tavern managers. During his adolescent education at the Maria-Magdalenen Gymnasium in nearby Breslau, Ehrlich began a lifelong friendship with Albert Ludwig Sigesmund Neisser. This connection is notable in part because Neisser would later discover the bacterial pathogen responsible for gonorrhea (and later still, the microorganism that causes leprosy), while Ehrlich's later contributions to medicine included arsphenamine, the first successful treatment for gonorrhea. Ehrlich's early education was also notable for the fact that he followed in the footsteps of his cousin, Carl Weigert, nine years his elder, who also attended Maria-Magdalenen Gymnasium and became fascinated with the emerging field of chemical dyes (in contrast to relatively rare and more expensive natural dyes such as indigo).

In the 19th and early 20th centuries, Germany reigned supreme in the science of creating and manufacturing dyes and other chemicals for use in the manufacturing of clothing, printed goods, and consumer products.[45] Weigert's seminal contributions to medicine included demonstrating that a set of dyes derived from coal tar could be used to stain certain human tissues (helping modernize the field of histology).[46] Carl Weiger's first report demonstrating the use of aniline to study human tissue was published in 1871, six years before Paul Ehrlich's first scientific publication.[47]

The chemical dyes and bacteria that captured the imaginations of Weigert and Ehrlich were quite the rage among late-19th-century German scientists.

Arguably, the fascination began with the discovery of a link between bacteria and disease as revealed by the Prussian scientist Theodor Albrecht Edwin Klebs, who used these dyes and a simple microscope to identify bacteria in the mucus of individuals suffering from pneumonia (and who is honored with the eponymous name of a major causative agent of pneumonia, *Klebsiella pneumoniae).* It seems that Klebs was a rather cantankerous character. His 1913 obituary in the prominent *Science* magazine attributes his impulsive and combative style to a genetic predisposition representing "the Slavic element in his composition."[48] Klebs's combative style likely also reflected the environment in which he trained. His principal mentor, Rudolf Ludwig Carl Virchow, was referred to as the "Pope of Medicine" and is remembered today not only for his many contributions to science but also for his public denigration of colleagues with whom he disagreed: he pinned the title of "ignoramus" upon Charles Darwin and publically branded his student, the prominent naturalist Ernst Haeckel, as a "fool."[49] Virchow's passion for socialist causes led him to politics, where he served more than a dozen years in the Reichstag and cofounded the Prussian liberal party. These experiences also revealed an audacious character, as evidenced by his well-publicized criticism of the Bismarck government, which led to his being challenged to a duel by the great statesman. In an apocryphal story, Virchow reportedly responded by agreeing to the duel only if he could choose the weapon: a Trichinella-loaded sausage.[50] As we will later see, this is not the only time that German wurst enters our story.

Returning to Klebs, the link between bacteria and pneumonia was intriguing but indirect. A contemporary of Virchow, the Silesian physician Carl Friedlander, built upon Klebs' work and revealed in 1882 that the cause-and-effect relationship between the bacterium and pneumonia was absolute.[51] This pronouncement was antithetical to many in the Prussian scientific establishment, who maintained a myth propagated since the times of ancient Greece that bad air (or *miasma*) was the cause of such diseases. Despite the growing criticism, Friedlander recruited a junior Danish scientist by the name of Hans Christian Gram to join his pathology laboratory in October 1883. Within two weeks, Gram identified a chemical dye, known as aniline gentian violet, that would selectively stain bacteria. By combining this "Gram stain" with a dye pioneered by Ehrlich, Gram would help dispel such outdated notions.

After graduating from the gymnasium, Ehrlich trained in medicine at multiple institutions throughout Germany, receiving his doctorate at nineteen years of age. Five years later, Ehrlich utilized aniline and other chemical dyes to stain the cellular components of blood. These studies revealed distinct populations of leukocytes, based on their ability to uptake or exclude various dyes. For example, one type of phagocyte was and still is referred to as an eosinophil for its ability to be selectively stained with the dye eosin.[52] Another population of cells, known as basophils, were so-named for their ability to take up basic dyes (*-phil* is an ancient Greek suffix meaning 'love' or 'affection,' as used in words such as *Anglophile*). Three years after publishing his study of blood cells, the junior Ehrlich attended an 1882 seminar given by the illustrious Robert Koch, which would change the lives of both researchers as well as the worlds of science and medicine forever.

Koch himself was a precocious child, who shocked his parents at the ripe age of five with the revelation he had taught himself to read.[53] After training in medicine, Koch became interested in understanding the causes of infectious diseases (which was still governed by superstitions such as "*miasma*," or bad air, as the basis of many diseases). Utilizing the still-emerging field of microscopy, Koch founded a new field: bacteriology. In 1876, Koch isolated a putrid organ from an anthrax-infected animal and isolated a bacterium, now known as *Bacillus anthracis*, thereby demonstrating the cause of anthrax.[54] Koch cultured the bacterium in the laboratory by isolating the exudate from infection onto a slice of potato to provide a source of nutrients for the microorganisms living within. After waiting a few days, Koch spied little white bumps on the potato, which contained millions of individual bacteria. He found the growth was optimal at body temperature (98.6°F or 37°C). Unfortunately, not all bacteria were amenable to being grown on a potato slice, and the vegetable medium was itself not ideal, as it tended to soften and disintegrate, particularly under the relatively warm conditions favored by bacteria.[55]

Fortunately for Koch, his studies to purify and then identify the biological agent that caused anthrax and other bacterial diseases were aided by a rather extraordinary team. One team member was Walther Hesse, a pathologist, who came from a large family of physicians (four of five brothers became doctors or dentists) in the east German region of Saxony. After obtaining medical training in Leipzig, Hesse served as a rural physician before volunteering for the Prussian army in 1867. This adventure led to his participation in the battles of Gravellote and St. Privat during the Franco-Prussian War of 1870–71. Hesse's frustration about the sanitation and logistics of the army hospitals revealed his early leanings towards academia.[56] Hesse briefly returned to eastern Germany after his discharge from the army in 1872. He tried to settle down as a physician in Dresden, but wanderlust led him to serve as a ship's surgeon on a German passenger ship (the New York Line). This decision also afforded Walther an opportunity to visit one of his brothers, Richard, who had established a successful dental practice in Brooklyn. During this time, Hesse became interested in the subject of seasickness (which engrossed him during his two-year stint on the passenger liner) and penned what has been described as the first objective scientific treatise on the subject. While Walther was on a North American layover in November 1872, Richard introduced him to New York society—specifically to the Eilshemius family, a family of successful Dutch importers. The eldest daughter, Angelina (Fanny), apparently caught Walther's eye. Their first meeting was cut short by Walther's need to return to his work, but Fanny was planning her Grand Tour of Europe (as was the custom with many wealthy Americans). While the trip was intended to focus on Switzerland, Fanny convinced her younger sister to join her on a diversion to Dresden in Saxony. Walther's return to Dresden (or more specifically the suburb of Zittau) might have been based in part on the fact that the medical community had learned of Walther's work on motion sickness and awarded him membership in their medical society. Impressed by the shimmering Saxon capital (described at the time as the Florence of the Elbe River), Fanny fell in love with Dresden and with Walther, and the two were married in Geneva the following summer (during the family's next excursion to Switzerland). Walther was later promoted to become the lead county physician in charge of Schwarzenberg im Erzegebirg, a mountain region near the Czech border. Mining for heavy metals such as uranium was the primary industry of the region. It was accompanied by myriad diseases, infectious and otherwise, associated with the poor working conditions and environmental damage of the heavily mined region. Such widespread malaise was not questioned and simply accepted by most doctors, both in Schwarzenberg and elsewhere in the world, at least until the arrival of Walther Hesse in Schwarzenberg.

Upon his arrival, Walther considered that the poor environmental conditions might cause an increase in airborne infectious agents and thereby contribute to the high disease burden in Schwarzenberg. This question drove Hesse to become interested in bacteriology and ultimately led him to take an academic sabbatical with Robert Koch. All the while, the well-educated and equally curious Fanny worked with Walther, often behind the scenes. Like her younger brother Louis Michel Eilshemius (who is now considered a leading figure of naïve art), Fanny had a talent for art. She became heavily involved in documenting Walther's work pictorially and joined him for his sabbatical.

During the Hesses' sabbatical, Koch was trying to identify ways to replace potatoes while culturing bacteria. One of his assistants, Julius Richard Petri, had created a small dish that still bears his name to this day, but what to put in it? An early experiment tried to utilize gelatin, but this material tended to liquefy in the warm and humid lab environments preferred by bacteria. These setbacks proved worrisome for Walther's attempts to identify bacteria in the air.

In addition to her artistic abilities, Fanny was an expert in making jellies and American-style puddings. Walther and Fanny realized that the solid consistency of these jellies, even on the warmest summer days, might provide an answer. Drawing from her upbringing in New York, Fanny had prepared her famous jellies using agar rather than gelatin. This East Asian culinary trick was learned from Fanny's childhood neighbor, a Dutch woman who had immigrated to New York from Java (at the time, a Dutch possession). Returning to work with a solution to the long-standing problem, Walther was feted by Koch, as this breakthrough greatly accelerated ongoing studies of tuberculosis and other infectious diseases (though neither Walther nor Fanny were credited for their contributions). Unperturbed by this oversight, Walther and Fanny returned to Dresden, where Walther continued his research in public health, with emphasis upon plague research. Upon his death in 1911, his laboratory had to be burned to the ground, since the magnitude of contamination was such that it was considered a public health threat.

While Petri and the Hesses were optimizing the technical necessities needed to culture bacteria, Koch was outlining the criteria that must be met to irrefutably link the cause of a disease with an identifiable bacterial pathogen. As spelled out by Koch in 1880, the criteria required that the responsible pathogen be present in every case of the disease; it could be isolated and cause disease in a healthy subject; it could again be isolated from the subject who had been infected. These simple rules created a much-needed standard for linking a microbial agent with disease. The prescience of Koch's postulates continues even today to overturn inaccurate conjecture about disease causes (including objective resolution of controversies linking HIV with AIDS, or Zika virus with microcephaly). Known as Koch's postulates, these assertions provided the cornerstone for understanding communicative diseases and utterly dispelled age-old superstitions, including miasma.

Given the impact and fame that accompanied Koch's discoveries, he was invited to give talks all throughout Europe. An 1882 seminar on the cause of tuberculosis was attended by a twenty-something Paul Ehrlich. Ehrlich was so impressed with Koch's demonstration of an assay to detect what is now known as *Mycobacterium tuberculosis* that he went straight back to his laboratory to test the idea. Within a day, Ehrlich had dramatically improved the assay and reported these results back to Koch. This interaction between a budding young scientist and his more senior (by more than a decade) counterpart began a professional and personal relationship that persisted throughout their lifetimes. Koch helped foster Ehrlich's career and offered him a position in his laboratory, a considerable improvement as Ehrlich up to this time had performed his studies in his sparsely-appointed private laboratory (akin to the foundations of the computer age arising from work in a handful of garages on the American West Coast a century later). The partnership to improve the tuberculosis assay became quite personal for Ehrlich, as he used the technique in 1885 to self-diagnose his own disease after suffering from a persistent cough. As was common among those who could afford it, Ehrlich recuperated in the warmer climes of Egypt. Following his recuperation, Koch asked his friend to join him in Berlin in 1891 as a founding faculty member of the newly created Institute of Infectious Diseases, now known as the Robert Koch Institute.

By the time of his 1891 promotion, Ehrlich had already begun a series of studies in mice with poisons such as ricin. As a brief aside, ricin is a powerful toxin found in the seeds of the castor oil plant. While it is by no means the most powerful natural toxin (far eclipsed by the lethality of botulinum or tetanus toxins), ricin has garnered considerable notoriety in the real and fictitious worlds of spy craft. The deserved fame is derived from KGB-sponsored assassination attempts using ricin against Bulgarian dissidents in 1978 London.[57] In the midst of the Cold War, Georgi Markov was a Bulgarian novelist who had defected to the West and frequently broadcast critical reports on the Bulgarian government on the BBC World Service and Radio Free Europe. Determine to rid themselves of the problem, the Bulgarian secret service enlisted a technology from the KGB and reportedly recruited an Italian-born smuggler by the name of Francesco Gullino (code name Picadilly) to successfully assassinate Markov. Gullino's weapon of choice was an umbrella with a hidden pneumatic device brilliantly designed to inject ricin into an unwitting victim. On September 7, the birthday of the Bulgarian leader and frequent target of Markov's vitriol, Todor Zhikov, Georgi was strolling across the Waterloo Bridge in London when Gullino briefly bumped him in the leg with a lethal dose of ricin. Within three days, Markov was dead. An investigation of his suspicious death began in earnest and revealed that just over a week prior, another Bulgarian dissident by the name of Vladimir Kostov had experienced a similar encounter but survived. In his memoir, Kostov writes,

> *There were crowds of people in the Metro corridors. A few seconds before stepping off the escalator, I felt a sharp pain in the small of my back, just above my waist. At the same moment, I heard a sound like the rattle of a stone hitting the ground. Natalya heard it, too, without suspecting that anything had happened to me. My first thought was that I had been struck by a stone slung with great force, as though from a catapult.*[58]

An autopsy on Markov revealed a small, ricin-coated pellet lodged in his thigh; a similar pellet was retrieved from Kostov's back. Moreover, the surviving Kostov was later found to have developed antibodies specific for anthrax, a coincidence that will return us to the story of Paul Ehrlich.

Just before Ehrlich's 1891 move to Berlin, he had begun a series of experiments to investigate if he could make mice resistant to poisons such as ricin. Starting at a low level of exposure, Ehrlich slowly increased the amount of toxin and found that within a month or two, the mice had become entirely tolerant to ricin. Ehrlich realized that this outcome represented an acquired immunity that resembled the situation in which vaccinated individuals could resist subsequent exposure to smallpox. He reasoned the body must respond to toxins (or microbial pathogens) by recognizing portions of the offending substance. He described these residues as "antigens" and the protective substances generated in the animals as "antibodies." In doing so, Ehrlich single-handedly initiated a science, today known as immunology, that objectively debunked superstitious and further mention of humoral factors—though not entirely.

Sera & Antisera

We will return to Ehrlich when we delve into the history of antibodies later in our story. For now, we will introduce some basic concepts that provide an overview of the system responsible for the acquired immunity as described by Ehrlich. Lymphocytes are a specialized subset of the larger group of leukocytes. Overall, lymphocytes come in two major flavors (and a few minor ones that are not essential to our story), popularly known as T cells and B cells.

Anatomists investigating a mysterious organ that rests atop the heart discovered it to be full of a type of cell rarely seen elsewhere in the body. Given their location in this organ, known as the thymus, the cells were thereafter known as T-cells. Throughout most of history, the butterfly-shaped thymus had developed a rather bad reputation. Returning to the venerable OED, the etymology of the name itself connotes a negative character. The appellation derives from the Greek word *thumos*, which translates as 'anger.' This may reflect the idea that palpitations of the heart arise from strong emotions originating from the organ overlaying the cardiac muscle. To some degree, the negative connotations associated with the thymus were somewhat prescient, as the thymus is the site of massive death, though the overall outcome of this carnage is unquestionably necessary for a healthy life.

A rather odd fact had perplexed physicians and scientists since the first identification of the thymus. Strangely for a human organ, the thymus is largest in babies and atrophies down to virtually nothing by early adulthood. This behavior was first noted by Galen, who, you may recall, was among the first to record the Antonine (smallpox) Plague in the 2nd century, and arguably was the most accomplished physician-scientist in the ancient

Greco-Roman world. In a newborn, the thymus can be larger than the heart, enveloping the vascular organ such that when an operation is needed to assist cardiac function in infants, most of the thymus must be cut away to allow the surgeons to access the heart. Whereas the heart continues to grow as the child matures, the thymus shrinks, relegated to the state of a "vestigial organ" (a seemingly nonfunctional structure such as the appendix) by the time a child reaches puberty.

The reason for this progressive loss of size is that the thymus is essentially a schoolhouse for the immune system. By the time a child reaches puberty, most, if not all, of the instruction has been completed. Within this strange and complex organ, immature T cells undergo a series of on-the-job learning activities designed to create an immune system that can distinguish self from others (described by immunologists as "non-self").

This instruction is essential because the human immune system is intensively powerful and must be tightly controlled. Thus, there are many safeguards to minimize the potential for unleashing the vast firepower of an uncontrolled immune system, as misdirected mobilization of the immune system can be, and often is, fatal. Examples of the consequences of not properly limiting the immune response range from the acute effects of something known as a cytokine storm (which can cause death in circumstances such as an allergic reaction to a bee sting or infection with the Ebola virus) to more chronic but no less deadly autoimmune diseases such as lupus or multiple sclerosis.[59] In each case, the pathology arises because the immune system is inappropriate or over-vigorous in attacking what is perceived as a hostile foreign invasion. To minimize such diseases, the thymus oversees a massive culling of all potential T cells, known as "central tolerance," that might otherwise turn against the body.

When these remarkable T cell receptors "sense" an invader, they can send a signal to the cell interior that causes a general mobilization of the body's host defense. The so-called activated T cells produce powerful cytokines that cause the cell (and others around it) to proliferate and go into "attack mode." Arguably, the most well-known of these T cells are those that display on their surface a molecule known as CD4. These CD4 cells, also known as "helper T cells" can be considered the "generals" of the immune system, which alert the body that it is under attack and then direct the counterattack until the pathogen has been eliminated. Once activated, these helper cells dump enormous amounts of powerful cytokines, known as interleukins, that can stimulate or synergize with other signals to amplify the power of an immune response. Without this help from CD4 T cells, the immune system is largely defanged.

The essential function of CD4 helper cells is now widely understood, since CD4 happens to be the target for the human immunodeficiency virus (HIV), the cause of acquired immune deficiency syndrome (AIDS). Indeed, the lethality of HIV/AIDS is based upon the fact that the virus seeks out and destroys CD4 cells. The brilliance of this strategy, at least from the perspective of the virus, is that HIV infection cripples the key regulator of the immune response, which might otherwise serve to detect and fight the invader. HIV is particularly crafty because it does not shut down the immune system all at once, as this could jeopardize its ability to jump to another unwitting victim before it uses up its host. Instead, HIV infection triggers a slow deterioration of CD4 T cells, and eventually the system collapses. However, the virus itself is usually not the direct cause of death but rather weakens the immune system such that other pathogens (often exotic fungal or bacterial diseases that are easily addressed by the immune system) convey the coup de grace.

Two other sets of T cell functions are also worth noting. First, some T cells can directly execute foreign invaders. These "killer T cells" are particularly useful for hunting down and eliminating tumors and virus-infected cells (as we will see in the next chapter), both of which had been normal but because of a diseased condition or infection are consequently perceived as "non-self." Such T-cells are referred to as cytotoxic, a modification of a Greek term meaning "cell-killing" and these cells evolved novel means to puncture the membrane of the disease-bearing cells (the biological equivalent of popping a balloon or puncturing a tire). Specifically, these cytotoxic T-cells deliver a complex and extremely lethal hit that simultaneously attacks and destroys both the DNA and proteins inside the targeted cells. Few cells, human or otherwise, can survive such devastation.

A second set of specialized T cells are known as memory cells. As the name implies, these cells arise during the resolution phase after a foreign threat has been eliminated. Carrying forward billions or trillions of cells that target one pathogen would be wildly inefficient, particularly as a countless number and breadth of invaders are encountered over a typical lifetime.

Instead of expending the energy to keep such a system surging at full speed, a subset of the most efficient and effective cells are placed in a sort of long-term storage. The "memory" cells are ready to pounce but kept in check until the antigen they recognized is encountered again. As an example of the efficiency of memory, the first time that a pathogen is encountered, the immune system may require days or weeks to recognize, mobilize, and eliminate the complex of "foreign" antigens that distinguish the potentially-harmful pathogen from the "self" molecules of the host. The presence of memory T cells that target the foreign antigens can allow the same disease challenge to be eliminated in days, if not hours. Consequently, memory cells convey the most important aspect of the human immune system: the ability to recognize and rapidly respond to repeat challenge by an antigen. This is a very constructive achievement if you risk becoming infected by a pathogenic microorganism but less appreciated by sufferers forced to endure annual allergic responses to tree pollen or molds.

A key to understanding the spectacular specificity of T cells and all immune function resides in one of the most bizarre and fascinating aspects of the human body: you were born with billions of T cells, and each one was subtly different from its siblings. The variations reside in a molecule known as the T cell receptor (TCR) complex. T cells recognize foreign invaders through a complex of molecules that comprise a TCR, which spans the cell membrane so that a sensing portion of the complex is localized on the outside of the cell. When stimulated by a foreign antigen, the TCR triggers a powerful signal and thereby prepares the T cell for combat. Importantly, this system is tightly regulated so that T cells often cannot be activated unintentionally. Instead, a series of specialized defense cells known as antigen presenting cells (which includes the neutrophils, eosinophils, and basophils described above, as well as an aggressive patrolling phagocyte cell known as a macrophage; a Greek term meaning 'large eater') possess specialized abilities to acutely stimulate T cells following an encounter with a foreign invader. As the name suggests, an antigen presenting cell is generally a cell that obsessively gropes its way around surrounding tissues, blood, lymph, and nodes of the body, always sampling its environment and gobbling up any debris it encounters. Consequently, these cells are known as phagocytes, a modification of a Greek term meaning "cells that eat." Any proteins ingested during this surveillance are broken into small bite-sized chunks (known as peptides) and nestled within a specialized protein complex on the cell surface. This complex is known as a major histocompatibility complex (MHC) molecule. High-resolution analyses of MHC molecules reveal their structure to look remarkably like a catcher's mitt, with the foreign peptide cushioned in the pocket.[60] This antigen-loaded catcher's mitt is the structure that interacts with the T cell receptors on T cells.

While the antigen presenting cell is sampling its environment, it is also performing an intricate ballet with T cells in its proximity. Usually the dance is quite quick and uneventful, only a unique combination of T cell, MHC, and "foreign" antigen will trigger a response. If the TCR and foreign peptide engage like two complementary pieces of a jigsaw puzzle, the T cell becomes activated. Otherwise, the T cell releases the antigen presenting cells and the sequence begins anew with a different pair of cells. On those rare occasions when the proteins "displayed" on MHC include bacteria or bits of material believed to be foreign and matching its specific tastes, the activated T cell enters a veritable frenzy and gains a license to kill and kill again.

The activation of a T helper cell initiates a cascade of events to rapidly and vigorously destroy perceived enemies. The activated T cell will immediately begin a process of rapid proliferation, increasing its numbers exponentially. In parallel, these activated T cells begin pumping out vast amounts of highly active interleukins meant to increase the growth and power of the T cell response as well as to elicit assistance from other leukocytes all throughout the body. A familiar sign that this is happening is the lethargy and fever often associated with infections such as influenza. These frenzied activities require an enormous investment of energy. For example, when the infection is localized (such as a wound), the site of infection can be particularly warm to the touch. Once the infection has been stemmed, it is essential for the body to halt the considerable energy investment, and the process is slowed. Most of the cells that participated in the fight are decommissioned. Many are killed, with their component proteins, sugars, and fats recycled for use in manufacturing future cells.

In a remarkable mechanism that balances such unsustainable energy usage with the need to avoid future dangers from encounters with the same pathogen, a subset of the T cells that successfully responded to the infection are put into a storage hall of fame. These memory T cells, which we met above, are equipped with the ability to respond even more vigorously and efficiently to the danger should it ever be encountered again. These memory cells provide a means to deploy learning of the past to ensure future recall. However, over the years, encounters with more and different types of antigens can lessen the readiness of some of these memory cells.

All of these interactions are based on an exquisite recognition of foreign antigens expressed within the groove of an MHC molecule to interact with and trigger a T cell that displays a particular T cell receptor targeting that antigen. However, the diversity of molecules that exist in nature is staggering and this presents a challenge to the T cell. The question boils down to this: How can T cells anticipate "foreign" antigens that the body has never encountered and thereby provide protection for new threats that may arise or evolve in the future?

Through a fascinating process that is unique to T cells during their time in the thymus (and B cells as they mature in the bone marrow through an analogous process), the genes for T cells undergo a spontaneous transformation that cuts out, rearranges and mutates multiple small pieces of DNA. These pieces eventually come back together in different combinations to create the unique region on the T cell receptor (TCR) that binds antigens. This complex process can be thought of as being analogous to having dozens of colors of Lego blocks in different piles. One or more block from each pile can be brought together to generate a final structure (e.g., a Lego building) with the colors arranged in an almost infinite number of combinations. Even more diversity derives from the fact these genes are rendered hypersusceptible to mutation as this process of building takes place. Extending the Lego analogy, a red block might mutate to become pink or orange. The result of this bizarre remodeling of T cell receptor DNA is that the resulting T cell receptor protein found on each T cell in the thymus differs slightly from the TCR on neighboring cells.

These subtle differences can allow a full complement of T cells in our body to detect, in theory, any molecule that might ever be encountered in a lifetime. Although such diversity is crucial for protecting our bodies against the wide variety of threats posed by an environment replete with bacteria, viruses and parasites, one can imagine that such extreme variability might also create opportunities to trigger "friendly fire" casualties were a T cell to recognize and become stimulated by "self" molecules or protein conformations. This potential hazard in turn explains the need for the massive culling associated with "central tolerance," a mechanism by which certain cells of the thymus can identify and eliminate any T cells with the potential to target "self" tissues before they can cause harm.

From the moment of birth, a child is bathed in an environment crowded with microbial pathogens, any of which could be fatal. Therefore, humans are born with a fully functioning immune system that is prepared for such assaults from the moment of birth. This capacity requires that the work of the thymus in educating the immune system begins in earnest prior to birth. Hence the thymus is one of the most metabolically active of all organs during gestation and shortly after a child first enters the world. As an understandable means to conserve energy and not have to constantly repeat the process, evolution allows T cells to survive for decades (unlike most cells). Consequently, the education overseen by the thymus is concluded prior to puberty. Like many teenagers, the immune cells seem to know everything by the beginning of high school. Having completed its mission, the thymus withers away.

As is well understood by anyone who has experienced an immune system gone astray, the power of host defenses has the potential to wreak havoc. Specifically, the same mechanisms that can mobilize the body to fight infections can cause extensive damage to the host itself when the system malfunctions. The outcomes range from relatively mild effects of seasonal allergies to extreme and sudden death in reaction to a bee sting or the chronic, debilitating, and sometimes fatal array of autoimmune disorders such as multiple sclerosis, rheumatoid arthritis, and lupus, to name but a few. Indeed, in looking at the raw power that the immune system can impart (think of how quickly a temperature can spike during a fever), it is rather remarkable that autoimmune misfiring is not more common. This fortunate feature is attributable to a series of safeguards that have co-evolved to mitigate the harmful effects as the firepower of the immune system has increased over the eons.

Dual Key Systems

Those who have viewed or read any of the genre of popular culture tropes focused on the use of nuclear weapons is at least generally familiar with the "dual key" or "two person" system. The idea is that no nuclear weapon can be launched by any single individual, since two keys (and locks) that are separated by a large distance (at least wider than a person's outstretched arms) must be turned in exact synchrony. If one key is turned before the other, the system is disabled. This safeguard generally works quite well in Hollywood films (and we must have faith in the real world as well) and is quite like the safeguards regulating the immune system.

An analysis of how T cells are activated in response to a perceived foreign attack reveals a surprisingly similar set of safeguards as those portrayed in the Matthew Broderick film *War Games* or any number of Tom Clancy novels. As we have seen, the T cell receptor is the primary signal that alerts a T cell to the presence of an infectious assault. This single key alone is insufficient to trigger the vigorous and energy-consuming activation of an immune response. As often occurs during a frenzied state of mind, decision making is often compromised by a need for action, any action. In the case of the immune system, this means killing something, and fast. The considerable power exerted by activated immune cells could thus wreak havoc upon innocent bystanders if left unchecked.

The immune system has evolved a series of "co-stimulatory" signals that must be relayed at the same time, for the exact same rationale behind the dual-key system: to prevent unwanted attacks. Although we discussed the exquisite ballet that governs how the T cell receptor dynamically interacts with the MHC molecule and the peptide therein, the dangers of an unintentional misfiring of the T cell receptors are so potentially catastrophic that a co-stimulatory interaction of another pair of molecular partners—one on the T cell and the other on the antigen presenting cells—must also engage within a predetermined time and location on the surface of the T cell. Absent this second set of activation signals, the T cell is irreversibly shut down through a process known as anergy.[61] Likewise, other series of precautions have evolved over time, including the use of certain processes known as checkpoints, which can actively prevent a frenzied T cell from activating and killing innocent bystanders during an attack meant to target foreign invaders.

Life is a complex process. At any given moment the body is multitasking in an infinitely complex manner. Energy and focus are constantly demanded by many of the trillions of cells in the body. With this in mind, it is important to understand that the evolution of a complex immune system is not arbitrary. It earnestly produced a defense against many disease-causing events. These diseases represent the wide array of pathogens that threaten not only individuals but can and do convey existential threats to entire species. As one example, while the world remained transfixed upon the terrible and highly publicized Ebola virus outbreak that began in 2013, considerably less attention was paid to the fact that the same virus had already killed more than 90 percent of mountain gorillas and is hurriedly pushing that species towards extinction.[62] Indeed, the only way to avoid the catastrophic loss of this noble ape may be active intervention in the form of a vaccine originally developed to manage the outbreak in people.[63] Before returning to the subjects of vaccines in detail, however, it is necessary to better understand adversaries such as Ebola virus and the myriad of other viral and bacterial pathogens confronting our species and others. Thus, we now turn our focus for the next two chapters to some of the many challenges presented by the microbial world.

4

The Wurst Way to Die

The germ theory of disease relates that organisms too small to be seen by the unaided eye can cause disease. This idea is not particularly new, yet it needed to be rediscovered on different occasions. Early evidence of this fact can be found in Roman writings. Marcus Tarentius Varro was a high-ranking equestrian soldier, farmer, and scholar who rose through the ranks to become a tribune of the Roman Republic. Despite being a supporter of Pompey in the Roman civil war, Varro was pardoned by none other than Julius Caesar himself (although later banished by Mark Antony and, later still, reinstated by Augustus). Casting aside his militant days, the older and wiser Varro settled into a life's work focused on scholarly pursuits. His noteworthy achievements included the first encyclopedia and comprehensive (albeit inaccurate) listing of all Roman consuls from Lucius Junius Brutus (who is credited with ending the monarchy and transitioning Rome to a republic) through the end of the Roman Republic with the rise of Augustus (during Varro's lifetime).

Like many early authors, his reputation if not the works themselves remain largely incomplete, with one exception. Varro's *Rerum Rusticarum Libri Tres* conveyed an encyclopedic overview of conventional Roman views on agriculture and farming.[1] Within this volume Varro expresses a word of caution for readers contemplating work in or around swamps, which have a long-understood link with diseases such as malaria. Varro counseled precaution in working near swamps, "because there are bred certain minute creatures which cannot be seen by the eyes, but which float in the air and enter the body through the mouth and nose and cause serious diseases."[2] It is presumed that Varro did not necessarily originate this idea but rather passed along a concept from others.

Sadly, this advice and understanding of the tiny microbial world would mirror the rise and fall of the Roman Empire, being lost for at least a millennium until being rediscovered in the 16th century by the Italian scholar Girolamo Fracastoro. Fracastoro's contributions to science include astronomy (for which he was honored by putting his name on a prominent crater on the moon), geography, and biology. Regarding disease, Girolamo proposed that there were small particles, which he termed as "seeds" or "spores," that could convey infection between and among people, even when there was not direct contact between them.[3] Although this is quite close to the modern definition of infectious disease, Fracastoro did not specify whether these spores were of a biological or chemical nature. In a more specific reference that rings truer to a 21st-century audience, he used a word translatable by the English *tinder* to describe the disease today know as syphilis (which coincidentally reflects the fact that the modern mobile phone application Tinder promotes "hook-ups" that can increase susceptibility to the same venereal disease).

Seeing is Believing

It is perhaps a bit glib and overly obvious to state that the definitive discovery of the microscopic world awaited the invention of the microscope. The key breakthrough in this regard arose from a most unusual source. Antoni van Leeuwenhoek was a drapery maker in mid-17th-century Delft, in the Dutch Republic. As a scion of an upper-middle-class family, Antoni combined an education with considerable curiosity and a variety of professional experiences and personal hobbies, which eventually ranged from accountancy to land surveying and glass blowing.

As a creator and purveyor of fine linen products, van Leeuwenhoek sought means to better observe the threads in his work, a need sated by experiences gained during his glassblowing hobby.[4] While melting a glass rod and pulling out a long string of glass, Van Leeuwenhoek realized that when he did it in just the right manner, the surface tension of the glass could create spherical glass beads. Moreover, Antoni realized smaller spheres of glass could magnify objects near it, acting as what we now refer to as a simple microscope. The draper proceeded to make more than two dozen such devices, improving his technique with every new version. These tools proved quite useful for observing very small objects such as a fraying thread, but Van Leeuwenhoek was not satisfied with the vocational applications of his invention. Instead, he began to look at many everyday objects in a different way.

In doing so, van Leeuwenhoek discovered a strange and fascinating microscopic universe all around him. Among his discoveries were muscle fibers (in meat), spermatozoa (within male ejaculate), and small animalcules (small animals) residing everywhere, including drinking water, raindrops, and saliva. We now know these animalcules to be the earliest representations of the microbial world, extending to the smallest of the observable species, the bacteria. His remarkable observations (and drawings based on his findings) were shared with his friend Reinier de Graff, a Delft anatomist, who touted van Leeuwenhoek's work to the Royal Society of London.[5] At first, the hobbyist van Leeuwenhoek demurred from the considerable interest conveyed by the Royal Society, but eventually he penned almost two hundred different reports on the microscopic world.

Unfortunately for the field, van Leeuwenhoek was not as forthcoming in detailing how he could generate such superb microscopes with high magnification and resolution. This reticence encouraged the indignation of the irascible Robert Hooke, an English polymath and serial provocateur, who is better known today for his contributions to physics, including the laws surrounding elasticity, advances in understanding the phenomena of gravity, and the introduction of the pendulum and watch spring. Based on reports from van Leeuwenhoek, Hooke became enraptured by the microscopic world. In January 1665 he published a treatise, *Micrographia*, which popularized the emerging field of microscopy.[6, 7] Despite the critical and widespread success of this book, Hooke remained frustrated that his microscopic observations could not surpass those of the comparatively pedestrian Dutch draper. Even in the face of haranguing from Hooke and others, Van Leeuwenhoek remained taciturn about his techniques, and the field of bacteriology nearly succumbed with Van Leeuwenhoek's death in 1723.

Progress in understanding the tiny world of bacteria was to remain largely dormant for more than a century until it was rejuvenated by German naturalist Christian Gottfried Ehrenberg.[8] Ehrenberg completed a doctoral dissertation on the subject of fungi in 1818. Only two years later, he agreed to embark on an expedition with his friend Wilhelm Hemprich to survey the Libyan desert to identify novel plant and animal specimens.[9] This campaign on Minutoli, which was underwritten by the Egyptian government, was to be led by the legendary Prussian explorer and former military general Heinrich Menu. The Nile Valley–based dynasty of Muhammad Ali was still evolving as an independent nation under nominal control of the dying Ottoman Empire. Ali had taken over Egypt after the withdrawal of the French, who had captured and occupied his homeland since an invasion by Napoleon Bonaparte. Looking to establish its sources of opportunity and avoid taking sides with either England or France, Ali looked to the seemingly disinterested Prussians to lead a scientific survey of his newly independent territories. This was the first in a series of events that would, and not for the last time, allow a bacterium to delineate Ehrenberg's career. The first defining event of the expedition began with an outbreak of typhus in Alexandria, which caused the team to divert to a study of the Upper Nile basin rather than the Sahara.

Typhus is a louse-borne bacterial pathogen that has quite a history of changing history. One of the first written descriptions of typhus was recorded during the Spanish siege of Granada in 1489. The delirium and rotting sores characteristic of the disease hastened the downfall of the Moorish civilization in Iberia in 1492. These victories heartened Spanish King Ferdinand and Queen Isabella to open their purses and sponsor an expedition by an Italian navigator from Genoa by the name of Christopher Columbus.[10]

Though the fall of Granada meant later funds could be redirected from the siege to exploration, the victory came at a high human cost, with roughly five sacrifices to typhus for each battlefield death. Three centuries later, typhus would again sway human history by decimating the ranks of Napoleon Bonaparte's army (a decade after leaving Egypt) during its retreat from Moscow in the winter of 1812, again claiming more French soldiers than the Russians had.[11, 12] The utter breakdown of public health in times of war allowed typhus repetitively to claim thousands of victims in virtually every conflict from the English civil war through the Second World War, including the young lives of Anne Frank and her sister Margot, who were held under pestilential conditions at the notorious Bergen-Belsen concentration camp.[13] The spread of typhus has only been managed with the 20th-century advent of dichlorodiphenyltrichloroethane (DDT), an insecticidal chemical developed by the Rockefeller Foundation, which, as we will see throughout the book, has waged a long war against insects.[14]

The impact of typhus on Ehrenberg was considerably more personal than the experiences we have noted, but they were defining nonetheless. As their exploration progressed from the Upper Nile to the Sinai Peninsula, Syria, and Lebanon, Ehrenberg and Hemprich collected samples of local flora and fauna until Hemprich was bitten by a lethal viper snake, known by the Latin name *Vipera bornmuelleri*, while exploring in Lebanon.[15] Although the bite was not immediately lethal, Hemprich was still far too weak from his bout with typhus. The snake toxin caused local bleeding, which was exacerbated by a factor that prevents clotting and tissue repair. Despite being severely injured, Hemprich refused to rest, and this decision proved lethal. Hemprich died on June 30, 1825, suffering from a fever on the Red Sea coast of modern-day Eritrea.[16]

The death of his friend Hemprich created a personal crisis for Ehrenberg, but his career and subsequent remembrance in history would be established by the expedition. Shortly after his return to Prussia, where there was widespread reporting on the expedition's success (beyond Hemprich's death), Ehrenberg was approached about a new adventure by Alexander von Humboldt.[17] Humboldt was the preeminent explorer of his day and had dreams of exploring Egypt. Two decades before Hemprich and Ehrenberg's survey, Humboldt had plotted to explore Egypt with Napoleon. However, this adventure had to be abandoned because of a tribal uprising against the French. The timing of this revolt could not have been worse, as Humboldt had been in the process of gathering the personnel and equipment needed for the expedition. All prepared but without finances or a destination, Humboldt had found himself interviewing with the Bourbon Spanish monarchy, which had been seeking a team to explore its possessions in South America. The change in plans had proved fortuitous, as Humboldt had explored the Orinoco River in Venezuela, Cuba, Ecuador, Peru, and Mexico. During this expedition, he'd become arguably the most famous science-based explorer and geographer of his time. No less than Thomas Jefferson invited Humboldt to visit the fledgling United States, toasting him as "the most scientific man of the age."[18]

What Humboldt recruited Ehrenberg for in 1829 was an expedition to Russia. The country was massive, but its rulers knew comparatively little about their domain.[19] The tsar was particularly interested in having Humboldt assemble a team to explore mining opportunities east of the Ural Mountains that might ultimately provide silver and platinum needed for coinage. The tsar's government provided an escort consisting of a company of Cossacks, who were employed both to ensure the protection of Humboldt and Ehrenberg's team as well as to prevent them from encountering (and later reporting upon) sensitive subjects such as the social conditions of the country's many serfs. Although the excursion was intended to venture no further than Tobolsk, a town in south central Russia (and the town where Tsar Nicholas II and his family would later be confined in the days following the Russian revolution), Humboldt and Ehrenberg pushed farther, eventually turning back at the Chinese border. A chance encounter with another bacterium almost cancelled the adventure midway through the expedition. Specifically, a local outbreak of another bacterium, this time anthrax, caused considerable unease within the party and led to discussion about cancelling the expedition. However, the sixty-year-old Humboldt pressed on, stating, "At my age, nothing should be postponed."[20]

With a treasure trove of samples obtained from his expeditions to Egypt and Russia, Ehrenberg settled into a domestic life spent staring at the water, rocks, flora, and fauna he had collected with Hemprich and Humboldt. Germany was and remains a dominant innovator in the field of optics and microscopy, and Ehrenberg had access to some of the finest instruments of his era. For the three decades after his return from Russia, Ehrenberg subjected his samples to microscopic analyses and discovered many "animalcules," including multiple species of bacteria and single-celled eukaryotic life known as protists. His work was widely followed, and it reinvigorated interest in the invisible world of microorganisms. Given the thousands of hours spent behind a microscope, it is quite fitting that Ehrenberg was honored in 1877 as the first recipient of the Van Leeuwenhoek Medal by the Royal Netherlands Academy of Arts and Sciences, a once-a-decade award that recognizes outstanding achievements in microscopy.[21]

Pasteurized

Arguably, the individual most responsible for our modern understanding of infectious disease and vaccines was raised in a setting and started his career in a manner that would not have predicted such success. Louis Pasteur was born on December 27, 1822 in the eastern town of Dole, France.[22] The glory days of the city, when it had served as the seat of Parliament for the region and host to its only university, had long passed after the government and academic institutions had been transferred to Besancon in 1422. Indeed, the city's major claim to fame arose a century and a half later with the notoriety of Gilles Garnier, a hermit who confessed to the actual or attempted kidnapping, murder, and cannibalization of at least a half dozen of Dole's youth.[23] This loathsome killer gave rise to the myth of the "wolfman" and unknowingly spawned a genre of horror fiction that remains popular until the present.

This unhopeful setting nonetheless gave rise to Louis Pasteur, the son of a poor tanner. His academic training did not suggest much promise, as fishing and art were the primary passions of the young Louis. He did manage to graduate from the local university (which, you may recall, had been relocated to Besancon), though he failed in his first science examination (and scored particularly low marks in chemistry). Despite these setbacks, Louis continued diligently to develop as an academician, though failing at his first attempt to gain entrance to the prestigious Ecole Normale Superieure in Paris.[24] Pasteur's tenacity paid off with his eventual matriculation. His graduation with high honors brought him to the attention of Professor Antoine Jerome Balard, who recruited Pasteur to study a chemical wonder known as "chirality." Pasteur's work showed that solid crystals of the same molecule can exist in two different states that polarize light in different ways. Pasteur's studies revealed a phenomenon that broadly distinguishes molecules as either right- or left-handed; the idea that biology had distinct left and right-handed features had the unintended consequence of influencing fiction writers such as Lewis Carroll's *Through the Looking Glass* and James Blish's 1971 Star Trek novella, *Spock Must Die.*[25, 26, 27]

As the professional life of Louis Pasteur was beginning to come together, tragedy devastated his personal life. Pasteur had married Marie Laurent in 1849, and the couple had gone on to have five children, four girls and a boy. As was all too common throughout the ages, the eldest daughter died of typhoid fever at the age of nine years. Sadly, the same disease would go on to kill their two-year-old daughter Camille in 1865 and twelve-year-old Cecile in 1866. This series of heartbreaks hardened Pasteur's resolve, and he thereafter dedicated his life to understanding the bases of infectious disease and the microorganisms responsible for his children's untimely deaths.

The chirality findings launched Pasteur into a career that balanced chemistry and biology. It was guided by the principle "Chance favors only the prepared mind," a popular saying that was uttered first by Louis in 1854.[28] This credo was put into effect during his early work on fermentation, when he explained the age-old basis of beer brewing: microscopic yeasts are responsible for transforming sugar into alcohol. He further applied this finding to wine (where yeasts are found on the skins of the grape) and his recognition that microorganisms cause food to spoil. In doing so, Pasteur debunked the antiquated theory of spontaneous generation (the sudden and frequent emergence of life from non-life).

From a practical standpoint, Pasteur translated this basic knowledge into practice by demonstrating that heating a liquid, such as milk, could kill the microorganisms within and thereby extend its shelf life and prevent food poisoning (indeed, products bearing the term *pasteurized* can still be found in virtually every food market today). Nonetheless, the adoption of pasteurization was slow, as most people, even in the scientific community, were skeptical about his ideas. This doubt is perhaps best captured in an 1860 editorial in the Paris-based newspaper *La Presse*, which announced, "I am afraid that the experiments you quote, M. Pasteur, will turn against you. The world into which you wish to take us is really too fantastic."[29] Having faced and overcome adversity during his early academic years served Pasteur well later in life, as he ultimately succeeded winning over the scientific community and the public with the same tenacity used to overcome his early academic failures and personal tragedies.

Despite these considerable contributions, Pasteur's most famous accomplishment was a test of his adage about chance favoring the prepared mind. During the summer of 1880, Pasteur and his 29-year-old assistant, Charles Chamberland, were performing a series of studies of the effects of anthrax on chickens.[30, 31] As Pasteur was heading out for a much-needed vacation, he handed Chamberland a jar of bacteria and instructed him to finish up the work with a final study. For reasons that are not entirely clear (perhaps because the boss was out of town or because Chamberland himself was preparing for his own holiday), Chamberland became distracted and forgot to initiate the study before he left for his own vacation. When he returned to work, he realized the oversight and decided to correct the situation by finishing the study, presumably before Pasteur returned. However, the injected chickens thrived and did not die. Moreover, when the same chickens were again injected with anthrax, this time with the full-potency version, they showed no signs of disease.

Pasteur was later informed of the results by a dejected Chamberland (who was preparing to throw out all the old samples). Pasteur became quite excited and told Chamberland to repeat the experiment, including the delay. Both scientists were amazed to see the same lack of toxicity when the inoculum was used to infect a new batch of chickens. Even more staggering, the inoculated chickens had become immune to subsequent infection with other and even more deadly anthrax cultures.

We now know that bacteria are just as mortal as you and I and the culture that Chamberland was using to infect chickens was dead and acted as a vaccine. Specifically, the vaccine consisted of a dead or severely weakened dose of bacteria that, rather than causing disease, sensitized the immune system to repel infection. Ever quick to translate his work into a real-world solution, Pasteur soon looked at a chemical methodology for weakening bacteria to treat anthrax spores rather than waiting long periods for the bacteria to weaken. He goal was to ensure the bacterium was so feeble that it could not cause disease yet intact enough to elicit a durable immune response.

The chemical of choice was soon identified as carbolic acid (also known as phenol), a chemical with a rather storied history. Carbolic acid occurs naturally in many forms of whisky. It conveys the distinct and highly sought-after taste associated with Scotch produced on the island of Islay, just west of the mainland border between the Highlands and Lowlands. Phenol is also the principle component of throat sprays such as Chloraseptic®, though its use is rather controversial, given that phenol is known to be a carcinogen. Returning to Pasteur, some of his fame arose from an announcement in the summer of 1881 that he'd used phenol to create an attenuated form of anthrax to elicit immunity against the disease. This finding captured the public's attention, and Pasteur named his new product a "vaccine" in honor of Edward Jenner's—or, more accurately, Blossom the cow's—earlier contributions to eradicating smallpox. As we will see in chapter 7, history has a way of simplifying stories to create visionary heroes (as we saw with Jenner), though the accurate background is far more interesting.

Hand Wringing and Washing

Coincident with Pasteur's first and failed attempt to gain admission to the Ecole Normale Superieure, and fifteen years after the conclusion of Ehrenberg and Humboldt's Russia expedition, a newly minted Hungarian physician graduated from the prestigious University of Vienna. Ignaz Semmelweis tried to make a go of being a specialist in internal medicine, but when he failed to find a proper position, he switched his specialization to obstetrics. In 1846, Semmelweis was appointed to the Vienna General Hospital, working under the direction of Dr. Johann Klein.[32, 33, 34] Klein had taken the helm at the hospital after the forced resignation of his predecessor, Johann Lucas Boer. Boer, it seems, was a bit of an iconoclast. In particular, he advocated minimal physician intervention in pregnancy, emphasizing the need for exercise and proper nutrition of the mother while arguing against the use of forceps and drugs during delivery.[35]

Infection had always been a major risk factor for pregnancy. A condition called puerperal sepsis was particularly problematic in birthing clinics of the mid-19th century. Known today by the rather self-explanatory name of "post-partum infection," puerperal sepsis is an infection of the reproductive tract that still claims far too many victims in the days following labor. Outrageously (to his peers, at least), Boer adhered to the controversial practice of hygiene, which maintained that puerperal sepsis could be spread among infected patients. In contrast, the accepted practices of the day included active intervention, frequent bleeding, and the use of dirty medical instruments such as forceps, which were rarely washed or even wiped down between procedures. Based upon his radical views, Boer was dismissed on "the pretext of insubordination" and was replaced by Johann Klein in 1823.

As an unfortunate coincidence, Klein's own reputation was largely based on his own advocacy of the emerging field of anatomic pathology. This practice centers upon the idea that physical examination of a diseased body, and of cadaver organs following death, can convey information that is useful for diagnosis and treatment of the living. The unfortunate bit is that Klein instructed the students and physicians in his clinic to interact with diseased specimens in the mortuary, with emphasis on cadavers with puerperal fever, as this was a real-world problem routinely faced by the physicians of the Vienna General Hospital. Thus, physicians would routinely travel back and forth between the morgue and the delivery suites (without proper handwashing or cleaning their examination tools, as we will see).

Another change instituted by Klein was necessitated by an ever-increasing acceleration in births, as mid-19th-century Vienna seems to have been a particularly fecund place. To address the demand, Klein separated the obstetrical practice into two clinics at the beginning of 1833.[36] A key event in the unfolding of our story arose because of a bureaucratic decision by the Kaiser's government, which segregated the roles of male and female obstetrical caregivers. Clinic One was to be staffed by medical students and physicians (male), while Clinic Two would be managed by midwives (female). Patients entering the hospital would be randomly assigned to one clinic or the other based on their day of admission, a practice that unintentionally created the basis for what we now recognize to be a randomized clinical trial.

Six years after the decision to staff the clinics in such a manner, a newly hired Semmelweis was puzzled to note substantial differences in the rates of puerperal fever between the two clinics. Specifically, the rates of sepsis were generally twice as high in the physician-run Clinic One as they were in the midwife-based Clinic Two. Such knowledge was apparently well known within the community, as expectant mothers actively avoided being admitted to Clinic One. Based on the desire to avoid Clinic One, some local women preferred street or home births to avoid the risks. Semmelweis later deduced that the likelihood of dying of puerperal sepsis was lower for women who gave birth on the street than it was for those cared for in Clinic

One. Despite this community knowledge, the mystery of why Clinic One was so dangerous remained unknown.[37]

In March of 1847, a tragic death saved the lives of countless women. Jakob Kolletschka was a 43-year-old Bohemian-born professor of pathology and a close friend of Ignaz Semmelweis.[38] While Kolletschka was using an autopsy specimen to instruct a team of students in the pathological diagnosis of puerperal fever, one of the students pricked Kolletschka's finger with a dirty examination tool. Within days, the pathologist began to manifest the classic symptoms of puerperal fever. He died shortly thereafter. The tragic event preoccupied Semmelweis, who struggled with the fact that the assumption of the day was that puerperal fever was only found in women.

As he sorted through his emotions and the evidence, Semmelweis realized the disease was clearly not unique to women and that something in the dead woman's body had caused Kolletschka's death. He incorrectly deduced that decaying or dead tissues from cadavers could kill the tissues of the living (the fallacy being that the dead tissues themselves, and not the invisible microorganisms in them, were the cause of disease). Armed with this new theory, Semmelweis realized that the practice of working with autopsy samples and then with patients had the effect of introducing "cadaverous particles" into the delivering mothers and causing disease. These deductions preceded Robert Koch's work by almost a half century. The fact that the midwives in Clinic Two practiced what we now know to be better (though still imperfect) hygiene and did not convey detritus from autopsy samples to the birth canal was responsible for the lower rates of puerperal fever.

Unfortunately, the medical community remained as intransigent as it had been with Johann Lucas Boer. Semmelweis's advocacy of proper hygiene led him into conflict with Klein, but fortunately for our protagonist, the senior doctor was on the verge of retirement.[39] Semmelweis was suspended for his improper encouragement of septic procedures, such as washing hands with chlorinated liquid, but he was reinstated by Klein's successor. The practices advocated by Semmelweiss were adopted more widely in Clinic One, and the rates of puerperal sepsis decreased accordingly. Nonetheless, Ignaz Semmelweis was informed in 1849 that his two-year contract would not be renewed, largely as a consequence of his past feud with Klein. Despite the positive results with Clinic One, Semmelweis became a leper within the Viennese medical community and was figuratively run out of town. In response, Ignaz increasingly lashed out at his critics. Even a return to his native Hungary did not resolve the situation, as Hungary was in the midst of an independence movement. The Viennese-trained Semmelweis was seen as a symbol of the Habsburg Empire and was shunned by his professional colleagues. Rejected by his own countrymen, Semmelweis began a long emotional decline, becoming obsessed with his theory for the last years of his life. Sadly, the perceptive physician died an angry soul in a Viennese asylum at the age of forty-seven. Ironically, the cause of his death was attributed to septic shock arising from the infection of wounds received after a severe beating by the guards. His contributions to medicine however lingered, and the name Semmelweis now adorns a Budapest medical school, as well as multiple women's clinics. In 2008, the memory of Semmelweis was commemorated with an Austrian coin.

Now that we have seen the inspiring and often sad history of the birth of bacteriology, our attention will turn to understanding how and why some of the smallest organisms rank among the deadliest.

No Stomach for Sausages

The Latin term *botulus* refers to the belly or womb. Its modern-day negative connotation reflects the notorious side effects too often experienced by diners following a filling meal of sausage-filled pig stomachs. The link between consumption of this delicacy and food poisoning was sufficiently prominent that Byzantine emperor Leo VI (also known as Leo the Wise for reasons that will be self-evident)[40] outlawed their manufacture during his reign in the 10th century. At the time, the Byzantines had a strong predilection for pig stomachs (*Blunzen*) filled with otherwise unused bits of meat combined with blood. The salted mixture of animal fats and blood congealed (or, more accurately stated, clotted) prior to cooking. As the inclusion of salt was known to delay spoilage, such prepared meats could linger in the butcher shops for many days, unlike fresh meat. Unknown to the butchers or their customers was the fact that this mixture often created an ideal incubator for the growth of bacteria that would feast upon the putrefying contents of a rich mixture of proteins, fat, and blood. The sickening aspects of the lethal brew could, however, be overcome with proper and thorough heating. Unfortunately, sufficient heat and time were not always achieved, resulting in a disease that gained the moniker of botulism based on this association with sausages.

Despite the best attempts by Leo VI and others to eliminate sausage making, the annual tradition of Oktoberfest continues to this day and is testament to the fact that native Germans and their diaspora have continued to relish sausages despite occasional outbreaks of botulism. Arguably, the worst wurst poisoning occurred in 1793 in the Wurttemberg village of Bad Wildbad. The mountainous region, tucked into the southwest corner of modern-day Germany near the French border, was known for sausage making. The location of Bad Wildbad made it the epicenter of upheaval in times dominated by the French Revolution, ever-changing borders, and civil unrest. As still occurs today, political and pecuniary turmoil joined hand in hand, and the locals increasingly sought ways to economize, which included increased emphasis upon sausage making. A few years before the dawn of the 19th century, thirteen residents of Bad Wildbad dined upon insufficiently cooked blood sausage.[41] All the diners became ill, and six died. Despite this calamity, the continued economic chaos in Wurttemberg throughout the Napoleonic years favored the continued use of *Blunzen,* and an epidemic of botulism persisted throughout the region.

An unintended consequence of this regional misfortune was that local universities became an epicenter of food poisoning research. A University of Tubingen professor with the daunting name of Johann Heinrich Ferdinand von Autenrieth was an early founder of the field of medical forensics (think *CSI Wurttemberg*).[42] As the scion of a relatively wealthy civil servant, the young Autenrieth enjoyed the benefit of being educated by some of the most important scientists of his day, including Georges Cuvier (the father of paleontology), Antonio Scarpa (who pioneered otolaryngology and cardiology research), and Johann Peter Frank (who developed the record-keeping approaches that Ignaz Semmelweiss used to relate puerperal sepsis with sanitary practices, as we saw above).[43] Within Wurttemberg, Autenrieth was renowned and widely feared for chastising housewives who undercooked blood sausages. The practice of undercooking was quite common, as overcooking blood sausages could rupture the pork stomach and allow the contents to spill out into the boiling water. Therefore, many housewives tended to err on the side of rarity, and the outcomes often were sickening.

Continuing the local tradition, an 1815 outbreak of botulism in the Wurttemberg town of Weisberg was documented by the local physician, Justinius Kerner. A report on the incident by the 29-year-old physician and part-time poet sufficiently impressed Kerner's former professor—Autenrieth—that it became his first published work, albeit scientific prose rather than poetry.[44, 45] Over the following years, Kerner amassed case histories of scores of botulism poisoning victims and began to experiment on extracts from improperly cooked blood sausages. Kerner's work with a variety of animals, from flies through cats, revealed that a poison secreted from the bacteria exerted its toxic effects upon the nervous system by altering the transmission of signals from nerve endings to the brain. To confirm such findings, he utilized himself as a guinea pig and recorded the effects of poisoning. These reckless actions garnered Kerner a sharp rebuke from Autenrieth, who, upon learning what Kerner had done, directed his fiery temper against his foolhardy former student. While Kerner went on to an internationally recognized career as a poet and author of medical treatises, one unfulfilled wish was to isolate the toxin responsible for botulism. This wrinkle to an otherwise flawless career was not simply a cosmetic desire to finish his interest in food poisoning; it was based on a belief that botulinum toxin might convey medical value. It would likely please but perhaps confuse Kerner to know that this substance (known as Botox®) is intentionally injected into the faces of millions of people today.

The steps towards discovering the botulinum toxin would require additional calamities. On December 14, 1895, in the small Belgian village of Ellezelles, a funeral was held for one of the town elders, the 87-year-old Antoine Creteur. As was the custom of the place and time, a local brass band with the name of Fanfare Les Amis Reunis was hired to play at the funeral. They were later feted with a dinner at La Rustic, the local inn.[46] The menu that evening at Le Rustic included smoked ham. Although smoking meats had been popular for years, the process tended to occur at a low temperature insufficient to kill all bacteria, spores, or toxins. These inadequacies were unfortunately realized by the ill-fated band members, all of whom became ill, with three dead. While such outbreaks had occurred innumerable times, this outbreak was unique, as the autopsies (and the remaining ham) were analyzed at the University of Ghent by pathologist Pierre Maria Van Ermengem at a time when the modern understanding of bacteria was just emerging. Ermengem had trained under Robert Koch in Berlin prior to his position in Ghent. As such, Ermengem was familiar with the means of culturing many types of bacteria, including the anaerobic bacterium that had infected the ham. As suggested by the name (the translation of the Latin *anaerobic* is 'without air'), these bacteria do not care for air and will not grow on petri dishes left in the open. Using approaches that deprived the cultures of oxygen, Ermengem could isolate a bacterium that fulfilled all of Koch's postulates. After a few name changes (as understanding of bacteriology increased), the organism came to be known as *Clostridium botulinum*.

The discovery of the toxin responsible for the nervous system paralysis caused by *Clostridium botulinum* required an additional three decades of research but was eventually attributed to a team at the University of California, San Francisco, led by Dr. Herman Sommer.[47, 48] Subsequent work within academia and germ warfare laboratories throughout the interwar years revealed purified botulinum toxin to be the most toxic substance known to man (a record that still holds true today).

Concern about the potential use of botulinum toxin propelled the United States to establish the U.S. Army Biological Warfare Laboratories at Camp (later Fort) Detrick in Frederick, Maryland, at the height of the Second World War to explore both the offensive and defensive needs of biological warfare.[49] Fort Detrick has remained to this day a center of biodefense (though not offense). Along with the Centers for Disease Control and Prevention, it plays a frontline role in countering biological threats, both natural and man-made.

Although there is no evidence that the United States ever deployed an offensive biological weapon during the Second World War or thereafter, much speculation remains that our British allies might have. Specifically, a biological agent has been implicated in the assassination of the man who is arguably the most nefarious character of the Second World War (excepting Hitler himself). Reinhard Tristan Eugen Heydrich was the progeny of the opera composer Richard Bruno Heydrich (the name Reinhard being the name of one of his father's favorite characters from his own opera, *Amen*, and Tristan being a tribute to Wagner's *Tristan and Isolde*).[50] Heydrich matured into a professional sailor and a handsome ladies' man, known for his many romantic liaisons, before settling down after a remarkably short two-week courtship and proposing to Lina von Osten. Lina had been an early adopter of Nazism. After Reynard was suddenly discharged from the navy in April 1931, Lina encouraged him to join the party. Heydrich then assisted Heinrich Himmler in establishing a counterintelligence division of the Schutzstaffel (better known as the SS). Heydrich was a hardworking and organized administrator, and his influence on Himmler and later Hitler grew quickly. Within a year, Reinhard was named as head of the Sicherheitdienst, Hitler's security service and later the dreaded Gestapo. Heydrich's malevolence included a variety of cloak-and-dagger operations to undermine, arrest, and eliminate "persons endangering German security." In this capacity Heydrich was a key instigator of, among other things, anti-Jewish pogroms such as *Kristallnacht* and the formation of the *Einsatzgruppen* tasked with the elimination of Jews in occupied territories. Later he was tapped to lead the Final Solution.

Among his various other responsibilities, Heydrich was the acting Reich protector of Bohemia and Moravia (formerly Czechoslovakia). In a biography of the fuehrer entitled *Adolf Hitler*, preeminent WWII historian John Toland suggests that Heydrich's efficiency in managing Bohemia and Moravia was such that these territories risked conversion from occupied territories to loyal members of the Third Reich. Such concerns prompted the exiled Czech government, domiciled in London, to attempt to assassinate Heydrich.[51]

On December 28, 1941, the British initiated Operation Anthropoid, involving the airdrop of two Czech soldiers, Jans Kubis and Jozef Gabcik, into occupied Czechoslovakia outside Prague. The assassins spent half a year analysing Heydrich's daily routines. On May 27, 1941, the two assassins were armed with specially modified grenades and Sten guns. The pair waited at a key hairpin intersection on Heydrich's daily commute. As the Mercedes convertible staff car slowed during its approach, the two soldiers attempted to open fire, but their Sten guns jammed. Rather than speed off, Heydrich unsheathed his sidearm and ordered the driver to halt so that he could shoot the assassins. As Heydrich took aim at Gabcik, Kubis lobbed the modified grenade into the Mercedes. The explosion wounded Heydrich, though an adrenaline rush allowed Heydrich to continue to pursue Gabcik for a few blocks before the Nazi leader's wounds caused him to collapse in the middle of the street.[52, 53]

Still conscious, Heydrich had convinced himself his injuries were suffered upon his exit from the Mercedes. He was aided by a local woman, who flagged down a truck that transported him to Bulokova Hospital. The hospital staff detected fragments from the grenade in his abdomen and lung and removed Heydrich's spleen. The protector's postoperative recovery was unremarkable, and a full recovery was anticipated. However, Heydrich's condition suddenly degraded a week later, and he was dead within a day. An autopsy by senior Wehrmacht pathologists indicated a cause of death by bacteria or poisons carried by bomb splinters. This speculation proved unremarkable until the 1982 publication of *A Higher Form of Killing*, by Robert Harris and Jeremy Paxman.[54] In a story that reads like a Hollywood thriller, the authors reveal that the head of the British biological warfare program, Paul Fildes, bragged about modifying the grenades used in the Heydrich assassination to include botulinum toxin. The boasts attributed to Fildes include, "[I] had a hand in his death" and "He was the first notch on my pistol." While the principals involved in the operation have long since expired, these accounts may eventually be proven or dismissed with the declassification of relevant British archives, which is expected within the next two decades.

The Spanish Strangler

The symptoms of diphtheria begin with a minor sore throat, followed shortly thereafter by a low but nagging fever. Over the next few days, the symptoms worsen and are accompanied by growing patches of gray and white material (looking to a modern-day person like a piece of spent chewing gum attached to the back of the mouth) that first coat the tonsils and then spread to other parts of the throat. The growth of these patches progressively narrows the vital passageway needed for the flow of oxygen to the lungs. These growing obstructions are compounded by such an enormous swelling of lymph nodes that the victim's skin around the neck can be stretched taut in a vain attempt to retain the infection within. The struggle to breathe is further complicated by a barking cough that closely resembled the sound of a baby's struggle with croup, a sound unforgettable to the countless parents who have sacrificed hours of sleep and sanity agonizing over that characteristic hollow, barking cough. Eventually, the struggle with the disease resolves, culminating either in a recovery or a lethal shutdown of the lungs and heart (due in part to myocarditis as the heart also became the target of assault). A recent estimate suggests that even with modern supporting care, the case-fatality rate (how many infected individuals succumb to a disease) for diphtheria is at least 5 percent (one in twenty) and perhaps as high as 20 percent, or one in five (with an impact equally tragic for adults over the age of forty).[55] To put that staggering statistic in perspective, roughly one in a thousand people infected die from diphtheria. The case-fatality rate for even the most devastating influenza outbreak in recent times, the so-called Spanish flu of 1918, was only 2.5 percent.[56]

The first historical record of diphtheria was logged in 1613, when Spain experienced an epidemic described as "*El Anos de los Garrotillos*" (the Year of Strangulations).[57] The strangler continued to remain a problem throughout the Western world thereafter, with occasional outbreaks, such as one in colonial New England that started in 1735 and remained a pandemic until at least 1740. Diphtheria has been known by many different things in many places, such as the Strangling Angel of Children, Boulogne Cough, the Bull Neck, Canker Ail, Malignant Croup (or angina or quinsies), and Gangrenous (or putrid or pestilential) sore throat. It was not until 1826 that the French physician Pierre Bretonneau invoked diphtheria, the Greek term for 'leather,' labeling the disease with the name it's still known by today.[58]

A half century after Bretonneau's moniker, the disease struck the family of Queen Victoria's daughter, Princess Alice. Alice had by then become matriarch of the Hessian court. She nursed her family to health and in doing so, allowed her bloodline to sustain multiple European royal families. Indeed, Alice is the great grandmother of the current Queen consort, Prince Philip.[59] Over the next few years, extraordinary progress would be made in assessing the identity of the killer and prosecuting the murderer to the fullest. In 1883, the Prussian-born pathologist Theodor Albrecht Edwin Klebs worked with another Prussian colleague, Friedrich August Johannes Loeffler, to identify the bacterium responsible for diphtheria, a pathogen that for a time bore the honorific title "Klebs-Loeffler bacillus" but is now known as *Corynebacterium diphtheriae* (though a prominent genus of opportunistic bacteria discovered by Klebs retains the name *Klebsiella*).[60]

Further understanding of the disease arose from studies utilizing new filtering technologies (a subject to be discussed in greater detail later in our story), which helped identify the toxin produced by the diphtheria bacteria that is responsible for its pathogenicity. This achievement represents yet another contribution to human medicine from Emile Roux, the Frenchman and cofounder of the Pasteur Institute. Working with the itinerant Swiss microbiologist Alexandre Yersin, Roux was able to isolate the toxin, which would provide a key step that would eventually help manage (though not entirely eradicate) the disease, as we will see in chapter 6. (Yersin later went on to identify the cause of the ancient disease known simply as "the plague," which today bears the eponyous moniker of *Yersinia pestis*.)

Too Much of a Good Thing

The toxins wielded by botulism and diphtheria are known as exotoxins, which simply means a toxin that is released from bacteria. These toxins can be insidious because even if the bacteria that have produced the molecules have been extirpated, say by antibiotics, the remaining toxins can convey deadly legacies unless neutralized (e.g., by high heat or chemicals). Another set of bacterial toxins, distinguished by the German scientist Richard Friedrich Johannes Pfeiffer, remain embedded within bacteria (live or dead). These endotoxins (meaning toxins from within) actively conspire with our host defenses in a manner that can be equally deadly.[61] Endotoxins are generally large molecules composed of a lipid (fat-based) core. The core inserts into the bacterial membrane and is linked to a large and outward-projecting sugar-based polysaccharide. These components serve the bacterium by providing structural integrity to the cell membrane. Throughout the evolution of the immune system, these telltale signs of bacteria have also provided an early-warning signal that alerts the host to the presence of unwanted bacterial invaders.

As anyone who owns a highly sensitive home alarm system knows, a strong defensive posture can often go awry. As our immune system evolved, the fine-tuning towards bacteria became so exquisite and overwhelming that misfiring could be deadly. Specifically, the immune system is so finely honed against endotoxins that the prevention can be worse than the disease. As one example, some endotoxins (appropriately named superantigens) can trigger a response so vigorous that up to 20 percent of all the T lymphocytes in the body are activated into a frenzy.[62] The call to action also alerts other cells of the innate and acquired immune systems to discharge their pathogen-fighting chemicals; the result of these cytokine storms includes a dramatic (and unwanted) opening of the capillary walls throughout the body.

This response is intended to open the blood vessels to facilitate the passage of host defense from the blood into tissues to fight foreign invaders. If an infection is limited to a site in the body, the increased blood vessel permeability will cause vital fluids to transit into tissues, which can cause localized swelling characteristic of acne, paper cuts, and other damaging events. However, when such a response occurs throughout the body, the loss of fluids undermines blood pressure and causes a deadly event known as septic shock. According to the Centers for Disease Control and Prevention, septic shock is the thirteenth-largest killer of Americans. The most common triggers are superantigen-based infections of the urinary tract, lung, or sites of catheter insertion.[63] More than 40 percent of septic shock occurrences are lethal, far outstripping many other events such as heart attacks (5 percent mortality), stroke (19 percent) or breast cancer (17 percent).

A dramatic reminder of the deadly nature of septic shock occurred in 1978. The sales of consumer products are notoriously humdrum and would hardly seem to provide the background for a dramatic tale of sudden death. In general, consumers tend to select a product and remain loyal for years to come. Such realization was problematic for the Procter and Gamble Company (P&G) of Cincinnati, Ohio, in the mid-1970s. P&G sought to enter the highly competitive and established field of tampons as an outsider. Tampon technology had remained remarkably unchanged for decades, and P&G executives believed the field was subject to potential disruption with new technologies. To this end, company researchers discovered that another consumer product, carboxymethylcellulose, might provide such a breakthrough.

Carboxymethylcellulose is a thickening agent that creates the consistencies of puddings, toothpastes, laxatives, and many other consumer products.[64] Its thickening properties came from its ability to absorb impressive amounts of water, and the compound was known to be safe. During the development of P&G's breakthrough tampon, researchers estimated that a single tampon could absorb all the fluids of a menstrual cycle without leaking. The experimental tampon could absorb twenty times its own weight. In doing so, it would expand to conform to the vaginal cavity, thereby providing a leak-proof barrier that would prevent embarrassing accidents.[65] The resulting tampon was therefore given the name Rely, and it was marketed with the catch phrase "It even absorbs the worry."[66]

P&G began test-marketing the new tampon in 1974 but ran into an unwanted snag in 1976, when the Food and Drug Administration (FDA) announced a redesignation of tampons as "medical devices," thereby increasing the requirements for objective data to verify safety and efficacy. Frustration turned to glee in the corporate suite of Procter & Gamble as company executives read the fine print of the new mandate, realizing that the new regulation provided a grandfather clause for existing products, which included Rely. The stars had aligned for P&G (consistent with their corporate logo), since this meant that their innovative new breakthrough would remain unchallenged for even longer, as future competitors would be required to meet the higher standards set by the FDA. The company launched Rely in August 1978. As part of their marketing strategy, they mailed forty-five million free samples to consumers throughout the United States.

Rely was a sensation, but the company and product suffered from its success. Many consumers were so impressed with the absorbency of the new product that they used it far longer than they had with conventional tampons. A representative example was documented in a 2011 report about a consumer, which stated:

> *It was 1980, and Styx was playing at the Cow Palace in Oakland. We took the BART from San Jose (or rather Fremont, near where we lived) and anticipated little bathroom access at the arena so the tampons offered a big "convenience" to her instead of waiting in the predictably long lines at the women's restroom. It was the first time she had used Rely, and the tampon worked amazingly well. But, she said, "I remember removing that Rely tampon after getting home late at night and wondering whether I had lost my virginity, that thing had gotten so huge. I stopped using them after that because of being too grossed out."*[67]

Shortly thereafter, epidemiologists throughout the United States noted increased occurrence of women with high fever, hypotension, and extreme fatigue. Other odd symptoms included the loss of skin on the palms of the hands and the soles of the feet. These symptoms rapidly progressed into coma, organ failure, and death within days.

In the closing days of 1979, CDC investigators were inundated with cases of a new disease, later named toxic shock syndrome, whose victims were primarily women. The cases could be tied to one another by the presence of a common bacterium, *Staphylococcus aureus.*[68] This particular bacterium is broadly found in the environment and resident on the body, as experienced by anyone who has had acne or minor food poisoning. Most Americans suffer minor food poisoning at least once a year, usually manifesting itself as a queasy stomach or light diarrhea. However, the extent and severity of *Staph*-based toxic shock in the late 1970s outbreak initially stymied CDC investigators. Within three months, strong detective work revealed a link with the use of tampons in general and the Rely brand in particular. Working backwards, we now know the powerful moisture absorbency of carboxymethylcellulose had unintentionally dried the normally moist environment of the vaginal cavity. The sudden drying of these tissues created small ulcerations in the normal microflora and thereby provided a foothold for foreign pathogens such as *Staph. aureus* to take hold. Compounding the problem, the increased viscosity of the fluids remaining in the parched environment conveyed advantageous growth conditions for the invading microorganisms. As the degree of infection increased, the woman's body sensed the foreign interlopers, and endotoxins present within the *Staph* bacterial membranes triggered host defenses to perceive a massive infection with bacteria. In response, these immune cells overreacted and caused blood vessels to leak, decreasing blood pressure and the ability to succor vital organs. In this complex but fundamentally primitive manner, the seemingly innovative and innocent idea of creating a more absorbent tampon unintentionally caused a devastating response.

Friend and Foe

The examples of *Staphylococcal* and *botulinum* infections may seem to portray bacteria as dire threats. However, this would unintentionally convey an oversimplified view of human interactions with bacteria. A more nuanced exposition must convey that our species, like many others, has developed an intricate relationship with microorganisms that conveys extraordinary benefits. A failure to recognize the complexity of these interactions could be shortsighted and disastrous. The recognition that humans, animals, plants, and even microorganisms work together in a larger ecosystem parallels the discovery of the microbial world itself.

An early pioneer in the field was the French microbiologist and ecologist Rene Dubos, who has been given credit by some for the well-known saying "Think globally, act locally."[69] (This saying was attributed to Dubos during a 1972 United Nations Conference on the Environment but has also been credited to the American ecologist David Ross Brower, who might have used it as a 1969 slogan for the Friends of the Earth, a nonprofit he founded.[70] Additionally, it has been attributed to the architect and designer R. Buckminster Fuller.[71] All these contenders might have been eclipsed a half century earlier by the Scottish biologist Patrick Geddes, who alluded to its use in 1915.[72] Regardless of its unclear parentage, this statement is quite prescient not only in reference to the environment (think water pollution or climate change) but also in reference to the microenvironment that exists everywhere we look. This sentiment was championed by Rene Dubos, whose range of thinking spanned from the very small (he was trained as a microbiologist) to the very large (he championed the causes of global environmentalism). Dubos trained in microbiology with the Canadian-American scientist Oswald Avery, who would later discover DNA and is widely regarded as the most deserving scientist never to receive the Nobel Prize. While studying the myriad bacteria and fungi in dirt, Dubos discovered a set of compounds that could kill other bacteria.[73] Some of this research led to the discovery of the antibiotic streptomycin. Dubos and Selman Abraham Waksman were awarded the 1948 Lasker Prize, the highest American scientific honor for the discovery of this key antibiotic (Waksman later received the 1952 Nobel Prize for the work).

An even more impactful outcome of these studies was the recognition that bacteria such as *Streptomyces* (from which the antibiotic was discovered) produce antibiotics to thwart the growth of other organisms in the environment. Dubos realized that the same dynamics that occurred in soil samples also occur throughout the body's surfaces, including the skin, mouth, nose, throat, and lining of the gastrointestinal system. Another prescient view was that the study of individual bacteria was insufficient to model the extraordinary dynamics among the different species. Consistent with this, surprisingly few bacteria within any given specimen (be it a soil sample or cheek swab) can be cultured and studied in the laboratory due in part to the fact that biology unforgivably defaults to a complexity that is largely beyond our ability to grasp. For example, some organisms may become so reliant upon interactions with others that their isolation disrupts their normal functioning or is irreversibly lethal.

Virtually every square micrometer of our body that interacts with the external world is host to a teeming scaffold of diverse species. Although the numbers can vary, the latest estimates from investigators at the Weizmann Institute indicate that an average 150-pound person is composed of forty trillion bacterial cells and thirty trillion human cells.[74] The complexity of the cohabitation is evidenced by the fact that the mouth alone serves as host to hundreds, if not thousands, of different bacterial species but the different parts of the mouth (e.g., the tongue, tonsils, soft palate, and saliva) each plays host to a different density and identity of microorganisms. Altogether, the combination of humans and their microbiome brethren has elicited a new category called "superorganism."[75]

These bacteria are not just passengers along for a ride. As we learn more about these coinhabitants, we appreciate their necessary contributions to an ever-increasing number of key functions needed for everyday human life. For starters, the act of recruiting benign microorganisms, particularly those that bring along their own defensive strategies as described above, provides a first line of defense against some of their more nefarious toxin-wielding or flesh-eating cousins. Alterations in the microbiome have been associated with a variety of diseases.[76] For example, a strong course of antibiotics can devastate the normal flora of the gut, rendering a person susceptible to particularly aggressive and deadly bacteria such as *Clostridium difficile*. Beyond this, the bacteria lining our gut, of which more than one thousand different species have been identified, contribute to our ability to digest certain sugars, synthesize essential vitamins, and regulate the metabolism.[77] Unsurprisingly, a poorly functioning microbiome is also associated with a variety of diseases of the digestive tract, ranging from malnutrition to obesity, as well as autoimmune and inflammatory diseases of the bowel.[78] A particularly interesting study from the laboratory of my colleague at Washington University in St. Louis, Dr. Jeffrey Gordon, shed light upon the interplay between the microbiome and obesity.[79] The gut bacteria from normal or overweight mice were isolated and transplanted into germ-free mice. Remarkably, the simple transfer of the bacteria from fat mice was sufficient to cause the germ-free mice to gain weight—despite the fact that their caloric intake was tightly regulated and unchanged. Stated another way, weight gain was dictated by the bacterial component of the "superorganism" and was unlinked from the number of calories consumed. Expanding the study further, the Gordon laboratory isolated fecal bacteria from sets of identical twins, where one was lean and the other obese. Much as was seen before, the mice receiving the bacteria from the obese twin became fatter even though the calories were strictly monitored.

Whereas the impact of the gut microbiome on physiological processes such as weight gain might be expected, given the known location and function of bacteria in digestion, much less obvious is evidence linking the microbiome with susceptibility to a variety of human diseases ranging from depression and schizophrenia to diabetes and rheumatoid arthritis. In a 2016 study from investigators at the Mayo Clinic, a link between the microbiome found on the patient's breast (skin) or in her mouth related to a diagnosis of breast cancer.[80]

Such emerging findings are consistent with a thesis advocated by Dr. Martin J. Blaser of the NYU School of Medicine. Blaser maintains that the widespread use of antibiotics has critically altered the composition of the microbiome in a manner that has increased susceptibility to obesity and a variety of diseases.[81] The livestock industry is well known for supplementing animal feed with antibiotics to increase the growth rate and weight gain of livestock. Such findings may not be unique to livestock as antibiotics alter the microbiome of virtually all human and non-human animals. Blaser's studies link one organism, *Heliobacter pylori*, whose absence in both mouse and man can cause greater susceptibility to obesity, gastrointestinal reflux disease (GERD), and esophageal cancer, as well as the incidence of a variety of inflammatory and even neurological disorders such as autism. Although this field of research remains quite preliminary, a May 2016 report at the International Meeting for Autism Research by investigators from the Texas Children's Microbiome Center revealed a potential link between the intestinal flora and autism, though the causality, if any, remains questionable as of press time.[82]

Such findings are consistent with the idea that the health and well-being of the human component of the exquisitely complex superorganism is highly dependent upon our constitutive microorganisms. Such understanding underlines the need not to regard the bacteria in our environment as implacable enemies to be destroyed. Rather, keeping our superorganisms healthy requires us to properly distinguish between friend and foe. With this in mind, we turn to another set of organisms that has traditionally and rightfully been viewed as man's deadliest enemy: viruses.

5
Spreading Like Viruses

Many children ask, Who is the king of the mountain and master of the planet? An alien looking at Earth from a distance (or at least at our social media) might conclude the dominant species is canine or feline in origin. After all, humans often work long hours to provide food, shelter, and healthcare for their pets, while the pets partake in constant leisure. From this our alien might conclude that companion animals are the dominant species. Other observers might maintain the dominant species is either a cockroach or bacterium, based on sheer numbers. An occasional bigot might instead nominate our own species. The current ecological epoch is known by many as the Anthropocene, due to the impact of humans on the planet (though it's not necessarily a positive attribution).[1]

If our alien were to analyze the planet with higher-resolution optics, a more defensible answer would be viruses. Even the most inaccessible bacteria, such as those found in volcanic trenches in the lowest depths of the Earth's oceans, find themselves infected by, and providing sustenance for, a multitude of viruses. By the time such a question is asked by a young child, the youngster has already been feasted upon by at least a few viral pathogens.

In order to evaluate the public health danger of an infectious agent, public health officials gauge its deadliness (i.e., what fraction of infected individuals succumb) and contagiousness (the number of additional people an infected person is likely to infect). A pathogen with high scores in both could truly be species-ending, but, fortunately, such infections are rare. It is nonetheless striking that infections caused by viruses are seven of the ten most contagious (rotavirus, measles, mumps, chicken pox, rhinovirus, smallpox, polio) and four of the five most deadly (rabies, HIV, Ebola, pandemic influenza). Compounding the dangers posed by viral pathogens, these infections are notoriously difficult to treat for reasons we will soon address. Before addressing such complexities, it will be useful to recount the history of viruses and how we have come to fear and respect these deadly pathogens.

A simplistic view of a virus is to consider it a highly evolved and still rapidly changing organism. A never-ending controversy surrounds the question of whether viruses are alive.[2] For example, I maintain they are, though my scientist spouse (and undoubtedly others) believes me a fool for believing such notions. Regardless of your position on the subject, viruses have evolved in such a way that they remain at the top of the food chain. One can argue convincingly that virtually any living cell on the planet serves as the sustenance for at least one (and often many) viruses. This dominant position has arisen as viruses have over time evolved to take advantage of the machinery of the hosts they infect. Indeed, the most sophisticated viruses are often those with the fewest number of genes, as they leave the heavy lifting to the unfortunate victims they infect. Furthermore, viruses have been actively participating in the evolution of all living structures. A landmark paper by the Australian biologist Philip John Livingstone Bell suggested that the nucleus, a subcellular structure that contains the DNA of all eukaryotic cells, represents an example of a mutually beneficial partnership, known as an endosymbiotic event. It can be viewed as a viral variation of one bacterium engulfing another, which led to the appearance of mitochondria.[3]

Through the Filter

The last time we encountered the 19th-century microbiologist Charles Chamberland, he had forgotten to perform a study on anthrax for his boss. After returning from vacation, he had injected chickens with old bacterial culture and unintentionally found that exposure to dead bacteria protected the chickens from future infection (and thus acted as a vaccine). Rather than sacking Chamberland for the oversight, the appreciative Louis Pasteur kept him on, particularly because it seems that Chamberland had quite the touch for inventing devices to simplify laboratory research. For example, one year before the fateful chicken experiment, Chamberland had championed a project that led to the modern autoclave, a device used by virtually every biologist and clinician to sterilize their tools. This innovative aptitude also led Chamberland to unintentionally facilitate the discovery of viruses.

In 1884, Chamberland crafted a rectangular porcelain box to filter water.[4] The rationale for this invention was based on increasing evidence from the Pasteur team that some bacteria secrete toxins into their environment—the very toxins that are responsible for their deadliness (as we observed with botulism). Chamberland and Pasteur realized a filtering mechanism would allow them to separate the relatively large bacteria from the much smaller toxin proteins. The resultant Chamberland-Pasteur filter (as it is still known today) was a hit among microbiologists endeavoring to understand how toxins functioned, but its real value was even more transformative.

At about the same time Chamberland was tinkering with the development of the autoclave, the German agricultural chemist Adolf Mayer was leading the Agricultural Experiment Station at Wageningen in the Netherlands.[5] Mayer's interest in chemistry might have come naturally, as he was the grandson of the great chemist Leopold Gmelin. Though this name hardly resonates today, among Gmelin's contributions to science was the discovery of potassium ferricyanide, whose distinctive hue is familiar to anyone who has seen architectural blueprints.[6]

While serving his duties at the Wageningen agricultural station, Adolf Mayer was approached by tobacco farmers complaining of a new disease that was affecting their crops. This strange new disease revealed itself through a patterned and not entirely unattractive mottling of tobacco leaves in a manner reminiscent of a mosaic. Early studies on the tobacco mosaic disease had revealed it could be transmitted via the sap, suggesting it was infectious in nature. Mayer presumed this to be the result of an undiscovered bacterium. Eager to make his name by discovering a new microorganism, Mayer subjected the samples to microscopic analysis, as most bacteria can readily be seen by eye with the assistance of these optical magnifiers. However,

Mayer failed to see any bacteria in the infectious cultures. A second possibility was that the disease was being transmitted by a toxin in the sap rather than the bacteria itself. By the summer of 1886, Mayer published this speculation and began using Chamberland's filters to identify the responsible toxin.[7] Consistent with his hypothesis, Mayer was excited to see that something in the filtered fluid could cause the disease. However, excitement turned to confusion with additional investigation. Specifically, the plants infected with the toxin could themselves transmit the disease to other plants. This did not make sense, since the toxins could not replicate themselves and thus the infectious capacity of a toxin would be self-limiting. Yet Mayer's findings revealed that the pathogenic agent could propagate itself indefinitely. After eliminating the possibility that such a bizarre finding was not simply a reflection of some sort of contamination, Mayer appreciated that he was confronting something never seen (or, more accurately, unseen).

By the time Mayer was performing his follow-up studies, concerns about the rapidly spreading tobacco disease had captured the attention of scientists around the world. In 1892, the Russian botanist Dmitri Iosifovich Ivanovsky was studying a tobacco disease in the Crimea and likewise described the cause to be an infectious agent capable of passing through Chamberland-Pasteur filters.[8] Unlike Mayer, Ivanovsky remained convinced until his death that the cause of the tobacco mosaic disease was a bacterium that was simply too small to be captured by the filter. Such small thinking was refuted by the Dutch microbiologist Martinus Beijerinck, who experienced the same experimental outcome in 1896 but was convinced it was caused by a new form of life.[9] Beijerinck named this causative agent a virus (the Latin term for 'poison') to distinguish this activity from both bacteria and their toxins. These new viral beings remained a matter of blind faith, as they were far too tiny to observe with the eye or even the most powerful microscopes of the age. In 1934, the American Wendell Meredith Stanley utilized the new technology of electron microscopy, which affords far greater resolution than conventional microscopy. In his life's work, Stanley described the isolation and purification of a virus. It consisted of a hollow tubelike protein structure, known as a capsid, loaded with an RNA-based genetic material (which is closely related to but distinct from the more familiar DNA-based material that controls heritability in people).[10]

Unknown to science at the turn of the 20th century, the odd assemblage denoted by the term *virus* was wholly foreign to the conventional definitions of infectious agents and, indeed, life itself. As viruses were structurally and physiologically distinct from any known biological entity, including bacteria, the process of deciphering the riddles posed by their discovery and questions about how they reproduce and cause disease would remain for a time in the realm of bacteriology. This ownership was based not only on the fact that microbiologists rank among the earliest pioneers of virology research; it was also based on a fortuitous series of findings that coincided with the discovery of viruses.

The Smallest Shall Lead Them

Viruses are everywhere and can devour virtually every type of living matter. Scientists have identified only a small fraction of the number of conventional living organisms on our planet. Such ignorance is compounded by many magnitudes when applied to our understanding of the diversity of viruses. The current consensus is that the number of different viruses (the virome) vastly exceeds the number and diversity of all known and unknown prokaryotic and eukaryotic life.[11] Whereas the bias of our own species often presumes the primary threat of viruses is focused upon humans, as evidenced by scourges such as Ebola, HIV, or influenza, viruses in general are just as troubling for bacteria as they are for any mammalian species.

The current renaissance in the discovery of new viruses reveals most humans (and indeed probably all vertebrates) cohabit with myriad viruses that do not cause disease and are either neutral or perhaps even helpful, much as we have already seen with bacteria. The human virome includes hundreds of species of recently discovered viruses known as anelloviridae.[12] These small, ring-shaped viruses can be found throughout the body, but most have not yet been linked with disease. It is not too far-fetched to imagine that they may be beneficial in day-to-day life. For one example of how viruses can be beneficial, we turn to one group of viruses that helped scientists to understand how viruses function in the early days after their discovery and how these same viruses may provide unimagined future benefit. Understanding the set of bacterial viruses generally known as bacteriophage provided the gateway for understanding viral pathogens in humans, so we will begin there.

In the same year Martinus Beijerinck was verifying the discovery of very small pathogens targeting tobacco plants, the English bacteriologist Ernest Hanbury Hankin contributed a landmark breakthrough that revolutionized our view of disease.[13] The setting for the discovery was quite murky. It seems Ernest became interested in the notorious fact that bathing in the fetid waters of the Ganges River in India was one of the most efficient means of self-induced illness. The "miasma" theory of disease still largely dominated the public view of infectious disease, and Hankin sought to dispel these ancient notions by demonstrating that bacteria were the cause of the intestinal and other maladies often arising from direct encounters with the Ganges.[14] Hankin himself had trained as a bacteriologist in London and was an early proponent of utilizing aniline dyes to visualize bacteria under the microscope. After training with both Robert Koch and Louis Pasteur, the young English scientist had gained the credentials to begin an independent career in India, where he was granted the impressive title of Chemical Examiner, Government Analyst and Bacteriologist for the United Provinces, Punjab and the Central Provinces.[15] Although this was certainly a promotion from his student days, Hankin's detractors in the United Kingdom portrayed his departure to the Raj as a desperate flight away from ignominy suffered at home. The venom directed at Ernest was not based upon his progressive views of infectious disease but because Hankin had also gained quite a reputation as an advocate for the use of animal dissection as a means to promote scientific discovery and education. This view was anathema to many contemporaries, thus earning the ire of the animal rights community, which often posted hostile newspaper articles and letters to the editor, some of which threatened physical violence to "vivisectors" such as Hankin.

The rivers of India were (and sadly some remain) a bountiful source of waterborne diseases. One of Hankin's early contributions to his new home was a demonstration that boiling the river water was sufficient to prevent the spread of cholera and other waterborne illnesses. As the practice spread and cholera rates correspondingly diminished, the life-saving contribution continued to be derided by his fringe animal rights enemies in Britain, who published the following in the June 26, 1896 edition of *The Zoophilist*:

> *Remarkable to state, a vivisector has made a beneficial—and common-sense discovery. The discoverer in this case is Mr Hankin an old antagonist of ours, whom we met in debate at Cambridge before he took ship and departed to serve as a bacteriologist in India, where he still remains.*[16]

In that same year of 1896, Hankin published an article in the annals of the Pasteur Institute that would forever link him into the history of viruses, though he did not realize this at the time.[17] In a rather obscure study buried in the institute proceedings, Hankin noted the presence of substances in the polluted waters of the Ganges and Jumna Rivers that thwarted the survival of the cholera bacterium. Hankin went on to contribute knowledge to a wide array of scientific thought (including zoology, microbiology, anthropology, political science, and architecture) until his death in 1939. Nonetheless, his most lasting impact would ultimately center upon the 1896 observations from the sacred but putrid waters of the Ganges.

Almost two decades after Hankin's key observations, another English bacteriologist, engrossed in studies to improve the manufacturing of smallpox vaccine, built upon Hankin's findings from India. In 1915 English bacteriologist Frederick Twort reported the discovery of a small agent that could pass through a Chamberland-Pasteur filter and was sufficient to kill staphylococcal bacteria.[18] Twort postulated this might be a virus or enzyme that targeted bacteria but favored the latter, thus denying him conclusive title to a seminal discovery. The key breakthrough was instead to be made by the itinerant French-Canadian microbiologist Felix d'Herelle, who was fueled by booze.

A Parisian newborn was christened in 1873 with the name Hubert Augustin Felix Haerens, but his name was changed, likely by customs officials, after his parents moved the family to Montreal. Returning to Paris at the young age of six after the death of his father, d'Herelle completed high school and trained himself in the sciences. A not-so-grand tour of Europe by bicycle initiated a lifelong wanderlust, which drove him to explore South America and Turkey, mostly by bicycle, by the age of twenty.[19] The fixation upon travel halted briefly as Felix met his future wife, Marie, in Anatolia and settled down for a few years. D'Herelle's nervous energies were then directed, both personally and professionally, towards a new obsession: the science of fermentation. D'Herelle studied fermentation through books and by building a home laboratory (effectively a distillery). Ultimately, the pursuit of understanding and improving the fermentation process would propel d'Herelle to return to Canada, where he had been offered a commission to study the fermentation and distillation of maple syrup into schnapps. Although maple schnapps can still be obtained for a price, the precious cost of pure maple syrup (as compared with wheat, barley, potatoes, and other inexpensive sources of sugars and starch for the bacteria and yeast involved in fermentation) soon rendered this practice largely obsolete.

Unperturbed by the failure to launch maple syrup–based liqueurs, d'Herelle continued traveling the world to further his preoccupation with distillation (and because his poor investment skills compelled a constant need for income). D'Herelle accepted a job offer from the Guatemalan government to establish a bacteriological laboratory at the General Hospital in 1901. His primary goal was to help find ways to prevent the fungal infections that were damaging the local coffee crops (a task that he eventually achieved by acidifying the soils). On the side, d'Herelle also embraced an opportunity to ferment whisky from bananas, a product positively compared with Canadian Club, a high-end whisky from another of d'Herelle's adopted countries.[20] Later moving to Mexico, d'Herelle again moved his family, this time to the Yucatan peninsula of Mexico, where he was commissioned by the government to develop a means to produce schnapps from the agave sisal. This plant was primarily harvested for its fiber but is related to the blue agave plant that is famously used to produce tequila. Using the "throwaway" material not useful for rope making, d'Herelle was able to develop a novel and exotic variant of schnapps that resembled the anise-flavored liquors (like ouzo) indigenous to the Mediterranean basin and Middle East.

Sisal-based schnapps would not ultimately subvert tequila as the principal liquor produced in Mexico, but it nevertheless became popular in d'Herelle's native (though seldom-seen) France. This fact led d'Herelle back to Paris, where he helped oversee manufacturing and again did side jobs both for the money and the love of science. The side job was a request from the Mexican government to combat a locust outbreak. In preparation, Felix volunteered for an unpaid stint at the Pasteur Institute, where d'Herelle investigated the idea of developing and deploying bacteria that could kill locusts. The resulting discovery of the *Bacillus thuringiensis* was tested in Mexico and later Argentina, where it helped quash the devastation wrought by the insects.

Expanding the idea of using one biological organism to kill another, d'Herelle became interested in ways to combat dysentery. War and disease have a long-standing relationship. As we have already seen, it is no coincidence that the Athenian plague coincided with the Peloponnesian War and the Antonine Plague was linked with the quelling of an uprising by the Parthians. As recently as the mid-19th century, some two thirds of the 620,000 Americans who lost their lives in the Civil War succumbed to disease rather than wounds. World War I holds the distinction of being the first conflict in which more people were killed by people than by microorganisms. This improvement in large part arose because of the recognition of infectious bacteria and viruses in general, and also because of efforts by investigators including Felix d'Herelle.[21]

The advent of motion pictures in the early 20th century gives modern viewers a semblance of the murky and muddy conditions that characterized trench warfare. A fact that may be less familiar is that the mazes of trenches in Western Europe spanned from the Belgian coast to Switzerland, collectively comprising a network more than twenty-five thousand miles in all.[22] Amidst these horrible and widespread conditions, d'Herelle began a quest to identify natural agents that could kill the *Shigella* bacterium, which had been discovered by the Japanese scientist Kiyohsi Shiga just a few years before in 1897.[23] *Shigella* infection causes a violent form of diarrhea that is better known as dysentery. This disease has a long linkage with warfare, claiming the lives of famous historical characters, including King John of England (the signer of the Magna Carta), the admiral Sir Frances Drake, the humanist Erasmus, and the explorer David Livingstone (of "Dr. Livingstone, I presume" fame). Given the damp, cramped, and unhygienic conditions of trenches during World War I, fear of dysentery was endemic.

At the peak of the bitter trench fighting in September 1917, d'Herelle announced the discovery of "an invisible, antagonistic microbe of the dysentery bacillus." Specifically, Felix described an unknown biological (as opposed to chemical) substance that feasted upon *Shigella* bacteria.[24] While the exact nature of the disease-killing activity of this seemingly magical substance was unknown at the time, d'Herelle initiated a new business to supply the French military with twelve million doses of the life-saving concoction as a medicine to combat dysentery.[25] We now know that the efficacy of the substance capable of selectively and efficiently killing the causative agent of dysentery came from a bacteriophage.

D'Herelle continued his work to identify additional bacteriophage beyond the Great War, resuming his globetrotting ways both to identify new phage and to market the phage products he had already pioneered. In the midst and immediate aftermath of the Great War, d'Herelle isolated and marketed phage therapies to treat cholera, typhus, and other septic diseases. These products were broadly embraced, and a growing fame led to honorary degrees and, ultimately, a professorship at Yale University, the receipt of the prestigious Leeuwenhoek Medal, and eight nominations for the Nobel Prize.[26, 27] Likewise, d'Herelle provided the inspiration for the lead character in *Arrowsmith*, a novel by Sinclair Lewis that won the 1926 Pulitzer Prize.[28] All these well-earned awards were amassed despite the fact that d'Herelle lacked formal credentials and any idea of the composition of the beneficial concoction that he was marketing as a medicine or how the new treatments functioned. We now know that the magical substance was composed of bacteriophage (also known simply as phage), viruses that would seek out and kill harmful bacteria.

The impact of using bacteriophage to kill bacteria as a means to treat infection is reflected by the experiences of another great war. As we will see, the Soviet leadership embraced phage therapy, which was heavily deployed to treat soldiers wounded during the brief but fierce Winter War with Finland in 1939–40. Likewise, the leadership of the German Wehrmacht included strong advocates of phage-based therapy, and medical kits containing vials of phage were issued to the Afrika Korps and other German units.[29] Indeed, there is rampant speculation amongst some historians that the German decision to divide their armies to include Georgia as an early target for the 1941 Operation Barbarossa was in large part driven by a desire to acquire the phage research and manufacturing facilities located outside Tbilisi. The Soviet capabilities targeted by the Wehrmacht provides a direct connection back to Felix d'Herelle and thereby allows us to return to his remarkable story.

While d'Herelle gained considerable fame and wealth throughout the Jazz Age of the 1920s, he and his remarkable products had been all but forgotten within a decade. Phage therapy was destined to lose its luster considering the growth first in sulfa-based medications and later in antibiotics such as penicillin (innovations that were largely focused upon the English-speaking democracies during the Second World War). The reasons for the rapid diminution of phage-based medicines include their unknown mechanism of action. Because of this, it was virtually impossible to assess how efficacious, if at all, a particular batch of phage drug might be. Unsurprisingly, this vulnerability was exploited by unscrupulous or low-quality manufacturers fixated upon quick and disproportionate profits from inferior products. Such irregularities increased consumer skepticism amidst the discovery and product launch of attractive new sulfa drugs and antibiotics introduced in the 1930s and 1940s, respectively.

Unlike the advances that characterized the scientific revolutions of the interwar era, phage therapy persisted in the Soviet Union, in no small measure because of Felix d'Herelle himself. In 1934, d'Herelle received an invitation from no less than Josef Stalin to join the Tbilisi Bacteriophage Institute. Stalin had become aware of d'Herelle's work conducted by an old friend from the Pasteur Institute, George Eliava, who was pioneering phage-based medicines in the Soviet Union. For a while, it seemed that d'Herelle might finally settle down once and for all. Again, fate intervened as d'Herelle learned that Eliava had established his institute over the head of the regional Soviet party head, Lavrenti Beria.[30] Eliava had also gone over Beria's head to petition Stalin directly to extend the invitation to d'Herelle. These relatively mild actions were particularly dangerous given the figure of Beria, who embraced a paranoid hatred of influence by foreigners such as Felix. Compounding this, Beria's major responsibility was not mediating his role as overseer of Soviet Georgia but rather serving as the head of state security—the notorious NKVD.

It is thus unsurprising that when Beria and the NKVD initiated a purge of intelligentsia suspected of corruption by foreign influence, Eliava and his family were among the first corralled. On the evening of January 22, 1937, the knock on the door was quickly followed by charges of treason and the execution of George Eliava. While d'Herelle had initiated the construction of an intended permanent home in Tbilisi, he was in France at the time of

Eliava's arrest and never returned to the USSR. Despite his considerable achievements and perhaps because of the decline of phage therapeutics, d'Herelle did settle down, but his fame and business declined, and d'Herelle died in 1949 in almost complete obscurity. However, his legacy may again be revived by a new wave of interest in phage technology as a potential solution for the emergence of antibiotic-resistant bacteria.[31]

Spreading Like Viruses

Despite declining therapeutic application of phage technology, increasing understanding of bacteriophage greatly accelerated the emerging new science of virology.[32] Arguably, the most famous bacteriophage is known as T4. This particular phage has been the experimental model of choice for multiple Nobel Prize laureates, including Frank Macfarlane Burnet, Andre Lwoff, Max Delbruck, Salvador Luria, Alfred Hershey, James D. Watson, and Francis Crick.[33,34] Despite these illustrious credentials, the discovery of T4 has largely been lost to time. The earliest reference was a 1943 paper from Salvador Luria, Max Delbrück, and Thomas F. Anderson of Vanderbilt University, who described an electron microscope analysis of bacteriophage that revealed sperm-shaped viral particles.[35] With technology improvements came greater resolution, and we now know that bacteriophage T4 looks like a cross between an Apollo landing craft and a mosquito. A hollow and mathematically precise icosahedral head contains the genetic material (in this case DNA) and sits atop a ring-laden collar, which in turn sprouts an array of tail fibers that look like the spindly legs of an otherworldly insect. Rather than being used for locomotion, the tail fibers are tasked with acting like fingers that probe their environment in a never-ending search to seek out new prey. The interaction with a new bacterial victim triggers a landing sequence (again reminiscent of the Apollo lander) as the phage docks upon its soon-to-be bacterial victim. Once firmly attached, the phage body plunges down upon the bacterial surface with sufficient force to compromise the bacterial cell wall. This allows it to inject its genetic material into the bacterium. Once the DNA gains access to the interior, it efficiently hijacks the bacterial machinery to force the production of thousands of new progeny that literally burst forth from its bacterial victim like the shocking scene from the 1979 Ridley Scott sci-fi classic *Alien*. This entire sequence of events take place within a mere thirty minutes. The bursting forth of progeny multiples the infection and thereby allows the phage to quickly destroy massive swarms of bacteria within a remarkably short time.

This extraordinary efficiency, which was utilized by d'Herelle to treat bacterial infections, reflects the viral trick of appropriating the machinery of the host rather than bringing along most of its own equipment. Using this strategy of utilizing the host cell to do most of the work required for reproduction, the virus gains an extraordinary efficiency, as it only needs to encode for a small number of genes, often restricted to those that compose the viral particle structure, key enzymes needed to replicate the viral genetic material, and, of course, the machinery needed to facilitate the hijacking process. From the standpoint of genetics, one might think of viruses as the most highly evolved creatures on earth as they exploit the intense labor and machinery of their hosts for their survival and reproduction.

We will conclude our interlude with bacteriophage by relating a story that sounds apocryphal but was verified by me and during my years as a predoctoral student. This experience verifies that viruses truly dominate the planet, including even our most hygienic environments. This story also relates to the practice of scientific transparency and reproducibility, key features of the modern research endeavor. For those unaccustomed to the academic research process, the general practice is that once a scientific manuscript has been published, it is incumbent upon the authors to share their unique tools with the wider scientific community so that the work can be reproduced and thus (presumably) verified by others. Obviously, such actions might not be palatable to investigators prone to secrecy, as this practice might surrender their "edge" to the competition.

The apocryphal story is set in the early 1980s, when scientists often communicated by a system known at the time as "letters" (translating this for younger readers, this approach is more commonly referred to as "snail mail"). As was the custom, the source for this story, a professor researching bacteriophage at Duke University, sent a letter to a colleague at a prestigious East Coast institution, requesting a sample of the material that had been published by the colleague in a recent scientific paper. The colleague, who had a reputation for secrecy, responded politely a few weeks later with a letter indicating that he would not be able to honor the request. However, the bait had been taken, as the bacteriophage was at Duke and already in the hands of the professor. The reason for this is that the ubiquitous nature of bacteriophage, which waft along with the wind (often aboard the bacteria they are ingesting), meant that the letter and the envelope used to decline the request for the phage were coated with phage (and bacteria). The Duke investigator simply extracted the desired phage (along with others used by the colleague, as well as those that naturally floated around Boston and Durham), and with a little effort, purified the phage and began using it for his studies, all within a week. Such opportunities have been all but lost today due to e-mail. The only types of viruses that might be isolated in the more modern form of messaging are made by man and not nature.

Common but Not Trivial

This process of taking over the host and rewiring its programming was first discovered with viruses that infect bacteria (phage) primarily because bacteria are readily studied and less complex than eukaryotic cells. Similar processes of hijacking and replication haves been repeated by many other virus types, and it is safe to conjecture that every species on earth has a unique array of viruses that feed upon them. It may also be defensible to conjecture that virtually every type of cell in the body might be host to its own set of viruses. For example, the array of viruses that infect a hair follicle are distinct from those that infect the liver or circulating cells. Worse still, many viruses can exploit to their advantage the fundamental immune systems that have coevolved to prevent or limit viruses. For example, HIV tends to feast upon a subset of T lymphocytes that happen to control the immune system. This clever strategy undermines the ability of the host to reject the infection but ultimately dooms the larger host to succumb to other infections. As snails, insects, and other animals shed or molt to cast off unneeded or worn-out body parts, the virus constantly seeks to spread itself beyond a first host and propagate within subsequent victims.

This chapter opened with a statement that many of the most contagious and deadly diseases arise from viruses. Among the most contagious are an array of what we refer to as "childhood" diseases (measles, German measles, mumps, and chicken pox). The next chapter will deal with these maladies as a demonstration of the extraordinary successes vaccines developed in the latter half of the 20th century. For now, we will focus our attention upon the common cold, one of the most contagious viruses known to science.

According to the Centers for Disease Control and Prevention (CDC), the average adult suffers from two to three colds per year. Worse still, the infectiousness of the common cold is relatively high, with each infected person in turn passing the disease on to six other people.[36] Each cold lasts a week or longer, and many adults absent themselves from work during this time to avoid infecting others. Thus, while mortality is rarely threatened, the economic impact of the common cold can entail billions of dollars in reduced productivity alone. Given the incidence, impact, and contagiousness, much early effort was focused upon identifying the cause of the common cold.

Early works revealed that the causative agent was not captured by Chamberland filters, so it was accurately suspected to be some sort of virus. In 1956, word emerged from Johns Hopkins University of the discovery of the causative agent.[37] The discoverer was a Baltimore scientist by the name of Winston Harvey Price, who had high aspirations for fame.[38] A native New Yorker and son of a wealthy physician, Price had become inspired by Sinclair Lewis's *Arrowsmith* (based on Felix d'Herelle's life). He started his career working with several prominent scientists and institutions, including Princeton University and the Rockefeller Institute, before landing a position at Johns Hopkins. Whether consciously or not, Price had begun down a path that aped both the real-life d'Herelle and the fictional Arrowsmith. For example, Price's early scientific work followed upon the bacteriophage discoveries pioneered by d'Herelle, but Price's notoriously short attention span meant that he did not contribute substantially to the field during its development in the late 1940s.

On the personal front, Price also emulated his literary hero. Like Arrowsmith, Price married first an unexceptional woman but later sought out the company of a more elite partner.[39] (In real life, Price and his wife divorced, whereas Arrowsmith's wife suffered a gruesome death from poisoning.) The second wife was the result of his passion for abstract art. Price was an avid collector and paid a substantial sum for a painting by the New York artist Grace Hartigan. Hartigan was a leading talent in the second-generation American abstract expressionist community and ran in the same circles as Jackson Pollock, Willem and Elaine de Kooning, and Mark Rothko. The newly married Hartigan invited Price to visit her gallery in 1959. Unexpectedly, the two initiated a torrid affair, and both divorced their current spouses and remarried within months.

By 1960, Price was making quite a name for himself, both in the scientific and artistic communities. Within the former group, Price advocated a belief that infectious agents were normally present in most people and the diseases caused by them were triggered by environmental factors such as getting a chill. In studies reportedly conducted on thousands of volunteers, he advocated that physiological stressors were sufficient to trigger a collapse of the immune system (as evidenced by decreasing amounts of antibodies circulating in the blood), thereby liberating latent viruses and rendering individuals susceptible to infection. Such ideas were recognized as groundbreaking by some and made for great headlines, being touted as evidence by the *Science News Letter* of January 8, 1955 as evidence that "Grandma seems to have been right. You may be able to catch cold by getting your feet wet and sitting in a draft."[40] Indeed, the article goes on to relate that Price had received prestigious and lucrative rewards such as the Theobald Smith award from Eli Lilly & Company.

Although Price was the subject of much attention, his own colleagues at Johns Hopkins felt he was superficial in his ideas and liberal with the interpretation of the data.[41] Such charges had plagued Price from his earliest school days. While Price was receiving acclaim for linking cold feet with susceptibility to getting sick, he was working to discover the cause of the common cold. Using nasal washings from patients with a cold, Price reported the isolation of a new virus, which he called the JH virus (for Johns Hopkins).[42] Like most of his studies, the results were a bit superficial and open to questioning. The discovery of the JH virus (whose name was later changed to rhinovirus—*rhino* being the Latin word for 'nose') meant that greater notoriety would be accompanied by greater scrutiny of Price's methods and interpretations. For example, a widely reported 1957 study reporting the discovery of a vaccine for the JH virus (and thus an end to suffering from the common cold) lacked substance. Indeed his claims were refuted both by his supervisor at Johns Hopkins and by the preeminent vaccine aficionado Maurice Hilleman (whom we will meet later in our story), who claimed Price's work to be "a complete fraud."[43] Price quietly left the field and went on to study other areas of medicine. The discovery of what we now know as the rhinovirus would ultimately represent a high-water point in terms of Price's scientific contributions.

Challenges of Treating Viral Infections

As a young academic embarking on a career of teaching, I was asked to deliver a lecture on the first day of my inaugural academic year as a professor. The lecture was on antiviral medicines for a medical school course on pharmacology. Excited by the prospect, I did a lot of research. While more than a handful of antiviral therapeutics had been developed in the years before the lecture, the first sentence of my first lecture remains committed to memory: "The number of viral infections that can be definitively cured by modern medicines can be estimated to be exactly zero."

Our introduction to the common cold virus provides a jumping-off point to understanding some challenges associated with the treatment of viral infections. A prominent hurdle entails the timing during which one can successfully intervene against a viral infection. With a bacterial infection, the clinical signs (fever, chills, and malaise) often reflect the body's attempt to attack and clear the infection while it remains relatively benign (to avoid more lasting damage or death). As we saw in the last chapter, the human immune response has evolved to recognize and vigorously react to certain triggers, such as the "superantigens" associated with the cell walls of many bacteria. Consequently, even a trace amount of residue from a bacterial pathogen might be sufficient to fully alert the immune system to begin taking corrective actions. The malaise associated with a bacterial infection, including redness, swelling, and fever, is often attributable as much to the defensive counterattack as to any damage caused by the pathogen itself. Such strategies evolved because, as we have seen, the toxins produced by some bacteria are among the most lethal poisons known to man. Over time, the balance between allowing a bacteria to gain a foothold and overreacting, thereby causing potential collateral damage to normal cells, tends to favor long-term survival over short-term disquiet.

The situation with viruses can be very different. In general, the early warning signs that trigger the host defenses against bacteria are not applicable to viruses. The symptoms of a viral infection occur later during infection and often reflect pathogen-mediated damage to the body rather than a vigorously responding immune response. Thus, the earliest symptoms of a viral infection often arise after the infection is already well entrenched. If the viral infection is particularly aggressive or acute, it may be too late to intervene successfully. In recent years, the public has gained greater understanding of this limitation through the marketing of products such as Zicam® and Cold-Eeze®, which are zinc-based products that can lessen the severity or duration of symptoms from the common cold. Such medicines are effective in decreasing the duration or severity of infection only if taken within hours after the first noticeable symptoms. Likewise, the treatment of shingles (a subject to which we will return) is effective if treatment with drugs such as acyclovir is initiated within the first forty-eight hours after the first signs of the blistering characteristic of the disease. One obvious problem compounding this limitation is that most people tend not to realize they are infected or respond accordingly until after the tight window in which intervention might be useful has already closed.

Therapeutic intervention with antivirals assumes that one has access to effective medicines. For most viruses, this is not the case. This raises a key question. Given the impact of antibacterial medicines such as penicillin, why are there comparatively few antivirals? The answer in part lies in the fact that the proteins of the virus-hijacked cell are recruited, or more accurately enslaved, by the virus to build new viruses. Therefore, if one targets these proteins with a drug, the consequences likely will include intolerable collateral damage to normal human cells. In contrast, most of the molecules needed to facilitate the growth of bacteria are quite foreign, due to extensive evolutionary changes that distinguish our prokaryote cousins from ourselves. These differences provide targets for new medicines. Thus, a key (and limitation) to developing safe and effective antivirals is to identify and distinguish the small number of molecules that are unique to the virus.

The Love Bug

The idea of successfully targeting a molecule that is unique to a virus is perhaps best exemplified by one of the most maligned (though not malignant) viral diseases. Attending high school in the early 1980s, I was subject to the requisite health education courses, the highlight of which was the discussion of sexually transmitted diseases. "The big three" killers were consistently portrayed as herpes, syphilis, and gonorrhea. All else paled in comparison. (Reports of a "gay cancer," later given the less derogatory moniker Acquired Immunodeficiency Syndrome, or AIDS, were sparsely heard on the media in the Midwest.) Even after the revelation of the magnitude of risk from AIDS, the perceived risk of acquiring herpes elicited the greatest fear among most peers in the late 1980s. Such fears were propelled by frequent mention of herpes, particularly in comedies (after all, the word *herpes* bounds off the lips with a droller sound than the more ominous undertones of *gonorrhea* or *syphilis*).

Despite its notorious reputation, genital herpes is mostly a benign disease. The family of viruses that cause herpes are broadly known as the Herpesviridae. The family members include HSV-1, the primary cause of small ulcerations in or near the mouth such as cold sores and aphthae (or canker sores, as they are known in North America). Individuals are infected with HSV-1 after contact with an infected person, who is actively shedding virus. The virus tends to infect the skin and nerve cells of the mouth. During an active infection, the virus will hijack and take over the cell it has infected, subverting the normal function of the cell machinery to redirect it to produce more viruses. As described at the top of the chapter, viruses have played a biological version of the old television show *Name That Tune*, where contestants would challenge one another to name a song with the fewest notes possible. Likewise, viruses seek to minimize the amount of nucleic acids and genes they convey as one means to remain efficient. For some of the most effective viruses, the genes are generally limited to some of the structural materials that compose their capsules and key enzymes needed to reproduce their genetic material. For the more mundane jobs, many viruses simply subvert the cell's native machinery to do their bidding.

Once a host cell has been successfully infected by HSV-1, it can choose from among three options. First, the virus can remain latent and simply "hang out" in the infected cell, remaining largely hidden (latent) from the immune system and not causing overt symptoms. Second, the virus can take over the cell, reproducing itself at a dawdling rate and budding out of the infected cells in a manner analogous to a slow-dripping faucet. In this case, the cell may remain alive but spew forth a constant low level of infectious particles that in turn can persistently spread infection to new cells and new hosts (if, for example, the mouths of two individuals happen to come in intimate contact). When feeling particularly aggressive, HSV-1 will completely take over a cell; after it has fully engorged the infected cell with new viral progeny, these burst open the cell in a process known as a lytic cycle. If the new progeny infect nearby cells and continue this aggressive approach, the damage will accumulate and become sufficient to destroy enough of the skin surface to form a painful ulcer, known as a canker or cold sore (depending on the location of the ulcer). How a virus chooses which of the three paths to take remains largely unknown to researchers, although preference for the lytic cycle is associated with certain stimuli, such as particular acidic foods, psychological stress, and immune suppression. Fortunately, the damage caused by HSV-1 tends to be rather uncomfortable but generally limited to ulcerations that heal within a week or two.

Despite its dreadful reputation, the genital form of Herpesviridae infection, HSV-2, tends to be remarkably similar and almost as benign as its HSV-1 cousin. Contrary to conventional wisdom, outbursts of HSV-2 are generally limited to localized, albeit uncomfortable, ulcerations on and near the genitals. Like HSV-1, these sores tend to heal and indeed may go completely unnoticed, as evidenced by a 1997 CDC report. This study revealed that while one in five adult Americans was infected with HSV-2, 80 percent were unaware of the fact.[44] The notable exception to the idea that genital herpes is relatively benign arises with the very young. When an infected mother delivers an infant vaginally, the infant may encounter an active sore during the vaginal birth. The relative weakness of the newborn immune system can allow the virus to run rampant and cause irreversible brain damage, respiratory or liver failure, and death. Fortunately, even the most extreme infections with Herpesviridae tend to respond favorably to a series of medicines developed in the early 1970s, with thanks to a gift from the briny depths.

Tectitethya crypta is a species of sponge found in the shallow waters of the Caribbean Sea.[45] These sponges are among the oldest multicellular creatures on earth, with origins in the period that immediately followed the Cryogenian period three quarters of a billion years ago. This period is known as "Snowball Earth" (or perhaps more accurately but less sexy from the standpoint of popularizing science, "Slushball Earth"), when the entire planet was encased in ice, snow, and magnitudes of dirty slush (not unlike a typical Boston January). Somewhat more recently, in 1950, a descendent of these early sponges was isolated from the waters off the Elliot Key in Florida. This specimen was shipped to New Haven, Connecticut, where Professor Werner Bergmann of Yale University dissolved a sample of the sponge in acetone (better known as finger nail polish remover).[46] The reason for doing this was not based upon an interest in aquatic creatures but because Yale was amidst a scientific and medical frenzy to develop new medicines for cancer. As elaborated in greater detail in *A Prescription for Change*, Yale was the epicenter in the postwar years for identifying compounds that could restrain the deadly disease.[47] The Yale study was disproportionately successful as it identified a series of molecules that would revolutionize cancer treatment, including the antimetabolites, cytarabine, vidarabine, and gemcitabine. An antimetabolite is generally a natural molecule that somewhat resembles an essential metabolite (or foodstuff) needed for the growth and survival of normal cells. As such, one can think of metabolites as the medicinal version of the Trojan horse supposedly invented by Odysseus and popularized by the ancient Greek poet Homer three millennia ago.

Cancer cells are often characterized by a literal feeding frenzy as they gobble up all nutrients that might be useful to support their rapid growth and metabolism. As such, the feeding of so-called antimetabolites that sufficiently resemble a potential foodstuff but are different enough to cause lethal indigestion can provide a means to target cancer cells. This basis of "selectivity" allows these drugs to be particularly noxious to cancer cells, while the fussier benign cells generally ignore antimetabolites. The strategy of discovering and deploying antimetabolites was exploited in the late 1940s and throughout the 1950s to identify new medicines. The need for ever more new medicines propelled the discovery of novel species across the planet (and under the seas), most of which were harvested to sample their antimetabolite activities.

Partnering for Cures

The concept of developing new medicines based on nature-derived (or natural) products was the hallmark of drug discovery since the earliest discoveries of botanicals such as the willow tree and the poppy, which ultimately led the way to aspirin and morphine, respectively. Arguably, the most prolific duo ever to harvest the potential of nature to target disease is the unlikely team of Gertrude ("Trudie") Elion and George Hitchings. According to the autobiography she composed to commemorate her 1988 Nobel Prize in Medicine (one of the most entertaining ever written for such an occasion), Gertrude Belle Elion was born in New York City "on a cold January night (in 1918) when the water pipes in our apartment froze and burst."[48] The precocious daughter of an immigrant family from Eastern Europe (whose fortune, like that of so many other families, was lost in the 1929 stock market crash), Gertrude's fertile mind was particularly drawn to thoughts of travel to distant lands though the family's limited financial means precluded such dalliances. Her father, Robert, had arrived in the United States from Lithuania at the age of twelve. He not only quickly assimilated into the new language and culture of his adopted land but also rose to become a successful dentist.[49] Likewise, Trudie's mother, Bertha Cohen, came to the United States from Poland at the age of fourteen and enrolled in night school, both to master the language and to gain skills needed for a career in the textile industry. The young and hardworking family expanded with the birth of Trudie's younger brother, Herbert, in 1924, and by the immigration of Bertha's family to the United States. Trudie excelled at her studies, skipping a grade on two separate occasions, thus allowing her to graduate from high school as a young adolescent of fifteen. Trudie was contemplating college versus the job market when her grandfather finally succumbed to an agonizing bout of stomach cancer. In her sorrow, Trudie proclaimed, "No one should suffer that much" and later recognized this to be a major turning point that started her down the path to develop new medicines for cancer.[50] Despite severe constraints on the family's financial resources that might have precluded further study, the family urged Trudie to enroll in college. Her high marks earned her a full scholarship at the all-girls Hunter College of the City University of New York (CUNY). Fascinated by the sciences and motivated by the fresh desire to alleviate the types of suffering she witnessed during the long, agonizing demise of her grandfather, she chose to major in chemistry rather than biology, due to her repulsion at the idea of having to perform research on animals.

The young graduate, who crossed the stage of Hunter College in 1937 to accept her bachelor's degree in chemistry, summa cum laude and Phi Beta Kappa, was nonetheless consistently rejected in her many attempts to apply her new knowledge.[51] Despite hard-earned credentials and a promising young intellect, the ongoing economic strains of the Great Depression aligned with the prevailing misogyny of the day to preclude Trudie from receiving offers of employment from the research laboratory jobs that had motivated her from the time of her grandfather's death. Women of that time (and for years to come) had essentially three career paths: secretary, schoolteacher, or nurse. Unqualified for the latter, Trudie enrolled in secretarial school but soon realized this path was not sufficiently fulfilling. She bounced among a bevy of part-time jobs (mostly as a secretary or teacher) and volunteered in a research laboratory at night, all the while saving the little money she earned to pay for the tuition needed to earn a master's degree in chemistry. Whereas her interest in chemistry had been rather unexceptional at the all-women's Hunter College, where she'd been one of seventy-five chemistry majors in her class, Trudie was the only woman enrolled in the master's program at New York University. During the two years needed to achieve the degree, Trudie lived a double life, serving during the days in the stereotypical role as a secretary and substitute high school teacher while at night she devoted herself to completing a master's thesis in advanced chemistry.

As she was embarking on her professional development, her personal life would first blossom and quickly wither. Within weeks after her graduation from Hunter College, Trudie fell in love with the man who would change her life forever, though perhaps not in the conventional way. The subject of her affection was a fellow CUNY alumnus by the name of Leonard Canter. Leonard had majored in statistics, and the two were well on the way towards planning their lives and settling down, a particularly impressive feat given the time commitments Trudie was juggling. On a spring day in 1941, Leonard began to develop chills and experience sudden weight loss and night sweats. Within days a shortness of breath was accompanied by pains in the chest and reflected the fact that a bacterial infection had taken hold of Leonard's failing heart. The resulting bacterial endocarditis likewise broke Trudie's heart with the death of the 21-year-old Leonard Canter on June 23, 1941. This tragedy renewed Elion's commitment to alleviate suffering through the development of new medicines. She would forever forsake additional romantic partnerships in deference to her love of both Leonard and science. (On a personal level, Trudie doted upon her brother's children and occasionally referred to them as "my children" rather than as her nieces and nephews.)

On the same day Leonard's heart stopped beating, the radio airways were filled with news that German tanks had begun pouring across the Soviet border, greatly expanding the war from a Western European conflict into a full-scale Second World War. As we know in hindsight, the global nature of the war would extend further with the inclusion of the United States and Japan; preparations for the looming war had already begun in Washington by the summer of 1941. The Selective Training and Service Act signed in September 1940 by Franklin Roosevelt disrupted the career aspirations of many American men and had the unintended effect of offering opportunities for Trudie. Her first work experiences as a chemist were admittedly menial as Trudie was hired by the Quaker Maid Company to perform a series of food quality tests such as testing the color of egg yolks destined to be used for mayonnaise production.[52] Despite the fact that she'd obtained a laboratory role, her desire to perform original research predominated, and she applied for and finally received a position as a research chemist at Johnson & Johnson in 1944. Again, excitement turned to frustration as the company disbanded that particular research team later that same year.

Within days after being furloughed from Johnson & Johnson, Gertrude was visiting her father's dental office when she noted that the anesthetic he routinely administered was produced by a local company just up the road in Tuckahoe, New York. By the end of that week, Trudie had interviewed with and accepted an offer by George Hitchings at Burroughs Wellcome, which was not a local company at all but an English-American pharmaceutical conglomerate that was well-known for developing innovative medicines.

George Hitchings, like Elion, was inspired to develop new medicines by a personal tragedy: the death of his father when George was only twelve.[53] The young Hitchings was also captivated by the life and work of Louis Pasteur. After obtaining his bachelor's and master's degrees from the University of Washington, Hitchings obtained a PhD at Harvard focused on the chemistry of purines, which comprise two of the four letters of the DNA alphabet. He later focused his research at the Wellcome Research Laboratories in Tuckahoe. Working largely alone at first, Hitchings hired Gertrude Elion based on the enthusiasm and intellect she displayed during her interview. This ignited a professional bond and partnership that would persist for many decades to come.

Hitchings was actively exploring concepts that would later provide the pair with the foundation for developing antimetabolite medicines in the fight against cancer. Trudie's early work identified molecules that were similar but different from the constituent components of DNA, especially the purine molecules that Hitchings had focused upon during his doctoral studies. To put this work in proper perspective, Elion and Hitchings' work predated the seminal discovery of the structure of DNA by James Watson and Francis Crick by three years and before the comparably groundbreaking research that would be started by Werner Bergmann at Yale a year later. Despite the fact that the Yale group gained the initial fame for discovering the impressive sponge that yielded such wonderful new options for cancer therapy, Elion and Hitchings went on to discover an even wider variety of new antimetabolites with applications for a broad set of diseases. The pair also realized that these same approaches might be useful to selectively targeting viruses. The dreaded (though not terribly dreadful) herpesviruses provided an early proof for these concepts.

The brilliance of viruses in avoiding drug-mediated extinction lies in their hijacking the machinery of their hosts (such as humans) to perform most of the work needed for their replication and further spread. The Burroughs Wellcome team realized the key word in this statement is *most*, and that the small number of essential proteins that viruses carry themselves (rather than stealing from their hosts) might provide targets for new medicines. In the case of herpesvirus, Elion and Hitchings exploited the fact that this family of viruses utilizes its own enzyme, known as thymidine kinase, to perform a function that is essential to allowing the virus to carry on its path of destruction. Humans also have their own version of this molecule, but, for reasons that are fortuitous for our species, the virus is biased towards its own version of thymidine kinase. By developing compounds that could selectively target the viral thymidine kinase, new medicines could theoretically be made to prevent herpes infection.

Theory became reality when Elion and Hitchings unveiled acyclovir (trade name: Zovirax) to the world in the mid-1970s. This revolutionary medicine provided a means to treat a wide array of herpesviruses, including not only oral and genital herpes but also those behind other well-known maladies such as chicken pox (and shingles), in addition to cytomegalovirus (which can cause extraordinarily painful infections of the eye) and Epstein-Barr virus (the cause of mononucleosis).

In these pioneering studies, Elion and Hitchings pioneered an approach that is still utilized today: exploiting the differences between humans and viruses to develop effective countermeasures. Hitchings went on to lead all research at Burroughs Wellcome from 1967 until 1976, and Elion led the Department of Experimental therapeutics through the early 1980s. The pair developed an unprecedented array of new medicines using their innovative approaches until and beyond Elion's retirement from Burroughs Wellcome in 1983. In retrospect, retirement seems to be a bit of an overstatement, given Elion's role in a plague that was coming to light just as she was stepping down as department head.

At roughly the same time Trudie was clinking champagne glasses and receiving heartfelt congratulations for her retirement from Burroughs Wellcome, a French laboratory at the Pasteur Institute was reporting on the identification of a virus implicated with the spread of a new disease that was quickly rising to epidemic and pandemic levels on both coasts of the United States (though focused at first in New York and San Francisco) as well as more sporadically throughout the rest of the United States, Western Europe, and Africa.[54] The so-called "gay cancer," or Gay-Related Immune Deficiency (GRID), underwent a series of name changes, finally gaining the less offensive and more familiar moniker of Acquired Immune Deficiency Syndrome (AIDS). Once the virus had been identified and the severity of the growing crisis was made known to the American people, a collective effort was made to identify molecules that could selectively target the virus with minimal collateral damage to normal cells. While the recognition of the slowness of the official response has been profiled in many outstanding reports, both old and new, within four years after the identification of HIV in 1983, the first new medicine targeting the deadly virus was approved by the FDA.[55]

Though technically "retired," Trudie was still at work in the laboratories of Burroughs Wellcome. A breakthrough in the early treatment of HIV/AIDS came from the discovery of the enzyme that the virus uses to reproduce its genetic material. This molecule, known as reverse transcriptase, was a particularly tempting target because it differs greatly from human enzymes, which provided an opportunity to develop medicines that could safely target the vital viral molecule. Trudie quickly mobilized her colleagues to screen for antimetabolites that might intervene against the reverse transcriptase molecule that was so essential for the new invader. Within four years, Trudie's team had spearheaded a scientific and medical campaign to expedite and gain FDA approval for another antimetabolite called zidovudine. The metabolic Trojan horse, also known as AZT, was carried by reverse transcriptase and incorporated into the viral genetic material, where it poisoned further viral nucleic acid synthesis, preventing its further spread.

Over the following few years, increasing understanding of how the virus infects host cells, hijacks their functions, and spreads to other victims allowed researchers to identify additional viral targets. One example is the HIV-1 protease, a protein encoded by the virus that effectively functions as a pair of scissors. It cuts and forms the lethal molecules of the virus in a manner analogous to the use of pattern shears in the deft hands of fashion designers as needed to craft intricate designs from formless pieces of fabric.[56]

At the same time, a committee of scientists in Sweden was secretly conferring to recognize Trudie and George with a Nobel Prize. Trudi never did earn a PhD in the conventional manner, but she'd been awarded no fewer than twenty-five honorary (but certainly earned) doctorate titles. Moreover, Trudie was invited to talks all over the planet, which finally sated her desire for travel. On a more practical level, while the discovery of acyclovir and AZT were landmark events, the value conveyed by medicines were, are, and will forever be limited by an evolutionary battle between viruses and man. While we occasionally win battles (such as the discoveries of the aforementioned therapeutics), we are unfortunately predestined to lose the war against HIV using conventional weapons such as AZT.

Collateral Damage, Insurgents, and Spider Holes

The militaristic-sounding name of this section reflects the fact the development of medicines and vaccines against viral diseases can be regarded in many ways as analogous to an arms race between two superpower rivals. Offensive and defensive measures are constantly deployed and modified by attackers and defenders. Tactics need to change constantly for both because defenders readily deploy countermeasures that are provided by evolution or technology. A virus has a particular advantage in this battle due to its inherent ability to rapidly and frequently mutate and evolve. As we have seen, propagating its genetic material (generally DNA or RNA) is one of the few activities for which viruses consistently rely upon their own machinery. To do so, the virus encodes for a molecule known as a polymerase, whose job it is to reproduce the virus's genetic material (DNA or RNA, depending on the virus). One example of such a "polymerase" is the reverse transcriptase (RT) targeted by AZT, which made up the first generation of antiretroviral drugs for HIV. This rather awkward nomenclature reflects the fact that the genetic material for HIV is RNA. Whereas most conventional human (and all eukaryotic life) uses DNA as a means to produce RNA through a process known as transcription, HIV does the opposite. To make copies of the virus RNA, it uses RT to create DNA (the reverse of the normal situation in humans) and using this DNA intermediary, the RT then transcribes the DNA into RNA. Hence the combination of the words "reverse" and "transcriptase". The advantaged for humans combating the virus is that RT tends to have a different set of restraints than human DNA polymerases and thus will utilize some Trojan-horse like molecules such as AZT, whereas human polymerases prefer not to do so.

What was not conveyed above was that AZT as a stand-alone drug provided only a temporary victory in the war against HIV/AIDS. Reverse transcriptase and evolution provided an escape route for HIV to avoid AZT. Specifically, the virus could mutate ever so slightly until it found, by random chance, a mutation that was resistant to the antiviral effects of AZT. Whereas other viruses that remained sensitive to AZT were efficiency dispatched, nature allowed the drug-resistant mutants to persist. Such mechanisms are well known, as we have seen with the widespread development of drug-resistant variants of bacteria such as methicillin-resistant staphylococcal aureus (MRSA) or multi-drug resistant tuberculosis (MDR-TB). However, nature has conferred particularly powerful abilities upon viruses such as HIV.

More accurately, evolution generally conveys upon viruses error-prone means to replicate their genetic material. To put this in perspective, think back upon the earliest publishing industry, which recalls images of monks with bald pates reproducing copies of canonical manuscripts by hand. Occasionally, one encounters significant errors, such as the dropping of one key word that caused the so-called "Wicked" Bible to instruct its followers,

"Thou shalt commit adultery."[57] However, such gross errors in humans, bacteria, and most other living creatures are subject to rigorous fact-checking, and errors are efficiently identified and removed.

A very different situation arises in viruses. Evolution has largely removed the constraints on error checking. Mistakes are thus actually encouraged as a means to adapt to ever-changing environments and the immune system. Whereas the error rate committed by human polymerases is something on the order of one mutation per every ten billion, the mistake rate for HIV is closer to one in two thousand.[58] Some mutations cause minor changes, but the majority are fatal. The result is the production of many viral particles that are doomed from the beginning. At first glance, this might seem like a very inefficient strategy, as most viral progeny erupting from an infected cell are dead or ineffective. However, recall that each infected cell may put forth thousands of viruses. Consequently, each virus sprouting from an infected cell is substantially different from its brothers and sisters. This additional genetic diversity is useful in their harsh reality, where the entire world is out to get them. While such changes evolved to allow viruses to adapt to constant attack from the immune system and outcompete other viruses to infect the next wave of victims, this same system was also rather useful in contending with man-made assaults from antiviral drugs. Unsurprisingly then, while AZT and other RT inhibitors carried the day shortly after their introduction in the mid-1980s, by the end of the decade these medicines were rapidly at risk of being consigned to obsolescence.

In recognition of this losing battle against HIV (and other viruses), humans relied upon their strengths in technology and growing knowledge of how viruses attack and defend. Within months after the first clinical studies with AZT, scientists recognized that drug-resistant HIV viruses were emerging via simple Darwinian selection. What was to be done? Investigators at Rockefeller University, working with scientists at Merck & Co., realized a combined assault upon HIV with multiple drugs, each attacking a different part of the HIV life cycle, might provide a challenge that would overwhelm the ability of the virus to parry several attacks at once.[59] Such an approach had already proven useful for targeting many cancers, another set of diseases characterized by a high rate of mutation/evolution. The approach was popularized and advanced by David Ho, a

Taiwanese-born American researcher who was among the first physicians to encounter HIV/AIDS while serving as an internal medicine resident at Cedars-Sinai Medical Center in 1981. By 1996, the combination of multiple different antiretroviral drugs showed impressive results, and Ho was named Man of the Year by *Time* magazine.[60] The impressive results of combination therapy, also known as AIDS Cocktail Treatment—a moniker that falsely conveys a convivial image—allowed a diagnosis of HIV infection to evolve from a death sentence to a long-term medical malady, comparable in some ways to a diagnosis of high cholesterol or hypertension.

The combination of different medicines to combat HIV/AIDS leads us to the next militaristic analogy, that of "collateral damage." A fundamental advantage of drugs like AZT is that they are more readily taken up and used by the viral polymerase (RT) than by the enzymes involved in the host (human) DNA replication. However, some of the drug is indeed utilized by human cells, and this so-called leakiness often results in the killing of these bystander cells. Such realities explain the military analogy to collateral damage, because many normal cells are killed or damaged by most antiviral drugs, and the consequences can be substantial. For example, the side effects of antimetabolites such as acyclovir and AZT often mimic those more closely associated with cancer chemotherapy: gastrointestinal distress, bone marrow suppression (e.g., anemia and lowered blood cell counts), and, in some cases, carcinogenicity.[61] Hence, the combination of multiple and different drugs is not without its own set of lifelong issues, though some improvements have been achieved by cutting back the dose of some of the components of day-to-day therapy. Despite these enhancements, the side-effect profile of antiviral therapy can decrease compliance. One example is a "drug holiday," or taking some time away from the medicine, perhaps at a time when one is on an actual holiday and thus at a time when one wants to feel their very best. Just a few days before David Ho was named *Time*'s Man of the Year, an article from a Stanford University team revealed the perils of drug holidays.[62] Drug holidays relieve the pressure on HIV; as the amount of drug tapers down (and is not replenished), the virus gets an opportunity to test new ways to avoid the medicines. When the holiday ends and drug compliance resumes, even a short period of days may be sufficient to allow some of the virus particles to become less sensitive to the drug. The Darwinian race restarts, and inevitably enough of the virus will become sufficiently resistant that it can subvert the effectiveness of the new medicines and continue its path of devastation. Unless drug holidays can be eliminated, which seems impossible, the idea that we can "cure" HIV/AIDS will inevitably be proven false.

As if our story weren't already sufficiently disheartening, another compounding problem is that some viruses have the frustrating habit of become "latent." One example is retroviruses such as HIV/AIDS. These viruses (of which HIV is but one member) have the annoying habit of incorporating themselves into the DNA of their hosts (you and me), where they can hide out for months, years, or even decades.[63] Sometimes these endogenous viruses (endogenous is a Greek term meaning "inside genes") remain within our genetic material for days, years and even generations. A recent estimate suggests that as much as 5–8 percent of human DNA is composed of proviruses, also known as endogenous retroviruses (or ERVs).[64] This knowledge should not be the source of disproportionate angst, as it is unlikely that innumerable viruses will spontaneously erupt. Indeed, most of these ERVs are ancient or no longer able to become infective. Quite the contrary—as our understanding, as well as the raw count, of ERVs has increased, we increasingly appreciate that the evolution of humans (and all other species) has been greatly assisted by ERVs, some of which can transfer genes among individuals and, in some cases, across otherwise disparate species.

Another form of viral latency is characterized by diseases such as genital and nongenital herpesvirus infections, including shingles. In these cases, the viral DNA does not incorporate into our chromosomes but instead coils upon itself, like a snake biting its tail, into a structure known as an episome. These episomes can remain quietly within the cell interior (known as the cytoplasm) and occasionally spin off progeny from time to time to test whether the conditions are right for the virus to reemerge en masse. As discussed earlier in the chapter, the reemergence is associated with the intermittent periods of latent and active infection associated with both genital and nongenital herpes, which tend to hide away in nervous system tissues. An even more dramatic example arises with shingles, a disease caused by herpes zoster virus, the same pathogen that causes the childhood disease known as chicken pox. Zoster hides away in the large dorsal root ganglia near the spinal cord. The persistent insurgent frequently tests the ability of the body's immune response to target the virus. Failed attempts by the virus to re-emerge and renew attacks upon the body occur constantly but are quashed by the immune system and rarely cause noticeable symptoms. Under conditions of a weakened immune system, the virus may evade detection long enough to break out (literally). An early raiding party may consist of relatively few virus particles, which form small bumps on the skin, hardly noticeable and not particularly problematic. As the disease progresses, the bumps become blisters that fill with pus and eventually break open, spreading the infection further, encompassing large patches of skin and causing painful eruptions that can last for weeks. These outbreaks fortunately tend to be contained to a ganglion (a single group of nerve bundles that radiate from the spine). Thus, shingles tends to occur in almost linear outbreaks that follow the major sets of nerves, remaining on one side of the body or the other. Given that the infection occurs within and near key nerve endings, shingles eruptions are particularly painful and rank among the most problematic issues for patients, as well as for people who care for those with immunosuppression, such as elderly individuals.

Fortunately for those with an intact immune system, the response to zoster virus can be boosted with a relatively new vaccine. This provides a segue to return from discussions of the immune system and its varied microbial enemies to the concept of vaccination. While the statement about drugs inevitably failing to cure viral infections remains unfortunately still relevant today, we do have one proven weapon that has demonstrated extraordinary ability to prevent or treat disease: vaccines. We now turn to these vital treatments and resume our discussion of vaccines with an examination first of passive and then of active vaccination.

6

A Sense of Humors

In the last two chapters, we began to introduce an array of microscopic pathogens capable of causing death and disease. The list is never-ending, as new pathogens are discovered at an accelerating rate and evolution ensures new pathogens will forever plague our species. New threats and old in the forms of exotic pathogens such as methicillin-resistant *Staphylococcal aureus* (MRSA), chikungunya, Ebola, Zika, and Marburg virus have grabbed headlines around the world and are the source of much anxiety. As the world becomes smaller, hotter, and more crowded, the conditions are set for encounters with ever more pathogens. We have also seen that the development of new therapeutic interventions against microbial threats is particularly challenging and often provides only temporary relief due to acquired drug resistance. Rapid-fire successes in developing antibacterial therapies from the 1930s onwards created a false sense of security that our species had conquered the microbial world, but this complacency has been shattered with the emergence of antibiotic-resistant "superbugs." Worse still, our ability to combat viral diseases has never met expectations and, for the reasons detailed in the prior chapter, likely never will.

In contrast to the shroud of doom that currently pervades anti-infective medicines, rare, extraordinary, and enduring successes in the war between men and microbes have been achieved by vaccines. These vaccines were mostly discovered by mobilizing the immune systems of mice and men to recognize foreign invaders. We now return to the subject of the immune system.

You may recall from chapter 3 that the topic of immunology is extraordinarily complex. One subject not addressed in the earlier discussion was a key element of our body's defenses: antibodies. The term *antibody* is widely known in popular culture, but it is not as well understood by many. These vital proteins are a key component of the milieu of biological substances that keep most people healthy and alive on a day-to-day basis. This chapter will focus on these magical substances that can alert the body to threats and then function as executioners. This subset of remarkable immune-triggering proteins conveys the ability to recognize virtually every molecule that has, can, or will exist in nature (and many others that do not).

Antibodies are the fulfillment of a concept known as a "magic bullet" as conceived originally by Paul Ehrlich. In 1878, the brilliant German scientist envisioned in his doctoral thesis hypothetical chemicals that could selectively target disease with exquisite fidelity and efficacy.[1] This dream has largely been realized for eons by the immune system. More recently it's been translated into modern medicine with the isolation or creation of these antibodies.[2]

As a means of understanding how and why antibodies and vaccines work, we will briefly recount a history of antibodies and distinguish how their isolation and use as medicines created an entirely new type of vaccine not envisioned by Edward Jenner (or even by Benjamin Jesty). Each one of us has the capacity to manufacture an extraordinary number (think of Carl Sagan's catch-line of "Billions and billions") of different antibodies. As we go through life, we encounter many potential threats to our well-being, including bacteria, viruses, fungi, and cancer cells. Antibodies provide one means the immune system has evolved to help distinguish and eliminate these potential threats (along with the cell-based treatments discussed above). Like the task facing T-cell, the complexity required to patrol for real and imagined invaders (bacteria, viruses, cancer cells, *etc.*) is enormous. Consequently, there must exist at least a comparable complexity in the ability of antibodies to recognize these manifold challenges.

The greatest strength and limitation of an individual antibody is that it can recognize only one small snippet of a chemical, cell, or protein. This site where the antibody binds its target is known as an epitope. One can only imagine the innumerable epitopes that exist in the natural world. As an analogy, there exist trillions upon trillions of different locks, yet the milieu of antibodies that each of us produces can serve as keys for each of these locks, and more.[3,4] Each of us contains the potential to recognize all potential types of foreign invaders, both real and imagined. The key word of this last sentence is *foreign*. As described previously, the body's defenses are extraordinarily powerful. If directed against the body itself, an immune attack can kill quickly (such as an allergic bee sting) or slowly (as occurs with multiple sclerosis). Accordingly, antibodies that bind to "self molecules" (the cells, proteins, and chemicals that compose our bodies) must be identified and eliminated to avoid self-inflicted damage. The need to maintain extraordinary diversity to recognize "foreign" threats while not imparting collateral damage is the challenge confronting the body's immune system. A sort of détente has been achieved through the process of evolution, which has devoted considerable energy and resources to the design and function of antibodies.

Antibodies are produced by B cells (more on these below). Each B cell in the body can create a different antibody, thus creating extraordinary diversity. How is this possible? At the most fundamental level, an antibody is a complex structure consisting of four different proteins, which are forged together in just the right way to form a structure with a striking resemblance to the letter *Y*. Each half of the *Y* is comprised of two proteins, a "heavy chain" and a smaller "light chain." The heavy and light chains come together to form a dimer, a scientific term meaning a combination of two molecules. In turn, two identical dimers assemble into a full-sized antibody. Two arms of this Y-shaped structure are thus identical to one another and are the sites where the antibody binds to its antigens or targets. The third leg of the letter-Y will be discussed below but first, it is essential to point out how these target-binding portions of an antibody are generated and why they are able to recognize an extraordinarily diverse range of different molecules.

One tip on each chain of the four different proteins is known as the hypervariable region, a name that reflects a remarkable feature of these proteins. Whereas the chemical structure of virtually all other proteins is constant and hard-wired within an individual's DNA, the hypervariable regions are intentionally capricious. Almost identical to the mechanism we described earlier to explain how T cell receptor diversity arises, the genes of antibody-producing cells (known as B-cells for their initial discovery in an avian organ known as a bursa) rearrange and mutate to allow for an extraordinary number of potential targeting specifidies. As a brief recall, this feature arises because a handful of the protein combinations in these tips can include any of the twenty different amino acids that exist in nature. By having a relatively small number of such hypervariable sites, nature creates extraordinary diversity. Through a combination of random mutation and intentional reorganization, each B cell in the body has a slightly different variable region, both for the heavy and light chains, from that of their brother or sister B cells. For example, each person is estimated to produce something like one million million (a one with 12 zeroes behind it) different versions of a single antibody molecule.[5]

By fashioning such variability into antibodies, nature has also created the potential for exquisite specificity and variability.[6] The enormous diversity can allow antibodies to recognize and interact with virtually any molecule found in nature. Indeed, the ability to recognize foreign molecules extends beyond what is found in nature. For example, exposure to man-made chemicals such as asbestos can elicit responses from certain antibodies in the body.

Once an antibody seeks out a molecule of interest, it can deploy a variety of different means to kill or neutralize its target. In some cases, the mere act of binding can disrupt the target. Antibodies, from the perspective of a cell, are rather large. The size of such very small things is measured using a term known as a Dalton (so-named to honor the memory of John Dalton, a 19th-century chemist). To put things in proper perspective, a typical hydrogen molecule weighs one Dalton, and a medicine such as aspirin weighs something like 180 Daltons. Compare these relatively small molecules with the comparatively heavy weight of an antibody at 150,000 Daltons, with some clocking in at more than 900,000 Daltons. As such, one can see why the sudden attachment of a large hunk of antibody onto a protein or sugar molecule can be sufficient to disrupt its function.

In other cases, an antibody binding to a foreign target can recruit other components of the immune system. Whereas we already discussed that two of the highly variable arms of a Y-shaped antibody bind the antigen, the third arm (known as a constant fragment, or Fc) can bind other components of the host defense system. One example is a structure know as an Fc receptor, which is conveniently found on the surface of many cell-killing cells of the immune system. If the antibody is bound to an antigen on a bacterium and then comes into contact with the Fc receptor on a macrophage (a killer cell), for example, the macrophage is triggered to engulf the antibody-antigen complex. If that complex happens to be a live bacterium—a preferred quarry of macrophages—then certain enzymes in the macrophage are alerted and go in for the kill, which generally shatters the intruder into many pieces. The macrophage then takes these pieces and shows them off to other immune cells (the T cells) to alert the wider immune system to the presence of the interloper.

The binding of an antibody to a foreign target can alternatively trigger a series of chemical reactions involving proteins that evolved many millennia before the first antibody. This collection of proteins, known as the "complement cascade," entails a version of the toxin-antitoxin system described in chapter 3 (though not to be confused with the use of the term *antitoxin* that will be used in this chapter). Returning to the example of an antibody bound to the surface of a bacterium, this binding might trigger the coordinated assembly of a structure that alerts the immune system early on to the presence of a foreign invader while also creating a hole in the membrane of the bacterium. This hole causes the bacterial entrails to leak out, thereby eliminating unwanted germs. Given this potent killing power, both the Fc-mediated engulfment and complement systems have multiple and ingenious safeguards (akin to the antitoxins described in chapter 3) to minimize the potential collateral damage that would inevitably occur were our own (human) cells to interact with a rogue antibody.

Behring Down on Disease

As we have seen, Robert Koch was a rare character in scientific history. He functioned as a focal point for understanding the body in general and deploying this knowledge to pioneer new medicines. Beyond his own extraordinary insights and personal gravitas, Koch surrounded himself with intellects as impactful as his own, including the ubiquitous Paul Ehrlich. A characteristic that drove Ehrlich to prominence is captured by the modern-day saying "He had a fire in his belly." Ehrlich was industrious, efficient, and tireless, always exploring new opportunities to glean new insights about health and disease. Fire also rose ceaselessly from his physical being. Despite lifelong repercussions from an early bout with tuberculosis, he smoked cigars constantly and was known to carry a box of cigars with him wherever he went (which might have contributed to the stroke that ultimately claimed him in August 1915).[7] Despite the constant plumes of smoke for which he was known, Ehrlich attracted and incubated extraordinary talent, including one colleague who came to Koch's Berlin Institute in 1888 and whose contributions came as close an anyone else's to ranking as important as Ehrlich's or even Koch's.

Adolf Emil Behring was born on the Ides of March 1854 in the Prussian town of Hansdorf (which was renamed Lawice and now lies in north central Poland).[8] As the fifth child (with eleven siblings) of a village school instructor in a sleepy corner of the Prussian state, Emil's prospects for obtaining an education beyond basic reading, writing, and math were not promising. However, the town minister helped the precocious Emil (he dropped his first name as a child) to gain acceptance at the regional gymnasium in nearby Hohenstein. This Prussian region had a rich tradition of military service, and Hohenstein itself had been the site of many battles since its founding in the 14th century. The many conflicts at this site included the Battle of Grunwald during the Teutonic War, and later attacks by Lithuanian, Teutonic, Swedish, Polish, and Russian invaders, the latter taking part in the Battle of Tannenberg in 1914. This battle took place a half century after Behring's birth but played a pivotal role in the Eastern Front in which German armies commanded by Generals von Hindenburg and Ludendorff would demolish Russian forces in the early days of the First World War.[9, 10] A rich marshal tradition coupled with severe family financial constraints led the young Emil to join the military and seek an advanced degree as a physician at the respected Army Medical College in Berlin. As is still the case in modern-day America, the "free" education came with a price—service in the Prussian military, which Behring began upon graduation in 1878.[11]

The newly credentialed physician was particularly interested in the nature of infectious diseases, publishing an early scientific paper investigating the antiseptic properties of iodoform, a disinfectant that was just starting to be used as an "antibiotic" to sanitize wounds.[12] An antibiotic in the literal sense refers to a compound that kills living things (not necessarily just bacteria). Iodoform was true to this definition and could kill virtually all cells it touched. As such, it wasn't used only as a disinfectant for wounds; it was also employed for decades as a means, however crude, to kill cancer cells. For example, iodoform was unsuccessfully deployed in 1907 to treat a fateful case of breast cancer in Linz, Austria. Rather than iodoform curing the patient, the patient, Klara Polzl Hitler, succumbed to an agonizing case of iodoform poisoning. This tragedy changed forever the life of her primary caregiver and son, a less beneficent man who shared a first name with Behring.[13] (The idea that the horrific experience of being a caregiver to his beloved mother as she was slowly poisoned by iodoform has been speculated as contributing to Hitler's anti-Semitic views. While Klara's physician, Eduard Bloch, was indeed Jewish, he had an unusually close relationship with Adolf for the rest of his life. Contrary to the aforementioned theory, Hitler provided Bloch with special protection from the Gestapo until Bloch's emigration to the United States in 1940.)

The antiseptic properties of iodoform were of great interest to a Prussian military establishment that understood that infectious diseases killed more soldiers than rifles and cannons did. Recognizing the opportunity to improve upon these odds, the Berlin military authorities directed Behring to train with Karl Binz in Bonn. In 1867, Binz had discovered the antimicrobial properties of quinine, a drug that would help limit the effects of malaria.[14] British mandarins quickly gained a reputation for seizing upon this new science. Specifically, the Anglo-Indian elites of the Indian Raj embraced the excuse for mixing gin with the quinine-containing tonic (meaning 'medically useful') water. Their excuse, albeit defensible, was that the quinine would protect against malaria.

After gaining additional insight on infectious diseases with Binz, Behring was again ordered by the military to move in 1888, this time back to Berlin, where he was to work under Robert Koch.[15] Behring remained with Koch until 1895, where he racked up an enormous number of achievements. Koch's laboratory was arguably the center of advanced biomedical research in infectious diseases at the time, surpassing even the vaunted Pasteur Institute, at least for a while. An interesting insight into how this period affected Behring, both in his personal and professional life, was compiled by the Nobel Foundation (which oversees the eponymous prize). Behring and his wife, Else, had a total of six children, and their godfathers now rank in the pantheon of great biomedical investigators. The godfather of his eldest son, Fritz, was Friedrich Loeffler, who would discover the cause of diphtheria and contributed to the eradication of the disease (as we will soon see). Likewise, the godfather of Hans, his third son, was Friedrich Althoff, who restructured biomedical research within Prussia (and later the greater nation of Germany) to create some of the earliest research institutes and hired the likes of Max Planck, Paul Ehrlich, Robert Koch, and, of course, Emil Behring. Emil Behring, Jr. counted among his godfathers Emile Roux, a cofounder of the Pasteur Institute and a key figure in the development of the medicine that stopped diphtheria in its tracks, and Elias Metchnikoff, who shared the 1908 Nobel Prize in Medicine with Paul Ehrlich "in recognition of their work on immunity."[16]

Beyond surrounding himself with influential and successful friends in Koch's laboratory, Emil Behring himself made extraordinary contributions to medical research. Nor did he work alone. A 34-year-old Emil Behring (these were the years before the honorific "von" had been conferred upon him) began studies in the laboratory of Robert Koch in 1888. By this time, a Japanese scientist, Kitasato Shibasaburo, had already been hard at work with Koch to isolate the agent responsible for tetanus, a bacterium now known as *Clostridium tetani*. Kitasato succeeded in doing so only months after Behring joined.[17] In that same year of 1888, the newly graduated Swiss scientist Alexandre Yersin was transitioning from his doctoral studies at the Ecole Normal Superieure (where he worked with Emile Roux) and spent a few months with Koch, Kitasato, and Behring in Berlin before joining the Pasteur Institute in early 1889. This transient encounter is important to our story, since Yersin's doctoral work with Roux was spent developing a rabies antiserum and investigators at the Pasteur Institute also had just discovered the toxin responsible for the deadly effects of diphtheria. Thus, the timely visit of Yersin to Berlin clearly resonated with Koch's team.

The year 1889 would also prove to be a pivotal one for Koch, Behring, and Kitasato. The team formed to begin a series of studies in which sublethal amounts of tetanus toxin were injected into rabbits. The immune system of these rabbits rightly recognized the tetanus toxin as a foreign substance and began to generate antibodies against it, which could be found at high levels in the serum. (*Serum* refers to a protein-rich substance that remains after whole blood has been subject to coagulation or centrifugation to remove its cellular components.) Behring and Kitasato then tested the sera from these rabbits and found that this material was sufficient to protect other rabbits from the otherwise deadly effects of tetanus.[18] The key point here is that the protected animals had not themselves had time to generate their own antibodies (which is known as active immunity) but rather utilized the antibodies found within the serum of previously immunized animals. This process of conferring protection was later known as "passive immunity." Within days after discovering the concept of passive immunity, Behring's team was joined by Paul Ehrlich, and the expanded group began to investigate whether they could obtain comparable results with diphtheria toxin.

The Berlin group successfully isolated diphtheria toxin (with guidance provided by Yersin) and used the material to immunize animals. Their goal was to ask if a passive vaccine might be used to treat diphtheria, which remained a major killer of children worldwide. Somewhere along the line (and this is a surprisingly controversial point among a small group of people who become quite passionate about such things), someone began referring to these immune sera as "antitoxins," since they could counter the effects of a bacterial toxin. By 1890, Kitasato and Behring used this term to describe the substances in sera from immunized animals that could confer protection upon others.[19] Since a similar procedure could be used to generate sera that recognized other types of molecules, the serum-based material that conveyed immunity became known by a more generic term: *antibodies*.

By late 1891, the Berlin team had isolated enough diphtheria antibodies to begin testing in people. The horrible destruction of diphtheria was ever present, and it did not take long to identify potential patients. On Christmas Eve, a very sick eight-year-old boy in a Berlin hospital was successfully treated with the antitoxin and survived an otherwise lethal infection.[20] Nine of eleven additional children would be saved over the next few months. The ability to save ten of twelve children contrasted dramatically with the fact that two thirds of untreated children in the same infirmary succumbed to the disease.[21]

As detailed above, Behring was credited with the discovery of diphtheria antitoxin and would later go on to receive many accolades. In contrast, the contributions of Kitasato and Ehrlich were minimized, especially by Behring himself, who did not credit their work in seminal manuscripts or in his later acceptance of the first ever Nobel Prize for Medicine. Likewise, Behring received a disproportionate amount of credit even though this work was built largely upon foundations established by their French rivals at the Pasteur Institute in Paris. The Berlin team was aware that Emile Roux had been developing a similar antiserum at the same time as Behring, and the Germans might have built upon his experiences. These facts rankled many scientists, who recognize the contributions of their colleagues and appreciate that such work rarely happens in a vacuum or has a single victor.[22] Ultimately, some degree of equity was achieved, as Ehrlich would receive a later Nobel Prize for work on immunity, though neither Kitasato nor Roux would ever receive the recognition each scientist rightfully deserved.

Within months, the idea of passive immunization to protect children against diphtheria caught on all around the globe. A key limitation was the availability of antisera. The use of small rabbits as a source of antisera to treat the thousands of children that suffered from diphtheria each year was clearly not viable. With encouragement from the venerable Robert Koch, Behring reached an agreement with Lucius & Bruening of Hoechst Germany (the company was later renamed Hoechst after a merger). Behring then helped scale up production of diphtheria and tetanus antisera using horses rather than rabbits.[23]

In 1895, Behring left Koch's team in Berlin and became a professor at Phillips University in Marburg. From a scientific standpoint, his work had peaked with the co-discovery of the diphtheria and tetanus antitoxins.[24] His later work at Marburg focused for a time on tuberculosis, and he announced the discovery of a breakthrough for tuberculosis antiserum that might lead to much-needed treatment for that disease. This announcement ultimately proved optimistic, and Behring later abandoned that project due to a lack of sufficient efficacy. Nonetheless, other less scrupulous manufacturers did sell such products, which fed upon an irrational exuberance for antisera in the early years of the 20th century. (The unregulated nature of this new industry is a subject to which we will soon return.)

In 1901, Behring received the inaugural Nobel Prize in Medicine (which coincided with the honorific "von" being added to his name thereafter) and invested his prize monies into a very different venture. In 1904, Behring founded the Behringwerke, a company focused on producing vaccines and antisera for a variety of infectious diseases (including the failed tuberculosis antisera).[25] The company continued to operate independently before being subsumed into the larger Hoechst (the company Behring had worked with to commercialize the first diphtheria antitoxins) in 1953. Although it has been subject to a variety of corporate mergers and spin-outs, the company still exists in the form of CSL Behring, a multibillion-dollar, multinational company with facilities in Behring's native Marburg. Fittingly, however, given the Prussian ancestry of its founder, its international headquarters has since been relocated to King of Prussia—a town in suburban Pennsylvania—rather than central Europe.

In an unexpected twist, Behring's name would again be indirectly linked with research into infectious diseases a half century after his death. In early August 1967, a handful of researchers developing a new poliovirus vaccine at the Behringwerke facility in Marburg reported feeling weak and lethargic, with accompanying headaches and fever.[26] These employees were sent home and no great concern was expressed, even after they reported back to the company physician with severe gastrointestinal distress. When the symptoms did not abate by the end of the week, the sickened workers were admitted to the local hospital under the assumption they were suffering from dysentery or perhaps even typhoid fever. Diagnostic tests failed to detect the microbes known to cause these diseases, but they did reveal that the livers of these patients were being systematically destroyed. Soon thereafter, some of the patients began to hemorrhage blood, and roughly a week after their first symptoms were felt, the first patients began to die. Worse still, the infection was not restricted to Marburg. Reports of the same constellation of symptoms arose from Frankfurt and Belgrade. In total, thirty-two people became infected, and seven died.

Within days, the medical and scientific establishments realized this was a new disease with no known etiology.[27] Various bacterial and viral causes of the disease were ruled out. As the blood from the infected patients could cause disease in guinea pigs, it was clear this was a new infectious disease. However, the responsible agent could not be seen by either conventional or highly sophisticated microscopes, and the disease arose before the routine isolation and sequencing of DNA that assists disease detectives today. Experts from around Germany and the world were called in to assist in the investigation. Eventually, an experimental electron microscopic technique developed by Dr. Dietrich Peters and Gerhard Muller revealed a strange, twisted viral structure unlike anything ever seen by man.

In parallel with the attempts to visualize this new virus (which was called Marburg, based on the location of the first reports of the disease), work was afoot to determine how the virus had found its way into Germany and Yugoslavia. It was understood a single Marburg-infected monkey must have triggered the crisis. After some extraordinary epidemiological detective work, it was determined the victims had all been involved in vaccine research and each had come into contact with monkeys days or hours before their symptoms began.[28] Strangely, the monkeys responsible for initiating the outbreaks in Belgrade, Frankfurt, and Marburg were each from a different source, so a single point of contamination seemed highly unlikely. As is often the case in detective work, the cause of the outbreak in retrospect resulted from a series of unlikely and highly unfortunate events with plot twists that would be unbelievable if ever inserted into even the worst Hollywood B-movie.

On June 5, 1967, the tiny nation of Israel was facing imminent invasion from much larger neighbors to the north, south, and east.[29] Though vastly outnumbered, the Israeli Defense Forces (IDF) launched an audacious series of preemptive airstrikes that utterly devastated the Egyptian, Jordanian, and Syrian air forces in quick succession. The fighting was particularly hot over the skies of Egypt, which bore the heaviest brunt of the blow from the Israeli jets.

The air combat over the Sinai and into the heart of Egypt caused most airlines to divert or cancel flights from Africa to Europe. Among the affected were multiple planes containing cargo of East African Grivet monkeys (*Cercopithecus aethiops*), a favorite of researchers and vaccine developers. Rather than being transported directly to Germany or Yugoslavia, as would have been routine, the planes were forced into a flight path far to the west and were then diverted to London.[30] A complicating factor was that London was amidst an airport strike that effectively shut down freight and passenger transfers. Consequently, many cages of monkeys sat for two days in an airport hangar. Complicating things further, at least two monkeys escaped from their cages, gaining free access to the warehouse before finally being recaptured. During this unanticipated and eventful stay in Britain, the monkeys were apparently in sufficiently close quarters that one infected monkey was able to transmit Marburg to others. At the time of infection, it was presumed that Marburg spread solely by direct contact. Consequently, the two temporarily liberated monkeys were implicated as the culprits responsible for transmitting the virus from cage to cage and thereby seeding the infections that would blossom simultaneously in three cities and two countries. As revealed by accidental events many years later in Reston, Virginia, and intentional events conducted by the Soviet, Russian, American, and British biowarfare experts, we now know that Marburg and its closest cousin, Ebola virus, retain the ability to spread infection via an airborne route as well.[31, 32, 33] Thus, the exact means by which the disease had spread remains a bit of a mystery.

Returning to the story of diphtheria, the improvements in antisera production, largely made possible through the intervention of Paul Ehrlich (though again, his contributions were largely ignored by Behring), facilitated widespread availability. The public health outcomes were dramatic. The Hôpital des Enfants Malades in Paris witnessed a 50 percent decline in the death rates of children with diphtheria.[34] Similar findings were reported throughout Europe. By the turn of the century, most major American cities had adopted the practice. However, the popularity of antiserum was destined to serve up other victims as triumph soon gave way to tragedy.

The St. Louis Incident

The early 20th century was a tough time to be a horse. This was a period of great transition, when a millennia-old reliance upon our four-footed equestrian helpers was supplanted by gasoline-powered mechanical contraptions. What to do with the excess animals? The discoveries out of Berlin and Paris provided one solution. Within two years after word spread about the outcomes attributed to Behring, major cities throughout the world mobilized to produce diphtheria antisera. The example of New York City is typical. Public health officials launched a series of studies to immunize virtually anything with four legs with the goal of developing life-saving antisera. Experiments were conducted with dogs, goats, horses, and sheep, but just as Behring had observed, the horse was found to be the most efficient factory for producing the much-needed diphtheria remedy.[35, 36] The city maintained a stable of a baker's dozen of these animals, primarily retired service animals, and periodically drew blood from the immunized animals to keep a steady reserve of the medication. These animals were extolled by the media and revered by the public as lifesavers, and the fodder and housing for these animals were underwritten by ample donations from a grateful public.

It may come as a surprise to many on either coast of the United States, but the city of St. Louis has a long tradition as a progressive innovator. A site just a few miles east of the modern city had served as the capital of the Native Mississippian culture, a civilization that spread from as far north as the Canadian border to Florida and from the Atlantic to the Red River valleys. Although the modern city of St. Louis was founded by French immigrants as a fur trading post in the mid-1600s, its population surged dramatically in the wake of the failed European socialist uprisings of 1848. Specifically, many German socialists relocated to St. Louis, and the city's population swelled eightfold from 1840 to 1860, reaching 160,000 (passing New Orleans) and eventually surpassing all other American cities except New York, Chicago, and Philadelphia by 1900. The socialist-led waves of immigration drove other innovations, such as the creation of urban parkland museums, all of which were launched and remain free to the public. For example, Forest Park in west central St. Louis is twice the size of Central Park in New York. This land served as the site of the 1904 World's Fair and the third ever Olympic Games, the first and only time the same city has served both roles at the same time.

Amidst the excitement associated with planning for the upcoming events, St. Louis's city fathers were eager to embrace the new antisera technology. At the same time New York was initiating their horse-based program, the St. Louis Health Commission established a factory and farm in Forest Park in 1894. True to its socialist roots, the city precluded private-sector companies from manufacturing or distributing the antitoxin, as the city intended to offer it for free to the poorest of the population.[37] The city also recruited a physician by the name of Armand Ravold, whose training was conducted in Paris and overseen by none other than Louis Pasteur, to lead the diphtheria antisera program. For more than five years, Dr. Ravold's program was the talk of the town and allayed parents' fears that the Spanish Strangler would come for their child.

On January 8, 1899, an article in the local newspaper, the *Post-Dispatch*, reported the stabling of a herd of horses in Forest Park. Some of the animals were dedicated solely to the production of diphtheria antitoxin.[38] The *Post-Dispatch* was a struggling local newssheet that had been put up for auction in 1878. It had been purchased by a Hungarian immigrant by the name of Joseph Pulitzer (and was led by a member of the Pulitzer family until 1995). Joseph was an entrepreneurial publisher, who later bought the *New York World* (also at an auction), founded an eponymous prize for reporting, and introduced the concept of sensational headlines, better known today as yellow journalism.[39] One of these headlines would highlight the story of a horse named Jim.

Photographs of the time reveal Jim was good-natured, powerful, and large. As a young colt, Jim pulled ambulances throughout town, but as the years went by, Jim slowed down and was retired to pull milk wagons (which don't require quite as much speed or stamina).[40, 41] Eventually, these wagons too became overly burdensome, and in 1898, Jim was sold to the St. Louis Health Commission to serve his remaining days as a living and breathing factory for the production of diphtheria antitoxin.[42]

Ever the vigilant worker, Jim reliably produced high quality product, contributing more than 30 quarts of serum over a few years.[43] Dr. Ravold bled Jim on September 30, 1901 and obtained what would become his last contributions to public health. Two day later, Jim's handlers noticed an anxious expression on his face. His tail was rigid and held out straight, the telltale signs of tetanus. Ironically, the local veterinarians did not have access to tetanus antisera, and the reliable old workhorse had to be euthanized. As word of this became known, Dr. Ravold instructed the elderly janitor, Henry Taylor, who happened to be nearby at the time, to discard the sera obtained two days earlier. It is not entirely clear what happened next, as Henry's later court testimony appears to have been coerced. However, it seems that rather than disposing of the serum, someone at the Health Commission elected to process the sample, most likely because sera collected the previous week had already been depleted and perhaps because a local diphtheria outbreak was starting to pick up steam.[44] The culprit is unlikely to have been Henry, whose position as a janitor is hardly consistent with making such a decision, but nonetheless, the elderly, African-American janitor became a classic fall guy. Likewise, it seems unlikely that the decision was made by Ravold, who staunchly defended Henry to the end but would also shoulder some of the blame.

The decision to proceed with the processing of Jim's antitoxin had a disastrous result. According to a news story in the *St. Louis Republic* (Pulitzer's cross-town rival) dated two weeks before Ravold bled Jim for the last time, Dr. R.C. Harris of Number 130S North Garrison Avenue returned to St. Louis after a much-needed vacation in the "northern resorts."[45] The soothing effect of this vacation must have been short-lived, as Harris was called on the evening of October 19 to the home of Bessie Baker, who was suffering from advanced diphtheria (recall that house calls were commonplace in that long-forgotten time).[46] Consistent with his recent training as a young physician, Dr. Harris administered the antitoxin. To be safe, he also injected her two siblings.

Four days later, Bessie's frantic parents called Dr. Harris back to their home. While the girl had been convalescing nicely from diphtheria, she displayed the classic symptoms of tetanus and was far beyond medical treatment.[47] She died the next day, on October 24, 1901. Adding to the parents' grief, both of Bessie's siblings were also ill with the same symptoms, and they too would be dead within days.

The same thing was happening all throughout St. Louis. In total, thirteen children died, each displaying the excruciatingly painful symptoms of a rigid spine and neck, broken only by waves of terrible spasms and an inability to swallow. As the damage from the tetanus toxin increased, the patients suffered from extreme hypertension, culminating in a slow and agonizing death. Although this story would grab headlines under any condition, it was a piece of raw meat for the voracious Joseph Pulitzer. His *Post-Dispatch* blared with headlines proclaiming the tragedy and further amplified the story nationwide via his ownership of the *New York World*.

Almost concurrent with Dr. Harris's treatment of Bessie and her siblings with diphtheria antitoxin, Camden, New Jersey, was experiencing a smallpox outbreak. An eight-year old girl succumbed to smallpox in early October 1901, triggering the local school board to enforce a law requiring vaccination for all students. The city hastily floated a contract for smallpox vaccine, and the bid was awarded to a relatively new and ambitious Philadelphia-based vaccine maker, H. K. Mulford. One factor working in favor of Mulford's bid for the work was that the company operated a relatively new antitoxin manufacturing plant in nearby Glenolden, Pennsylvania, fifteen miles away from Camden.[48] However, the production of an antisera made in horses was quite different from a smallpox vaccine produced in cows. Because of growing concerns about a burgeoning smallpox outbreak, time was of the essence, and Mulford was told to switch from the production of horse antitoxins to cow-based smallpox vaccine.[49] Widespread vaccination of students with the Mulford smallpox vaccine began within a week.

Shortly after the first student was diagnosed with smallpox in nearby New Jersey, the smallpox outbreak spread to Philadelphia's Pennsylvania Hospital. Panic ensued. As the Mulford vaccine was already being manufactured for Camden, some of this vaccine was diverted to assist with the Philadelphia outbreak as well.

On November 1, a sixteen-year old Camden student, William Brower, died of tetanus. Just three weeks earlier, William had been vaccinated for smallpox. Within days, ten Camden schoolmates or patients at the Pennsylvania Hospital likewise were sickened and died with the same symptoms.[50] A subsequent investigation implicated Mulford in general and cross-contamination between their horse-based antitoxin and cow-based vaccine manufacturing activities in the same Glenolden plant. In a manner remarkably reminiscent of present-day accusations of "fake news," Mulford's defense opted for an aggressive offense and accused its various business rivals of fraud and slander. Allegations and counter-allegations flew in all directions among Mulford and rival manufacturers Parke-Davis and H. M. Alexander. Once again, Joseph Pulitzer found his story, and the finger pointing made it even more salacious. The public became outraged. Likewise, powerful editorials in the *Journal of the American Medical Association* and the mainstream *New York Times*, among others, advocated strongly for government oversight of vaccine manufacturing.[51]

As the stories emerging from St. Louis and New Jersey became sensationalized throughout the United States, change came from a rather unexpected source. The Medical Society of the District of Columbia, founded in 1817, was and remains an advocacy organization focused on public health. Given its proximity to the Capitol, its members cited the twin tragedies in Missouri and New Jersey as reasons why federal intervention was necessary.

The Medical Society challenged local District of Columbia health commissioners to draft legislation to regulate the production of vaccines (both active and passive vaccines). Propelled by popular outrage, the bill was expanded to include the entire nation and sailed through Congress. The Biologics Control Act was signed by President Theodore Roosevelt on July 1, 1902.[52]

This legislation was particularly impactful given the rapid rise of the vaccine and antisera industry. In a few short years between their discovery and the passage of the Biologics Control Act, the diphtheria and tetanus antitoxins developed by Kitasato and Behring had given way to an enormous number of different antisera products. As might be expected of an unregulated industry, the claims made by some manufacturers often went far beyond what could be delivered. Many were outright snake oils. The products sold included antisera to combat a range of indications, such as tuberculosis, and were based on speculation or worse (recall that Behring himself had abandoned work on the tuberculosis antisera).[53] The new law required that processes for manufacturing of all serum-based products (known as biologicals) must be approved and licensed by the federal government, which would also oversee quality control. Given the double impact of events in St. Louis and New Jersey, the intervention imposed by the 1902 act was embraced by the medical community as well as ethical manufacturers. Many municipal- or home-based manufacturing operations folded due to an inability to conform to the new guidelines.

Oversight for the licensing of serum-derived therapeutics was tasked to a relatively small team known as the Hygienic Laboratory. As described in greater detail in *A Prescription for Change*, this federal service had its roots in a one-man operation in an obscure corner of the Marine Hospital Service on Staten Island, New York.[54] Given its new regulatory responsibilities, the Hygienic Laboratory gained more and more responsibility, later moving to Washington, where it was rebranded as the National Institute of Health. As its impact increased, the name became a plural and the organization evolved into the modern National Institutes of Health, or NIH, in 1930, following passage of the Ransdell Act.[55] The NIH, rather than the Food and Drugs Administration (FDA), continued overseeing vaccines and antisera for most of their history. In 1948, vaccine oversight within NIH was transferred from the Division of Biologics Control to the newly created National

Microbiological Institute (which came to be known as the National Institute of Allergy and Infectious Diseases, or NIAID).[56]

A more dramatic change occurred when biologics oversight was transferred in 1972 from the NIH to the FDA, which formed the Center for Drugs and Biologics. The center was again reorganized in 1987 to form the Centers for Biologics Evaluation and Research (CBER) and its counterpart, the Center for Drug Evaluation and Research (CDER). A final change arose in 2002 when oversight of recombinant monoclonal antibodies and some other biologicals was transferred from CBER to CDER, adding these technologies to the portfolio of more conventional medicines that it had historically overseen.

Serum & Sickness

The need for oversight was not limited to risks arising from poor manufacturing processes. Oversight was also necessary because of the inherent nature of the new drugs themselves. Antisera derived from nonhuman species (horses and cows) are inevitably seen as foreign by a chauvinistic human immune system. Shortly after the Berlin team's antitoxins began to be adopted by pediatricians around the world, this fact was recognized by two investigators working in the world's leading pediatric center in Austria.

Clemens von Pirquet (unlike Behring, *von* proclaimed that Clemens entered the world as part of the landed aristocracy) was born in 1874 near Vienna. Consistent with his noble parentage, he attended the finest schools that the Austro-Hungarian Empire could provide.[57] Originally entering college as a step towards a career in the priesthood, Pirquet changed his mind, deciding to undergo medical training at the University of Graz.[58] The aspiring doctor had fallen in love with the sciences of bacteriology and immunology (fields that had barely begun by this time) while training at the Universitäts Kinderklinic (University Children's Clinic), a Viennese clinic led by the legendary figure Theodor Escherich.[59] Escherich's name may sound vaguely familiar to contemporary ears today based on his earlier discovery of an eponymous bacterium now known as *Escherichia coli* (*E. coli*). This bug is a major component of the gut microbiome and helps us digest our food. Occasionally, the organism makes the headlines for darker reasons. It was the cause of a particularly nasty form of food poisoning that plagued the Jack in the Box restaurant chain in 1993. A variant (O157:H7) sickened at least seven hundred diners, many permanently, and killed four.[60]

After receiving training from Escherich, Pirquet graduated in 1900 and started a stint as a military surgeon. He later returned to his hometown of Vienna, taking a job at the Children's Clinic. In 1902, Pirquet began mentoring a promising new Hungarian medical student by the name of Bela Schick. This event was life-changing for both, as it triggered a lifelong friendship and professional partnership between the two (who were only separated by three years of age) that would greatly advance our understanding of the immune system. Pirquet and Schick's early collaboration centered on the fact that a sizeable fraction of children treated with diphtheria antiserum tended to display a constellation of symptoms one to three weeks after receiving their first injection.[61] These children often demonstrated fever, malaise, hives, itching, and severe joint pain, which was frequently accompanied by swelling of lymph nodes throughout the body. By comparing these clinical symptoms with the outcomes from basic research studies with animals, the two investigators realized that animal-derived sera were triggering an exaggerated immune response. Worse still, the response was amplified following each exposure to an antiserum (any antiserum), potentially culminating in death. A prominent example of this "serum sickness" reverberated at the core of the Germanic medical community when the two-year-old son of a prominent researcher, Paul Langerhans (who had earlier identified the key component of the pancreas that would later be shown to produce insulin), died within minutes of receiving diphtheria antitoxin.[62]

Over time, Schick and Pirquet advanced understanding of this disorder, realizing the human immune system (and that of other mammals) often perceives and overreacts during its assault upon molecules perceived as foreign. This defense mechanism evolved in part because viruses have the annoying ability to pick up portions of their host DNA and incorporate it into their genes. This process can function to provide a type of shield meant to confuse the immune system. Therefore, the immune system developed a countermeasure to recognize material from foreign species (e.g., cows and horses) and even from other people (the reason for the rejection of transplants). As we have seen, these reactions themselves can be responsible for many diseases. As the process was elucidated, Schick and Pirquet realized that these exaggerated defensive measures contributed to diseases beyond serum sickness and included more mundane (but still troubling) issues such as responses to animal and plant productions. The team was the first to term these diseases as "allergies" and appreciated that a variety of different stimuli such as seasonal pollens, foods, insect poisons, or even man-made chemicals could likewise trigger responses akin to that observed with serum sickness.

The recognition of the causes and effects of serum sickness motivated a series of efforts to minimize the potential for damage. Early ideas included purifying the efficacious horse (or other species) proteins away from all other horse proteins within the antitoxins. This concept led investigators to evaluate the proteins in serum that were responsible for the beneficial effects. We now know these to be large protein structures known as "antibodies," but this had yet to be discovered during the lifetimes of Schick and Pirquet.

Almost a half century later, within months after the Germans invaded Poland, sparking the Second World War, the American professor Edwin Joseph Cohn began to research ways to treat shock on the battlefield.[63] As a slight diversion, we will briefly introduce the concept of shock due to its key role in our story. More accurately known as circulatory shock, this malady is a physiological response that occurs when the volume of blood (or, more accurately, the presence of blood proteins in the body) drops as a result of bleeding. The body responds with a vigorous drop in blood pressure and heart rate, which causes rapid degradation of vital organs such as the kidney, invariably leading to death unless rapid and dramatic intervention, such as blood replacement, begins.

One treatment for circulatory shock is intravenous replenishment of blood. However, such resources may be precious on the battlefield or forward aid stations. Cohn knew that reconstitution of volume alone (e.g., with a saline-based solution) was insufficient to counter shock, as fluids alone do not contain the vital proteins sensed by the body. In analyzing the different components of blood, Cohn isolated these important proteins from blood using ethanol (alcohol), then separated them into different classes of proteins utilizing a laboratory-based machine he had invented. Cohn was quite proud of this accomplishment. During his scientific talks, he would often take some of his own blood and place it in the machine at the beginning of the talk. He would then conclude his presentation (about an hour later) by showing the audience the fractionated products that the machine had separated in the meantime. This approach unexpectedly created a spectacle during a 1951 seminar at the Instituto Tecnico in Lisbon, when the blood clogged a key valve and the piping burst, bathing the audience in his blood.[64]

Cohn's process of physical separation (known as "fractionation") revealed five different groupings of proteins. Many of these went on to provide essential treatments used during the war and thereafter. For example, the proteins of the fifth fraction largely consisted of albumin, a prominent serum protein that could be mixed with fluids and injected into soldiers to prevent shock. Other fractions contained fibrin, thrombin, and other proteins involved in blood coagulation. Later, these purified proteins were used to halt bleeding, both on the battlefield and for the rare individuals who suffer from clotting disorders such as hemophilia. Of particular interest to our story are a series of proteins found in the second and third fractions. These molecules, collectively known as gamma globulins, comprise what we today refer to as antibodies.

The modern field of American academic medicine has from its beginnings been heavily influenced by a prominent scientist with the surname of Janeway. The Janeways first came to New England in the 16th century after fleeing from religious persecution in France and soon spread throughout the English colonies.[65] Reverend Jacob Jones Janeway was vice president of Rutgers College in New Brunswick, and his son, George Jacob Janeway, was a prominent New Jersey physician. George's son, Edward Gamaliel Janeway, served as the health commissioner of New York in the late 19th century, while his son, Theodore Caldwell Janeway, was the first full professor at an innovative new medical school in Baltimore by the name of Johns Hopkins. Theodore's grandson, Charles Janeway (Jr.), was one of the world's most brilliant and creative immunologists in the heyday of the field during the third quarter of the 20th century.[66] In between Charles Jr. and Theodore was Charles Sr., who also served as a founder of modern immunology.

Charles Janeway Sr. started his research career in the months after Edwin Cohn demonstrated the feasibility of separating blood into its various protein components. His work with serum sickness confirmed that the use of horse serum–derived albumin was not feasible, in part due to serum sickness. However, his work soon turned to utilizing the same techniques on human donor blood in early 1941. Charles's work helped optimize the amounts and types of blood products that could impede shock, and the albumin-based products rode along in the Allied landing craft during D-Day three years later.

In parallel with his studies on albumin, Janeway was evaluating the products of Fractions II and III and isolated the aforementioned "gamma globulins."[67] Working with a medical intern by the name of Fred Rosen, Janeway demonstrated that these gamma globulins could help protect children who otherwise lacked a stable immune system from infection. The benefactors of this finding included premature infants and children suffering from chronic infections, both of whom have underdeveloped or insufficient host defense mechanisms. Over time, these gamma globulins were understood to be the same protective antibodies that Kitasato and Behring had generated many years before.

Having discovered these "antibodies," the question now turned to how and where these interesting molecules were produced in the body. Throughout this book we return to Pasteur's quote "Chance favors the prepared mind" even though it has become a bit of a cliché. A prominent example is seen with the question of how antibodies are produced in the body. In the mid-1950s, an Ohio State University poultry scientist by the name of Bruce Glick was studying an organ near the anus of the chicken known for years as the bursa of Fabricius. The organ had been first described by Fabricius ab Aquapendente, a 17th-century Italian anatomist. However, the function of this small bud of tissues had remained unknown.[68] Glick noted the bursa grew rapidly in the first three weeks after hatching but atrophied thereafter, much as we have seen with the thymus. To get at the function of this anatomical site, Glick surgically removed the bursa from adult birds but did not note any particularly interesting effects.[69] Months later, a graduate student by the name of Timothy Chang asked Glick for some chickens to prepare antibodies against *Salmonella* bacteria. All that Glick could provide were some older chickens, whose bursas had been removed months before.

Despite multiple attempts, Chang could not produce antibodies in those chickens lacking a bursa.[70] Moreover, the animals that he injected often died of *Salmonella* even though he was using a weakened strain that should not have killed the chickens. Follow-on studies by Chang and Glick demonstrated that the bursa plays a key role in nurturing the cells that produce antibodies. Moreover, much as had been seen with the thymus, the bursa of Fabricius was only needed for this function in the first weeks of life. Removal of the bursa in an adult chicken did not impair its ability to make antibodies, since the job of the bursa had already been completed. Given the origin of these cells in the bursa, the antibody-producing cells have thereafter been known as B cells. (The human equivalent of the bursa of Fabricius is not itself a distinct organ but is found in the bone marrow.)

Over time, concerns about serum sickness and other immunological responses against the horse-produced components of many antitoxins and antisera drove new research for alternative medicines. Whenever possible, human sources of B cell–derived antisera (gamma globulins) were procured. A prominent example is intravenous immunoglobulins (IVIG), which provides a source of immunoglobulins from healthy individuals that protect people with immune system deficiencies (such as those receiving a bone marrow transplant or following certain chemotherapies). More focused products included antibodies derived from people who had been exposed to (and successfully fought off) infectious diseases caused by cytomegalovirus (CMV) or respiratory syncytial virus (RSV). However, the breadth of isolating such medications is necessarily limited by the ability to find appropriate donors. In considering the much-needed antibodies to neutralize bacterial toxins, it is obviously unethical and impractical to generate human antisera against powerful poisons such as botulinum toxin or various snake and spider venoms. Many of these are therefore still generated in nonhuman species (such as horses).

A key feature that distinguishes these first-in-class antisera from a new generation that would follow is that these animal- or human-derived sera are polyclonal in nature. This term refers to the fact that a wide variety of different antibodies is elicited during a typical immune response, and the complexity increases even further when one considers that any given batch of product likely contains antibodies from hundreds, or even thousands, of different donors. Thus, these sera may target a wide variety of different sites on a "foreign" molecule. A strength of this approach, as appreciated by nature, is that the larger the number of sites that is targeted (known as epitopes), the more likely that a particular antibody binding site will be useful in targeting a perceived foreigner. The disadvantage is that the number of antibodies that can target any given epitope will necessarily be diluted by all the other antibodies targeting other epitopes. Even a particularly useful antibody will also be subject to some degree of dilution. Since not all the antibodies produced are equally efficacious, much effort was placed into emphasizing only the most potent antibodies. Even more ideally, it would have been useful to identify a single antibody that works better than all the rest and then produce only this most superior molecule. However, such pipe dreams remained as such until the comparatively recent revelation of a powerful new technology.

Biotechnology advances in the latter half of the 20th century utterly changed the prospects of making highly selective and safe antibody-based therapeutics. In the early 1970s, two scientists from the University of Cambridge (United Kingdom) developed a technique that would allow investigators to immunize animals and then isolate individual antibodies of interest. Georges Köhler and Cesar Milstein developed an elegant approach to fuse normal B lymphocytes and cancer cells using a derivative of the antifreeze routinely used to prevent overheating in automobile radiators. The resulting cells recall an ancient Greek mythological creature known as a chimera (a hybrid animal composed of parts from other animals such as the Sphinx), and these fused cells were known as hybridomas. The advantage of a hybridoma is that a single cell (known as a clone) can produce a single antibody that is not subject to the mutation and gene rearrangements that can alter an antibody in normal B cells. Unlike conventional immune cells, these hybridomas could also continue to grow indefinitely (reflecting the fact they are partially derived from cancer cells).[71] As such, these hybridomas could produce a single antibody that recognized a single epitope. This breakthrough is today known as a monoclonal antibody (meaning all the antibodies are derived from a single clone). Beyond providing a more consistent product that does not vary from donor to donor, this technology yields other advantages: genes from these monoclonal antibodies can be isolated and engineered to modify their behavior, including improvements in how well they function and, when necessary, ways to improve their safety, ability to be administered in the body, or even the efficiency by which they can be manufactured.

Over the past three decades, the field of monoclonal antibody research has evolved quickly. Among the most important contributions, investigators such as Greg Winter (also of Cambridge University) demonstrated that antibodies isolated from mice could be subject to genetic manipulation such that most of their mouse-derived structure is replaced by human sequences. In doing so, these products look to the immune system as if they are human and thus are not subject to rejection. Neither do they trigger adverse reactions such as serum sickness.[72] A variation on this theme was the creation of laboratory mice that possessed human immunoglobulin genes and therefore also could avoid rejection.[73] Greg Winter also reenters the story with improvements that arose from the use of genetically-modified bacteriophage.[74] In this case, he cleverly inserted certain portions of immunoglobulin genes into bacteriophage, which could be used to infect *E. coli* bacteria in order to facilitate the creation of antibodies within hours (rather than the months that are otherwise needed to endure the immunization and isolation of antibodies from mice). Such biotechnology advances in monoclonal antibodies have revolutionized many fields of medicine, especially oncology, and have begun to do the same in our wars against infectious diseases. For now, we will conclude the chapter with a brief return to Schick and Pirquet to assess the work they did following their initial descriptions of serum sickness.

Diagnosis and Disease

Eclipsing even the identification and causation of serum sickness, Schick and Pirquet's biggest contribution was the recognition that the same basic causes that lead to allergic responses could be used to predict whether an individual had been exposed to tuberculosis or other diseases. Tuberculosis has long plagued man and can insidiously remain dormant for long periods of time, not revealing who is infected until the symptoms have progressed to a point where intervention is challenging, if not impossible. Schick and Pirquet developed a skin test to assess whether an individual may be harboring tuberculosis. This test was optimized by the French researcher Charles Mantoux in 1908.[75]

Separately, Schick and Pirquet emigrated to the United States. Pirquet accepted a position at Johns Hopkins in 1909 but suddenly changed his mind and abandoned the position to sail back to Europe within a year.[76,77] This timing was fortuitous, as the Austrian had returned just in time to take over as head of the Vienna Children's Clinic after Escherich unexpectedly died in 1911. However, in another sense, this rather rash decision was a sad prelude to a spate of erratic behavior by Pirquet that worsened with age. Over the next few years, Pirquet became delusional and violent, as evidenced by his sudden decision to leap towards and vault out of a second-story conference room in the middle of a business meeting. His personal life was also troubled by a marriage that left him estranged from his wealthy relatives, causing further stresses, both personal and financial. Worse still, his wife began displaying her own signs of mental distress. Tragically, the bodies of Pirquet and his wife were discovered in their home on February 28, 1929, the victims of a double suicide.

The story with Schick is far more cheering. The native Hungarian emigrated to the United States in 1923 and critically contributed to the public health of his adopted country by spearheading campaigns to eliminate diphtheria. Using the same approach he and Pirquet (and Mantoux) had earlier used to develop the tuberculin test, Schick developed a test to assess exposure to diphtheria. The test was widely deployed to help eradicate the disease.[78] Whereas 100,000 Americans suffered from diphtheria in 1927, the disease was virtually eliminated within a few years. This grand outcome was in no small part due to Schick's contributions, both in terms of his test and in leading an advertising campaign (in partnership with the Metropolitan Life Insurance Company) that distributed more than 85 million pieces of literature advising parents about immunization and other ways to keep their children safe from diphtheria.[79]

With this history in mind, we now turn our sights to active immunotherapy, or, as they are better known, vaccines.

7

Lost in Translation

It is widely known the royal families of Europe are quite inbred, as evidenced by occasional maladies such as hemophilia and porphyria, which have made for rather colorful storytelling over the centuries. Despite and occasionally because of these relationships, European peace was significantly and consistently disrupted over disputes pertaining to royal succession. One such disagreement forever changed the course of history.

As the new year of 1870 dawned in Europe, things were looking up for the relatively new state of Prussia (a territory once ruled by Teutonic knights that remained a paper lion until the 17th century). Just four years before, Kaiser Wilhelm I had appointed Otto von Bismarck as prime minister, and this brilliant, albeit ruthless, strategist had begun a conquest to raise the minor kingdom into a major power.[1] The kingdom's power had grown following a sharp but decisive war with the fading Austrian-Habsburg Empire in that same year of 1866, which had netted the Prussian kingdom additional lands and sway over still-independent territories in its backyard.

Portrait of Lady Mary Wortley Montagu, wife of the British ambassador to the Ottoman Empire and progressive thinker. Without the knowledge of her husband, she directed the embassy's doctor to subject her son, Edward (pictured) to variolation and later became famous for advocating the practice in the United Kingdom.

ABOVE: A photograph of the hide of "Blossom," whose minor infection with cowpox provided the material expanded by Edward Jenner into the first true vaccine that would save billions of lives. The hide hangs on the wall of St. George's at the University of London (although its authenticity is suspect) and Blossom's horn can be found at the Edward Jenner Museum in Berkeley, England. BELOW: "The Wonderful Effects of the New Innoculation." An 1802 cartoon from the Anti-Vaccine Society promoting the idea that cowpox immunization would cause its recipients to acquire bovine characteristics.

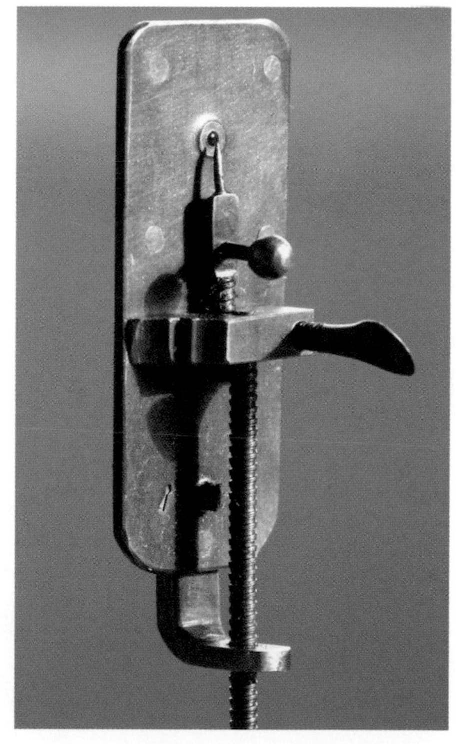

ABOVE: Portrait of Guy de Chauliac, a low-born peasant who rose to become the physician of multiple Popes and the first to describe the rather unpleasant manifestations of "pus," which helped revolutionize medicine. RIGHT: Simple but sophisticated. Shown is one of the first microscopes produced by Antonie van Leeuwenhoek. Although seemingly inauspicious, the manufacturing process needed to craft the lens (shown towards the top of the device) went to the grave with its inventor, and required more than a century to be reproduced so that microscopy could become commonplace.

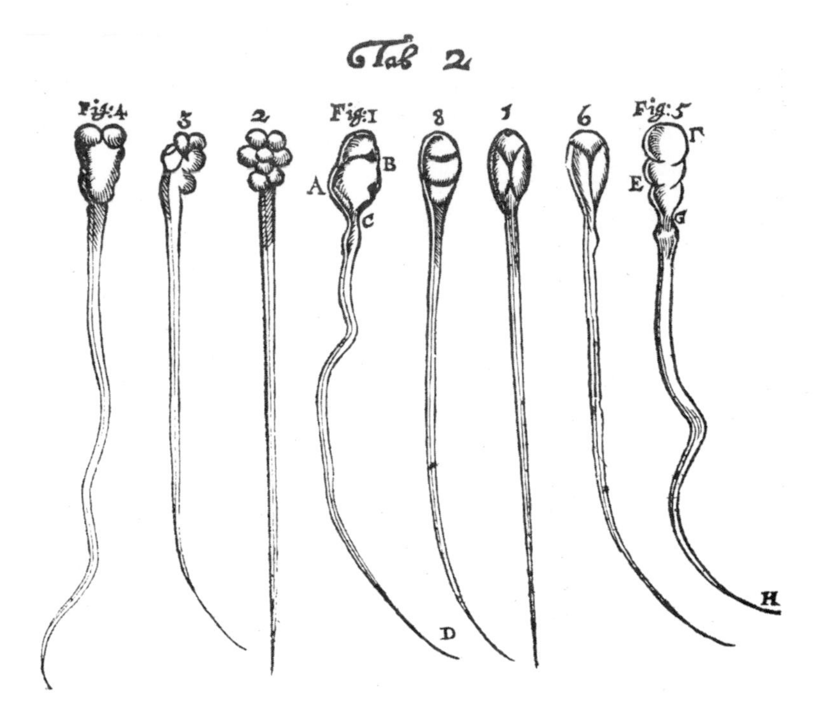

ABOVE: A whole new world. Using his simple microscope, van Leeuwenhoek catalogued the unseen, including his hand-drawn pictures of a sample of sperm that revealed the seeds of life. BELOW: A photograph of Louis Pasteur, who successfully tested the first vaccine against rabies and went on to direct or inspire the creation of a remarkable number of vaccines in the late nineteenth century.

TOP LEFT: A photograph of Joseph Meister in 1885, months after he was bitten by a rapid dog and successfully treated by Louis Pasteur and his Paris team. Meister remained employed as a caretaker at the Pasteur Institute until his tragic death during the German occupation of Paris in 1940. BOTTOM LEFT: A 1905 photograph of Felix d'Herelle reveals the intensity of the individual, who would later discover the first viruses that infect bacteria and went on to become an entrepreneur intent upon commercializing his discoveries around the world. ABOVE RIGHT: A dignified Emil von Behring, around the time of his receipt of the first Nobel Prize in Medicine, discovered antisera (later known to be comprised of antibodies) and led a team of Berlin-based scientists that rivaled their French competitors in Paris.

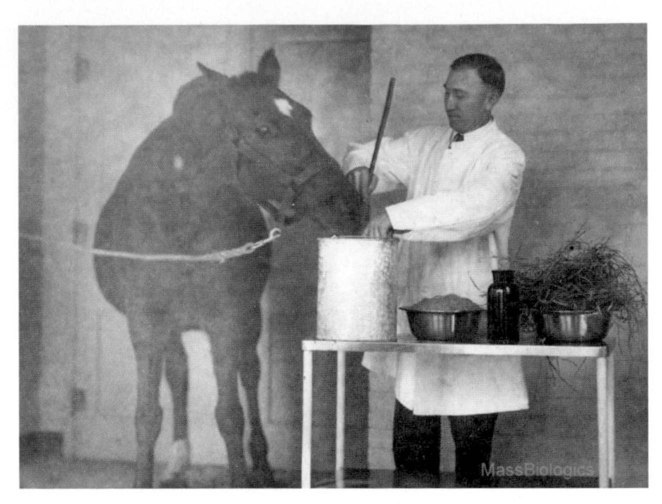

ABOVE: Though no authenticated image of "Jim the milkhorse" exists, this equine hero helped produced gallons of diphtheria antitoxin before himself succumbing to tetanus. The resultant distribution of contaminated antisera in St. Louis, and a related incident in New Jersey, triggered the enactment of legislation to regulate vaccine and antisera in the United States. The horse pictured here is another diphtheria antitoxin provider. BELOW: One of history's more forgotten heroes and overshadowed by Louis Pasteur, Gaston Ramon critically contributed to the discovery of vaccines for diphtheria and tetanus.

Born a preacher's daughter in Wheaton, IL, Peark Kendrick partnered with Grace Elderling and Loney Gordon to discover a life-saving vaccine against whooping cough.

The investigative journalist Brian Deer, of the *Sunday Times* of London, began his career as a vaccine skeptic but pioneered investigation that led to the revelation of greed-based frauds that had tainted the reputations of both the DTP and MMR vaccines, schemes that continue to threaten the lives of millions worldwide.

In 1868, the Spanish "Glorious Revolution" had deposed and exiled Queen Isabella II as the first step in a process of liberalization that would ensure instability and political uncertainty on the Iberian Peninsula for the next century.[2] The provisional government of Spain had lacked leadership and had soon been seeking a new monarch with hereditary ties to the Spanish crown. One candidate had been Leopold, Prince of Hohenzollern, whose wife had also been a princess of Portugal. However, Leopold's potential ascendency to the throne of Spain had been strongly opposed by the French Napoleon III, who'd threatened war as he'd feared contending with Hohenzollern rivals to the east (Prussia) and west (Spain).

While negotiations among the different candidates by the Spanish Cortes Generales had been underway, the French had continued to pressure Prussia against any consideration of contending for the Spanish throne, culminating in a seemingly innocent encounter on July 13, 1870. During a stroll on his summer vacation in the resort spa in Bad Ems, Kaiser Wilhelm was confronted by Vincent, Count Benedetti, the French ambassador to Prussia.[3] Benedetti, working on behalf of Antoine Alfred Agenor, Duke of Gramont and French minister of foreign affairs under Emperor Napoleon III, challenged the Kaiser to renounce all current and future claims to the Spanish throne. During their walk together, the duke implied that a Prussian failure to abandon aspirations for the Spanish crown could trigger war. The Kaiser politely demurred from answering and shared the story with his secretary, Heinrich Abeken, who in turn relayed the account to Otto von Bismarck.

Bismarck, the cunning architect of Realpolitik, comprehended the propaganda value of the encounter at the resort and crafted an accurate but stilted account of the Ems encounter into a dispatch that was shared with the foreign office and later published for all the world to see. In relaying the encounter in what has become known as the "Ems telegram," Bismarck utilized language that was crafty enough that a translation into French would reveal an insult to the French people by the Kaiser. This is not the only time that Franco-German translation skills will cause chaos in our story. Taking the bait, the French emperor took offense at the translation error, triggering a parliamentary referendum for war and mobilization of the military. Incensed, the French people demanded the parliament declare war, and their wishes were fulfilled on July 16.

The war was a complete rout and utter humiliation for France. A siege of Metz by the Prussians was decisive, as was a failed attempt by the French to relieve their embattled fortress. The relieving forces were utterly decimated at the Battle at Sedan in September (the future site of future misery in the Great War). Indeed, the Sedan debacle witnessed the capture of Napoleon III himself, further compounding the humiliation and triggering the creation of the French Third Republic. This third attempt at democracy was to be ended by another Germanic empire, then, later, by an equally successful invasion by Hitler's Third Reich in 1940.

In January 1871, the Prussians were besieging Paris, and France ultimately capitulated. For a time, the government dissolved into disarray, known as the Paris Commune.[4] This brief period was characterized by radicals taking to the streets and erecting barricades, triggering subsequent events including further inspiring a middle-aged Russian writer by the name of Karl Marx. Marx referred to the Paris Commune as the first example of the "dictatorship of the proletariat," which was a profound compliment (despite the modern views of the word *dictatorship*)[5] While the turmoil of the Paris Commune was pushed aside after a bloody week of street and barricade battles with the French regular army, its short existence inspired a young Vladimir Ilyich Ulyanov (better known as Lenin) to formulate the theory it might have succeeded had the Commune government been more heavy-handed and centralized.[6]

On the other hand, the Prussians succeeded in unifying many German-speaking territories into a single nation. They seized critical territories in the Alsace and Lorraine (which would breed future contention in the coming two world wars). The German star was ascendant and, likewise, German science was beginning to challenge French science led by the legendary Louis Pasteur.

A Marvelous Rivalry

While the battles between the powerful western European nations were fought, the proud nationalist Pasteur was in the south of France performing research on diseases of silkworms (a task not as menial as it might seem, given the economic value of the silk industry).[7] In this task, Pasteur was assisted scientifically, as was often the case, by his wife, Marie, who helped him grow the silkworms and tabulate the findings. Despite his geographical isolation, Louis was closely monitoring the political situation, particularly since his son had been drafted into the army. Like many Frenchmen of his age, Pasteur was deeply and irreversibly pained by the humiliation suffered at the hands of the Prussians, and he would retain his prejudice against Germans throughout his remaining days. Meanwhile Robert Koch was still more than a decade from presenting his postulates and was serving as a surgeon in the Prussian army (though nowhere near the front lines). As a mirror image of Pasteur's feelings of humiliation, Koch bathed in the national pride his fellow Prussians had achieved over the despised French.

The looming rivalry between the French, led by Pasteur, and the Germans, championed by the much younger Robert Koch, was not entirely nationalistic in nature. The rigid Koch, reflecting stereotypical Prussian attitudes, maintained that infectious organisms, such as bacteria and viruses, retained their character and behavior under all circumstances. This idea was applied by Koch to all bacteria discovered by himself or his team throughout a remarkable three-decade run from 1877 through 1906. In a mere thirty years, these organisms included the causes of anthrax, *Staph* and *Strep* infections, venereal diseases (syphilis and gonorrhea), typhoid fever, tuberculosis, cholera, diphtheria, tetanus, pneumonia, meningitis, food poisoning (including *Salmonella*, botulism, and *E. coli*), gangrene, dysentery, and paratyphoid.[8]

Like many scientists, Koch's proudest achievement likely was his first: the discovery of the organism that caused anthrax. A few years after the Franco-Prussian War, Koch had been appointed as the ranking medical officer for Wöllstein (modern-day Wolsztyn, Poland), a sleepy municipality near the region Koch had served in during the war.[9] The local dependence upon agriculture in general and cattle in particular led Koch to research the causes of a disease that had plagued domesticated animals from time immemorial. The etymology of the name of this disease, anthrax, reflects the Greek term for 'coal' and captures the black skin lesions associated with the disease. However, while many modern summaries of the history of anthrax credit Koch for the discovery of the causative organism in 1875, this feat was actually accomplished in 1850 by the physicians Casimir Davaine and Pierre Francois Olive Rayer.[10] These Frenchmen had likewise contributed to agricultural microbiology by discovering the bacterial cause of glanders, a fatal disease of horses and donkeys (and occasionally people). Also, Davaine and Rayer's work was heavily influenced by and directly supported by Louis Pasteur himself. Without question, Koch certainly popularized the discovery of the bacterium that caused the disease and dramatically increased knowledge of anthrax, including the deadly transmission. In particular, Koch help described the bacteria's insidious use of spores, tough particles that encapsulated the bacterium to protect it from harsh environmental conditions that would be lethal to bacteria exposed to the elements. These same spores were also more aerodynamic and allowed the infectious agent to spread among animals. Consequently, anthrax spores provided a combination of sanctuary from harm and a propulsion system that rendered it one of the most lethal dangers to agriculture (and as the American public witnessed in 2002, a favored tool of bioterrorists).

On the French side of the border, Pasteur and his team (and later Institute) were in the process of identifying the causes of some key diseases such as pertussis and rabies.[11] However, the French rivals to Koch were more intent upon developing new ways to treat or prevent such diseases altogether. To accomplish this goal, Pasteur focused on harnessing the immune system in a manner that could recognize and remember foreign pathogens (or the toxins they use to wreak havoc on the body). While the Germans, led by Koch and Behring, pioneered the use of antitoxins and passive immunotherapy, the French were developing new generations of active immunity, or, as they are better known today, vaccines. In the end, the nationalistic antagonisms, compounded by the difference in languages and the very different scientific beliefs, worked to the advantage of the rest of the world and future generations. Indeed, the mixture of German identification of infectious agents and ever-improving vaccine technologies served all of society. Although the stories behind this rivalry are cringe-worthy at times, the personal antagonisms between Koch and Pasteur resulted in one of the most productive rivalries in history.

The rivalry between Koch and Pasteur began in earnest in 1878, when the more senior and highly regarded Pasteur began studying anthrax. Anthrax was a deep passion of Koch's, and the Prussian took particular affront, as he believed the Frenchman was applying a fundamentally flawed and positively reckless approach to the study of his bacterium.[12] Specifically, Pasteur believed in the process of attenuation, a weakening of pathogens that could render them non-infectious while retaining the ability to elicit an immune response. The potential existence of such plasticity in microbial character was anathema to Koch's strongly held view that organisms remained constant under all circumstances. That Pasteur would contaminate his beloved anthrax with such a lurid approach created considerable resentments.

And contaminate he would. Pasteur proceeded to contaminate anthrax bacteria and its spores with a variety of different chemicals to determine if any of these might perturb its ability to cause disease. Pasteur was indeed successful, though the Frenchman did so by blatantly stealing the ideas of another Gallic scientist.

Jean Joseph Henri Toussaint was the son of a carpenter and a seamstress, who earned a degree as a veterinarian from the Ecole Nationale Veterinaire in Lyon in 1869.[13] Shuttling between the southern French cities of Toulouse and Lyon, Toussaint conducted studies to earn a PhD on the subject of anthrax, which he completed in 1879, receiving 100,000 francs after winning the prestigious Breant Prize for his work on the subject.

In 1880, Toussaint began asking if anthrax might be weakened (attenuated) such that the bacteria were no longer dangerous but able to elicit protection from future infection.[14] Within a year, he had successfully developed a rather complicated process that included treating the blood of infected animals to remove the clotting factors and then heating the sample to 55°C for ten minutes in the presence of small amounts of phenol (which you may recall as the active ingredient found in Chloraseptic sore throat spray). By July 12 of that year, Toussaint had detailed his approach in a sealed envelope and safeguarded this secret with the venerable French Academy of Sciences. This might seem an almost paranoid response—from a man who was beginning to suffer the effects of a neurodegenerative disease—but it turns out to have been a prescient step. Toussaint then organized a unique public demonstration of his findings in August by exposing vaccinated sheep to an otherwise lethal dose of anthrax.[15] The sheep survived and the sealed envelope revealed how the accomplishment had been achieved. These facts were presented to the regional Society for the Advancement of Science in Reims later that month. Toussaint further refined the phenol-based approach to improve his anthrax vaccine (for both large stock animals and for fowl). He then turned his attention to applying a similar strategy to the development of a tuberculosis vaccine.

Sadly, the next few years were not kind to Toussaint. The progression of his disease accelerated and his intellect failed, killing him in 1890 at the age of forty-three. At the same time, Pasteur was encroaching upon his work, albeit surreptitiously. In May 1881, it was Pasteur's turn to conduct his own public anthrax vaccine experiment.[16, 17] Pasteur announced that his vaccine consisted of anthrax bacteria that had simply been left open to the air for an extended period. He explained to a watching scientific community that time spent exposed to the elements had acted to attenuate the pathogenicity of the bacteria while not impeding their ability to elicit a protective immune response.[18] This outcome was the result of yet another fortuitous accident that Pasteur chanced upon. Earlier in 1880, Pasteur's associate, Emile Roux, had left some anthrax out in the air for many days before its intended use to invoke disease. However, the bacteria had become weakened and thereby helped promote the early concept of attenuation. Pasteur conducted a study, overseen by the French Academy of Sciences, at a farm in Seine-et-Marn, near Paris.[19] Pasteur invited skeptical colleagues to view the immunization procedure and then infect sheep with anthrax, both of which worked as planned. All the vaccinated sheep survived, and the unfortunate few who had remained unvaccinated did not. This outcome, and its implications beyond anthrax to create additional vaccines for human diseases, enthralled the public and vaulted Pasteur to even greater fame.

What Pasteur failed to reveal in the days after his success in the farm field was publically exposed a half century later in the memoirs of one of his assistants, published in 1938.[20] These documents revealed a fraud conducted by an icon of the French—and, indeed, the international—scientific community. Indeed, Pasteur's intentional deception was even more shocking as its revelation came from his own nephew and implicated additional scientific legends in the conspiracy. Adrien Loir was the son of Marie Pasteur's sister and had been given a job at the Pasteur Institute around the time of the fateful anthrax vaccine study. In his memoirs, Loir revealed that rather than simply exposing the anthrax to air as claimed, Pasteur had used a subtle modification of Toussaint's approach using potassium dichromate in place of phenol.[21] Pasteur's approach had been borrowed from work by Charles Chamberland and Emile Roux, who both were more convinced that chemical attenuation was far superior to letting the elements do their damage over time. Given Louis' dominance at the institute, these preeminent scientists were instructed not to reveal the truth until or unless a more effective air- and time-based attenuation approach was developed (which Pasteur remained convinced was a superior approach). Pasteur clearly remained aware of the significance of this deception, as his last will and testament instructed Chamberland and Roux to maintain their silence. Both scientists continued to maintain the deception, though Chamberland would later admit to the duplicity as an old man while composing his final memoirs.

Although Koch remained unaware of the potassium dichromate controversy throughout, he was nonetheless critical of Pasteur and his work. On one level, Koch took personal offense at the fact Pasteur's work on anthrax referred to the causative organism as "bacteridia," an arcane name dating back to Davaine, rather than the more specific name conferred by Koch, *Bacillus anthracis*.[22,23] As we have seen, Koch actively opposed the concept of attenuation and instead focused on killing the bacterium or its unique toxin. Indeed, this was the approach later favored by Behring in the generation of antitoxins meant to neutralize the toxins of anthrax, diphtheria, and other pathogens. Within months after Pasteur's highly publicized 1881 experiments with anthrax vaccines, Koch and two of his Berlin-based students, Georg Gaffky and Friedrich Loeffler, published a series of critical attacks on Pasteur, with accusations of sloppy technique and data.[24] Pasteur responded to the critiques in a September 1882 talk at the International Congress of Hygiene and Demography held in Geneva.

Reminiscent of the translation challenge behind the "Ems Telegram" that helped launch the 1870 Franco-Prussian War, the talk was translated to Koch (who did not speak French). The interpreter misheard Pasteur's use of the phrase "*recueil allemande*," which refers to "a collection of German writing" (referring to the critiques from Koch, Gaffky, and Loeffler).[25] Instead, the interpreter mistakenly heard "*orgeuil allemande*," which translates into "German arrogance." Koch was properly incensed, particularly as Pasteur's composure was unsurprisingly calm (and interpreted as haughty) after unknowingly lobbing the slur. It is not clear what happened in the moments following, but clearly, the mistranslation was not corrected. We know this because Koch published a blistering attack against Pasteur in 1882, calling his data "useless" and questioning his credentials.[26] Koch also made it clear that he viewed Pasteur's Geneva talk to be a personal attack on his character and career. Pasteur, presumably still confused at his Teutonic colleague's anger, struck back and defended himself in subsequent papers.

The professional rivalry raged on as both the Pasteur and Koch camps sent separate teams of investigators to Alexandria, Egypt, in August 1883 amidst an outbreak of cholera, a well-known disease with an unknown cause. The French delegation was led by Emile Roux and included Isidore Strau, Edmond Nocard, and Louis Thullier.[27] The latter was one of Pasteur's rising stars; despite being only twenty-seven, he'd helped in the creation and testing of the anthrax vaccines. Despite arriving before the Germans, the French team had no success in identifying the cause of cholera. However, they were more successful in isolating the organism than intended, because Thullier was struck down and killed by the disease on September 19, 1883. The grief-struck team packed their bags and left Alexandria to return Thullier's body to Paris.

Meanwhile, the German team, consisting of Koch, Georg Gaffky, and Bernhard Fischer, performed many autopsies on patients killed by cholera and were eventually able to isolate the responsible organism.[28] Already on the road, the team then proceeded to Calcutta, where additional cholera samples were isolated. By early January 1884, still in India, Koch announced the isolation of a new bacterium, *Vibrio cholerae*, with the odd shape "a little bent, like a comma."[29] Koch's updates on his work in Egypt and India were being closely followed by the German press, which applauded his findings. Ironically, the conclusive evidence that the organism isolated by Koch was responsible for cholera failed Koch's own postulates (for technical reasons having to do with how the organism propagates). Thus, while Koch was ultimately proved correct, international skepticism, perpetuated by the rival French team and by their English allies, reigned supreme until the preponderance of scientific thought eventually swung in favor of Koch.[30]

An investigator at the Pasteur Institute in Paris would follow up on Koch's discovery of the cholera bacterium and utilize Pasteur's attenuation techniques to develop a vaccine. Waldemar Mordecai Haffkine was born to a Jewish schoolmaster in Imperial Russia (modern Ukraine). He was trained by Elias Metchnikoff, a Russian immunologist who later shared the 1908 Nobel Prize with Paul Ehrlich.[31] Following the assassination of Tsar Alexander II in 1881 and the resulting pogroms that hunted down many in the "Jewish intelligentsia," Haffkine joined the resistance but was soon thereafter injured and jailed, released only after the venerable Metchnikoff vouched for him.[32] In 1888, Haffkine emigrated to Switzerland and followed his mentor Metchnikoff to the Pasteur Institute in Paris. Neither scientist ever returned to Russia, and Haffkine found the only job he could in Paris: as a librarian in the Pasteur Institute.

Haffkine's talents were soon liberated from the book stacks, and he initiated a campaign of chemical attenuation to develop a cholera vaccine.[33] As seems to have happened far too many times in a more naïve age, Haffkine first tested his vaccine on himself on July 18, 1892. He survived this lethal experiment and reported on his findings a mere twelve days later in a scientific report to the English Linnaean Society.[34] Still questioning whether Koch's bacterium had been the organism responsible for cholera, neither Pasteur nor Metchnikoff was convinced of the finding. Unperturbed, Haffkine was convinced his new vaccine held promise and was determined to test his vaccine in people.

By the time of, and perhaps due to, word of Haffkine's key study, the British biomedical community had begun to accept that *Vibrio cholera* was the organism responsible for the disease and that Haffkine's discovery might indeed hold value. As further incentive, cholera was a particular problem in the Indian colonies, and the British were highly motivated to investigate potential vaccines. As often happens, chance played a major role in Haffkine's future. The British ambassador to France, Lord Frederick Dufferin, happened to be a former viceroy of India and took particular interest in Haffkine's work (despite the fact that most of the French scientific establishment was still dubious about *Vibrio cholerae* and its links to the disease).[35]

Arriving in England, Haffkine met Ernest Hanbury Hankin. You may recall that Hankin was the English doctor who would later advance medical science by discovering the presence of an unknown substance in the polluted waters of the Ganges. We now know this to be the first description of a bacteriophage. Haffkine and Hankin initiated a lifelong friendship, and Hankin later followed Haffkine to India, where they also collaborated on a series of scientific partnerships.

Consistent with his passion to alleviate the scourge of cholera, Hankin served as Haffkine's fifth guinea pig to test the safety of the attenuated cholera vaccine (after Haffkine himself and three Russian exile friends). Hankin also helped Haffkine by communicating the encouraging findings to the prestigious *British Medical Journal,* which published his findings mere weeks before Haffkine set off to initiate a much more ambitious campaign of clinical trials in India.[36]

The trip was made possible because Lord Dufferin organized a meeting between Haffkine and Lord John Kimberley, the British secretary of state for India. Upon learning of the potential for a vaccine to prevent one of the most common diseases in the Raj, Kimberley embraced the idea and gave Haffkine access to all parts of the subcontinent. However, the British mandarin would only allow Haffkine to perform his studies under the condition that all subjects consent to participation in experiments (a relative rarity a half century before the development of the Nuremberg Code in the wake of Nazi atrocities during the Second World War). Haffkine was also forced to support his study with his own resources or donations. Undeterred, the adventurous Haffkine lobbied wealthy donors around the United Kingdom. Upon securing just enough funding to make the trip (but not for the clinical studies of the vaccine), Haffkine set out to confirm his belief in the cholera vaccine.

Haffkine arrived in Calcutta (now Kolkata) in March 1893. It was a monsoon-soaked metropolis that provided favorable conditions for the spread of a waterborne pathogen such as cholera. Unfortunately, the local authorities were not particularly enamored with the idea of a heavily accented Franco-Russian scientist conducting a series of dangerous medical experiments on their civilian population.[37] Fortunately for Haffkine, Hankin had just returned to the subcontinent to resume his role as the head of the bacteriological laboratory at Agra in Uttar Pradesh. He asked his friend to conduct the cholera trials almost a thousand miles away, but this site was in the drier conditions of north central India (not particularly know for outbreaks of cholera). Nonetheless, Haffkine recognized the opportunity to leverage the support provided by Hankin's offer and began the (voluntary) immunization of ten thousand British and Indian soldiers (at the government's expense rather than his own).[38, 39]

Luck worked against Haffkine for a time, as the monsoon rains did not bring cholera to Uttar Pradesh that year.[40] Although Haffkine was disappointed, a high-profile British official he'd met shortly after his arrival in India asked him to return to Calcutta and assist in evaluating an ongoing cholera outbreak. The Scottish-born Sir William John Ritchie Simpson was the lead public health officer for Calcutta and a pioneer in the emerging field of tropical medicine.[41] In appreciation for Haffkine's assistance in identifying the source of a cholera outbreak in his mandate, Simpson allowed Haffkine to initiate studies of his vaccine.

The human clinical trials began in earnest in the spring of 1894. Within months, the results of Haffkine's Calcutta experiment were sufficiently impressive to expand the number of subjects and for Simpson to identify additional resources to underwrite the studies. Amidst this exciting progress, Haffkine was struck down by malaria and was forced to convalesce in Europe for a time. Consequently, the work was continued by Simpson in India while Haffkine convalesced in England. As word of success spread locally, Indian and colonial English subjects concerned about the spreading cholera outbreak volunteered en masse, and the number of subjects soon swelled beyond forty thousand.[42]

We will return to Haffkine later in this chapter, but it will be necessary to briefly summarize the rapid-fire development of vaccines, particularly in France during the final years of the 19th century and the early decades of the 20th.

Mad About Rabies

Pierre Victor Galtier was a veterinarian. Born in 1846 to poor parents, he was effectively orphaned by the age of seven.[43] The young Pierre nonetheless demonstrated sufficient intellect to overcome a troubled start: he ran away from the nuns at the orphanage but became valedictorian upon his graduation from the Ecole Veterinaire de Lyon in 1873.[44]

After taking a position as a veterinarian and marrying the boss's daughter, he joined his alma mater in Lyon in 1876, eventually rising to lead the departments of pathology and internal medicine.[45] By 1879, Galtier published a thesis on rabies, which declared that the causative pathogen could be isolated from dogs and transmitted to rabbits. He also demonstrated that while changing species, the virus lost much of its vigor. What Galtier found (and which we now know happens at the level of variability in the DNA molecule) was that the ability to productively infect a new species produced a range of variants—essentially mutants—that would be less well-adapted to causing disease in humans. Such understanding led Galtier to consider the potential to develop an attenuated pathogen as a vaccine. Many contemporaries did not believe rabies was caused by a virus; they presumed this to be a bacterium that had yet to be isolated or, perhaps, a bacterial toxin. Likewise, the response of German scientists was largely skeptical, as many adhered to the idea that the behavior and tenacity of microscopic organisms was fixed.

Despite these handicaps, Pierre was determined to identify a way to treat or prevent the disease. Their desperate need for such a treatment is illustrated by examples put forward in an English 1878 treatise on the subject written for physicians. A primary recommendation for treatment of rabies was a procedure known today as "cupping."[46] Specifically, a cup would be heated with cigar smoke and then placed over the wound. As the temperature cooled in the cup, a bit of suction was applied, which was intended to raise the virus to the surface of the wound. Then, a physician had a choice. He could either use a red-hot poker to scorch the wound, or, lacking an iron poker or fire, he could rely upon the ubiquitous presence of guns (back in the day when guns were more prevalent). In the latter case, a bullet was to be dismantled and then sprinkled into the wound. The doctor would then light a match and drop it into the wound. The resulting explosion might sometimes be sufficient to kill the infectious material, though the efficacy of this treatment was doubtful. In other words, rabies was predictably a death sentence.

To help avoid such desperate (and largely ineffective measures), Pierre published a manuscript in 1880 that provided an overview of what was known about the disease and its cause.[47] This publication clearly inspired Louis Pasteur to take up work on a rabies vaccine, but the memory of Galtier and his key contributions have largely faded over the years. This oversight is perplexing, as Pierre continued his studies in animals, demonstrating in 1881 that the direct injection of the attenuated rabies virus into the blood could protect sheep from rabies. Despite a successful demonstration of the feasibility of a rabies vaccine, the relatively obscure country veterinarian was overshadowed by the dominance of the Parisian scientist.

Much as we saw with Toussaint, Pasteur took or was given credit for many ideas and outcomes that were more accurately attributable to Galtier. In 1882 Emile Roux, then an outstanding student in the Pasteur laboratory, began studies on rabies; his dissertation frequently cited the inspirations and background provided by Galtier.[48] In contrast, Pasteur, in most of his published works and public talks, was dismissive of Galtier and his work. This lack of attribution literally pained the more junior investigator, who was sensitive to Pasteur's lack of embrace and fearful of the possibility that Pasteur might emerge as an unfriendly rival.[49] Although Roux would essentially copy Galtier's studies and try to cite his predecessor's work, Roux's work was inappropriately hailed as the key breakthrough, particularly given Pasteur's profile and advocacy for Roux.

In July 1885, Roux and Pasteur at last conducted a key study that Galtier had not. Despite lacking medical credentials, Pasteur tested his vaccine in a person. A nine-year-old boy by the name of Joseph Meister had been mauled by a rabid dog and was undoubtedly doomed to a slow, painful death from rabies (thankfully, he had not been subjected to the gunpowder approach). Master Meister's only hope was the experimental vaccine being developed by Roux and Pasteur. The idea of treating the child was particularly dangerous, since Pasteur lacked a medical degree and could therefore be subject to legal prosecution.[50] However, the child was otherwise doomed, and Pasteur agonized about whether to conduct such an ethically- and legally-fraught experiment. Prodded by his pediatrician colleague, Jacques-Joseph Grancher, Pasteur relented.

Starting two and a half days after the mauling, the child was injected a total of twelve times with a preparation of rabies virus isolated from the spinal column of infected rabbits (recalling that the virus lost much of its killing potential when transferred from a dog to a rabbit).[51, 52] The viruses within the spinal cords had been further attenuated by drying the material for two weeks prior to its use as a vaccine. Meister survived. Eternally grateful to Pasteur, he served at the Pasteur Institute for the rest of his life.

As an aside, there's an apocryphal story about Meister ultimately meeting his end at the hands of Nazi guards (or suicide in a different version) while protecting the tomb of Pasteur during the German occupation of Paris, but alas, this story is even more heartbreaking. Meister sent his family out of Paris as the German military columns approached, only to receive word that his family was killed during their escape. The distraught Meister then committed suicide by asphyxiation with a gas furnace. In a tragic plot twist, his body was found later that day, when his family safely returned to their apartment in Paris.[53]

Upon learning of Pasteur's experience with the rabies vaccine and the ethical implications of using Meister as a human guinea pig, Koch was perhaps unsurprisingly dismissive of his rival's success.[54,55] Upon reflection, Koch conceded the implications of the discovery and later instructed his Berlin team to utilize the same techniques to develop a vaccine that could be distributed in Germany. However, the acrimony felt by Koch towards Pasteur was still sufficient in 1893 that he begged out of attending a seventieth birthday celebration, citing personal reasons (in his defense, he did end his twenty-six-year marriage with his wife that year). Three years later, Pasteur would be dead and Koch's feelings themselves attenuated over time. By 1904, Koch felt sufficiently comfortable to visit the Paris headquarters of the Pasteur Institute, where he was warmly received.[56]

Toxoids

According to the website ScienceHeroes.com, the French scientist Gaston Ramon is estimated to have saved the lives of sixty million people.[57] To put this in perspective, such an outcome equates to roughly the entire population of the United Kingdom or the number of people who voted for Donald Trump in the 2016 election (though three million less than his rival). Moreover, sixty million people is more than ten times the number of people saved by the extraordinary contributions of Gertrude Elion, who, as we saw, developed a toolbox of new medicines that included some of the most frequently used cancer and HIV/AIDS drugs.[58]

Despite this impressive feat, the name Gaston Ramon remains almost entirely unknown to modern ears, perhaps in part because his achievements never earned him the notoriety of a Nobel Prize. However, this was not for lack of trying, as Ramon holds the dubious record of being nominated on 155 separate occasions for a Nobel Prize, more than any person before or since.[59]

Gaston Ramon was the son of a baker, born in 1886 in the small town of Bellechaume near Paris.[60] In 1906, he enrolled in the local veterinary school, l'École Vétérinaire d'Alfort, located in the Parisian suburbs. It was one of four veterinary schools in France. Gaston's choice of Alfort was well timed and placed, as its director, Henri Valee, had deep connections with the Pasteur Institute. These contacts would serve Ramon well and in turn help revitalize the Pasteur Institute following the death of its eponymous and larger-than-life leader.

The Pasteur Institute was a government-backed research organization officially launched in 1888 to honor its founder.[61] The principals list at the new institute was a who's who of prominent scientists we have met throughout our story: Emile Roux, Charles Chamberland, Elie Metchnikoff, Alexandre Yersin, Jacques-Joseph Grancher, and Emile Duclaux.[62] With the death of the great man on September 28, 1895, the eldest of the group, Duclaux (who was a mere forty-seven, while all the others were in or barely out of their thirties), was elected director to replace Pasteur. Duclaux had a long history with Pasteur, accompanying him during his studies of silkworms during the Franco-Prussian War, then leading the internationally recognized *Annales de l'Institute Pasteur*—a periodical that remains a leading source of microbiology research—from its beginnings. During his tenure as the leader of the high-profile institute, Duclaux took up the cause of Alfred Dreyfus.

Dreyfus was a Jewish officer of the artillery. Based on dubious evidence, he was wrongfully convicted in 1895 of selling secrets of new artillery innovations to the despised Germans.[63] Given the magnitude of the crime, Dreyfus was publically humiliated before being exiled to a life of imprisonment on the notorious and squalid Devil's Island off the coast of French Guiana. Over the following months, evidence was unearthed by Lt. Colonel Georges Picquart revealing that the actual traitor was Major Ferdinand Walsin Esterhazy, but Picquart was silenced and assigned to a post in the southern deserts of Tunisia. As word of the cover-up leaked to the press, it became clear that Dreyfus's conviction reflected widespread anti-Semitism in the French High Command, and Dreyfus's cause was taken up by the famous author Emile Zola. Joined by many prominent citizens, including Emil Duclaux, these dissenters eventually gained a pardon for Dreyfus in 1899. He was fully exonerated in 1906.

After Duclaux's death in 1904, the reigns of the Pasteur Institute were handed over to Emile Roux.[64] Roux, you may recall from the previous chapter, had led Pasteur's efforts to develop an antitoxin for diphtheria. Roux also led the team that developed Pasteur's rabies vaccine and the expedition that sought to isolate cholera in Alexandria (until the death of Louis Thullier). In 1901, Emile Roux recruited Henri Vallee, who had just rejoined his alma mater veterinary school in Alfort (and would soon rise to lead the institution). Roux and Vallee shared research interests in tuberculosis and foot-and-mouth disease, both of which are problems for both livestock and humans.

In 1911, Vallee introduced Roux to one of his most promising students, a newly credentialed veterinarian by the name of Gaston Ramon.[65] An impressed Roux hired Gaston and tasked him with oversight of the production of a growing number of different horse-derived antisera, including antibodies targeting diphtheria and tetanus. As the Great War became entrenched in the north and western portions of France, disease outbreaks inevitably followed. As its part of the war effort, the Pasteur Institute was tasked with meeting the ever-increasing demand for antisera. Beyond simply manufacturing more material, Ramon was challenged with identifying ways to preserve these vital medicines. Once they left the institute, the sera were highly susceptible to contamination from bacterial contaminants, which ironically feasted on the proteins in the antitoxins as a food source. Ramon's solution was to package the antisera with small amounts of formalin, the active ingredient of embalming fluid. Gaston and other at the Pasteur Institute had routinely used formaldehyde to sterilize the glass tubes and flasks, and he considered that including small amounts of this chemical in tubes of antitoxin might likewise suppress the growth of contaminating organisms.[66] This simple idea worked and allowed the institute to increase the storage and transport of the antitoxins they produced while preserving their efficacy.

By the end of the war, Roux and Ramon's relationship had expanded beyond merely the professional when Gaston married Marthe Momont, Roux's niece.[67] In the following years, Ramon improved the ability to purify antibodies away from other animal proteins in sera to minimize the untoward effects of serum sickness. The early years of the Jazz Age also witnessed his epiphany that the same procedures he had used to preserve antitoxins might be used to inactivate the toxins themselves and thereby create a vaccine. By 1923, Gaston had subjected diphtheria toxin (the same material he routinely used to immunize horses) to a combination of formalin and elevated temperature and found that the modified toxin, which Gaston called "l'anatoxine," retained the ability to elicit an immune response but did not cause disease.[68]

As often is the case in science, Gaston did not work in isolation. Credit for discovery of the diphtheria vaccine arose simultaneously from Gaston's work and that of an English researcher by the name of Alexander Thomas Glenny. In 1899, at the age of seventeen, Glenny began working in the laboratories of the Wellcome Physiological Research Laboratories and simultaneously working towards a college degree at the University of London.[69] The company was producing large amounts of diphtheria antitoxin, which required large amounts of toxin that would be injected into horses to produce the antisera. The bacteria were cultured in large clay barrels and were decontaminated between batches using a formaldehyde solution, as was the standard practice of the day. One day in 1904, Glenny noted that a batch of bacteria was unable to infect the horses. Upon investigating the cause, he realized this batch had been improperly rinsed.[70, 71, 72] Glenny thereby surmised that the residual formaldehyde must have compromised the bacteria, much as Gaston would realize two decades later. However, Glenny did not follow up on this finding until 1923 (by coincidence, the same year as Gaston). Working with another researcher, Barbara Hopkins, Glenny returned to this finding and demonstrated that the formalin-inactivated diphtheria toxin could elicit an immune response. He referred to this neutralized toxin as a "toxoid."[73]

The ensuing row raised concerns that did not quite rise to the level of the Pasteur-Koch feud, but it did create some uncomfortable moments. The key question centered upon who should get credit for the discovery. Glenny only conveyed the 1904 finding in private conversations (one of which appeared in the book *A History of Immunization*, first published in 1965).[74] Likewise, it is not clear whether Glenny was suddenly motivated to investigate the question of formalin treatment of diphtheria toxin only after he heard rumors of Gaston's success.[75] Regardless, time has ruled a sort of split decision in that Gaston Ramon is generally acknowledged to be the discoverer of the vaccine, while Glenny's nomenclature of "toxoid" (rather than Gaston's "anatoxine") has prevailed. Nonetheless, the Nobel committees are notoriously shy about courting such controversies, which begs the question of whether the issue of Glenny's contributions might explain why, despite 155 nominations, Ramon never received the prize.

The discovery of a diphtheria "toxoid" provided a much-needed vaccine to replace the passive horse-derived antitoxins, particularly given the propensity of the latter to cause serum sickness. Stated another way, the vaccine allowed by a toxoid could prevent the disease from arising in the first place, while the antitoxin was administered after the disease had already gained a foothold.

The introduction of the vaccine also eliminated the need to rush antitoxin to sites of infection. The most notable example of this need occurred in 1925, when an outbreak in Alaska triggered a desperate race from the port town of Seward to Nome in time to prevent diphtheria from wiping out Nome's population.[76] This "Great Race of Mercy" is memorialized annually by the Iditarod Trail Sled Dog Race. Indeed, the antitoxin market, at least for diphtheria and tetanus, has largely disappeared.

Soon after the introduction of the diphtheria vaccines, the Spanish Strangler, which had routinely infected 200,000 children and killed 15,000 per year in the United States alone, was effectively eliminated.[77] For example, in the twelve years between 2004 and 2015, only two cases were diagnosed in the United States.[78]

Despite a success that on its own would have memorialized Gaston for all time, he was not yet done. Using similar approaches to inactivate tetanus toxin, he helped optimize a tetanus toxoid as a vaccine in 1925. The original discovery of a tetanus anatoxin (toxoid) was inspired by Gaston's work with diphtheria and was first published by Pierre Descombey, Ramon's colleague from the Pasteur Institute. Additional studies with Gaston and Christian Zoeller demonstrated the clinical efficacy of tetanus toxoids in 1926.[79] Appreciating the benefits of tetanus prophylaxis, the United States military was an early adopter, requiring recruits to be vaccinated against tetanus from 1940 onwards, just in time to help increase American preparedness for entry into the Second World War.

Diseases of the Body and the Mind

In an unusual twist, the success of the tetanus toxoid vaccine brings us back to the remarkable figure of Waldemar Haffkine. When we last left the Russian-born, French-trained, and English-supported researcher in India, he had demonstrated the safety and efficacy of his cholera vaccine during a December 1895 talk in London.[80] Wildly feted by the English intelligentsia, he could have settled into a comfortable academic position, but he opted to return to India weeks later.

The timing was propitious, as the Black Death had returned to Asia. By the time of his return, a wave of plague had overtaken Asia. As we have seen, bubonic plague is an awful disease made all the worse by a periodic tendency to break out en masse. Fortunately for the continuance of our species, there have only been three such plague pandemics.

The first pandemic was the so-called Plague of Justinian, so named for the sitting emperor of the Eastern Roman or Byzantine Empire. Likely entering the empire on the backs of rats succored by the grain of ships plying the trade route between Alexandria and Constantinople, the plague took hold of the empire by the year 541. This clash with the plague, which infected an estimated one in seven humans on the planet within a year, killed between twenty-five and fifty million people. A detailed history of the plague and its aftermath is captured by the outstanding book *Justinian's Flea* by the late William Rosen.[81] The disease cut down an attempted resurrection of the Roman Empire after Justinian's legendary general, Belisarius, reclaimed and reunited much of the former greatness of the earlier days of the empire. However, the massive societal and infrastructural havoc wreaked upon the empire during the vicious outbreak of the Plague of Justinian left it even more susceptible to enemies on every front, including Goths in the north and west, Lombards in the central regions (eventually taking Rome), and Arabs (later Muslims) to the east and south.

Not to be outdone by its prior achievements, the plague returned in force eight hundred years later, to be memorialized in Western literature as the Black Death. This is the pandemic experienced by Guy de Chauliac, which engulfed virtually all of Europe (and much of Asia), dispatching an estimated 100 million victims, almost one quarter of all humans alive at the time. The societal impact of this second plague was even greater than the first and included the elimination of most feudal societies (since a ruling class denuded of many workers could not count on the remaining few to carry on as before).[82]

Based on these two earlier experiences, fears of the rapidly accelerating third pandemic were manifest. Starting as a natural infection in the Yunnan Province of southwest China in 1850, the pandemic was accelerated, as is so often the case, by strife and warfare.[83] Specifically, the Panthay Rebellion, which began in earnest in 1856, represented a revolt by the Muslim Hui people, who had experienced state-sanctioned ethnic and religious persecution from the ruling Qing Dynasty.[84] Following a massacre of Muslims in Kunming, the restive Hui rose up against the Qing, triggering a seventeen-year revolt that witnessed the death of up to a million people and the temporary rise of a sultanate.

Though quashed by the Qing rulers, the disorder caused by the revolt helped facilitate the spread of bubonic plague. Canton experienced its first cases in 1894 and nearby Hong Kong suffered from 100,000 deaths within a few short weeks. The crisis in Hong Kong led the Pasteur Institute to dispatch the head of its local affiliate in French Indochina, a Swiss-born Frenchman by the name of Alexandre Yersin, to Hong Kong. In a shameful display of nationalism, Yersin was barred from using the facilities of nearby English hospitals but nonetheless commandeered a small ramshackle workshop, where he succeeded in isolating the bacterium responsible for the plague, now known as *Yersinia pestis.*[85] Kitasato Shibasaburo, who had returned to Tokyo after completing work with Robert Koch, might have received the distinction of discovering the bacterium, which he did independent of Yersin, a mere few days later.[86] The two appropriately did share the limelight for a time before Kitasato's contributions were unfortunately forgotten by future generations.

By the early autumn of 1896, the plague had arrived in India, not via its extensive land border with China but the same way it had entered Constantinople, London, Rome, and other cities during previous outbreaks: by sea.[87] The primary trading ports of the Raj began reporting cases in quick succession: Bombay, Pune, Karachi, and Calcutta. The terrified British rulers reacted with quarantines, embargoes, and travel bans. Waldemar Haffkine was recruited into the Indian Civil Service and tasked with developing a vaccine using the same principles he had used to develop the cholera vaccine. Continuing the risky practice of testing his vaccine on himself, Haffkine experienced a fever and general malaise but otherwise deemed his new vaccine safe enough to test on others.

His first "volunteers" included prisoners held at the royal prison in Bombay, where the natural form of *Yersinia pestis* had begun eating through the prison population.[88] Positive results from this trial, combined with increasing anxiety about the spreading plague, boosted demand for this new vaccine. Unlike the early days of his cholera vaccine studies, Haffkine's biggest problem was not in finding volunteers to test his plague vaccine but in meeting the urgent demands of a frightened public.

Haffkine's diligent efforts saved countless lives, stopping the third pandemic in its tracks before it could cause the civilization-altering devastation rivaling that of the Plague of Justinian or the Black Death. His works were recognized with many awards, including British citizenship and being named a Companion of the Order of the Indian Empire by the empress of India herself, Queen Victoria. All such honors would soon be conveniently forgotten and pushed to the side in the wake of what became known as the "Little Dreyfus Affair."

After scoring his second major coup with the discovery and implementation of the plague vaccine, Haffkine was soon back in the laboratory developing or refining new medicines. To meet the growing demand during the panic caused by the plague, Haffkine had gained considerable efficiencies by omitting a step in which the vaccine was treated with carbolic acid.[89] He did so after learning that this improved procedure had proven successful, as performed by former colleagues at the Pasteur Institute in Paris. Indeed, the experience in India largely revealed that the modified vaccine was as safe and efficacious as the original breakthrough.

On October 30, 1902, 107 patients in the Punjab village of Mulkowal were immunized with a batch of plague vaccine (as were many others across the country).[90,91] However, something had gone terribly wrong and nineteen died, all showing distinctive signs of tetanus (lockjaw) infection. After some quick detective work, all the deaths were linked back to one bottle of vaccine, labeled 53N. As the patients immunized with vaccines from all other bottles remained unharmed, container 53N was identified as the source of the tetanus. Further examination revealed 53N had been manufactured in Bombay almost a month before, and its history was closely tracked.

An inquest very quickly concluded that Haffkine's failure to include carbolic acid in the manufacture of the vaccine was responsible for the deaths.[92,93] Haffkine was put on what effectively was administrative leave. He retired to London to regroup and was later summarily fired. For two years, he pleaded his innocence and provided rationale that the contamination must have occurred in the village at the time of injection and not during manufacturing. While his evidence was strong, including the fact that all other bottles manufactured before or since had been safe and effective, these protests fell on deaf ears. The inquest panel needed to placate the outrage of those demanding to know who was responsible for nineteen excruciating deaths from tetanus.

Haffkine endured exile and unemployment for four years, continually pleading for the release of the commission's findings. These were finally released to the *Gazette of India* in Calcutta on December 1, 1906. Buried in the documents was exculpatory evidence in Haffkine's defense, including the fact that the bottle was, by standard procedure, sniffed by the vaccinators to ensure that no major contamination had occurred (a pungent odor would betray the presence of a contaminating bacterium, such as that which causes tetanus).[94] Most damning was the revelation that the technician administering the vaccines had dropped his forceps in the dirt just prior to using them to remove the stopper from bottle 53N. Clearly, simply wiping the tool with a cloth was insufficient to prevent massive contamination of the bottle with tetanus bacteria, which thrives in the soil.

Reaction from the Indian and British medical societies was loud and damning, led largely by the respected scientist Ronald Ross.[95] Comparisons with the French Dreyfus Affair were adopted by the press and propelled both by the fact that evidence was ignored or buried (especially the dirty forceps) and because Haffkine was a high-profile Zionist.[96] His arrest by the tsar's police for defending his community against the anti-Jewish pogroms had led to his exile from his mother country. Moreover, Haffkine had remained a vocal advocate for resettling Russian Jewish refugees in the British-administered territory of Palestine. As support for Haffkine grew, his tribulations gained the moniker "Little Dreyfus Affair."[97]

A campaign to exonerate Haffkine was supported by a high-profile group of British scientists, who published an open letter in the *Times* (London) on July 29, 1907. Their prominence, combined with a full revelation of the findings of the commission, prompted a belated and halfhearted act of contrition from the British India Office, which reinstated Haffkine (though the Raj officials utterly refused to offer an apology). However, the damage had been done, and the consequences of the dirty forceps used to open bottle 53N continued to cloud Haffkine's reputation for the rest of his short career. At the age of fifty-five, Waldemar retired and gave up on the British altogether, moving to Paris to live with his sister.

The fact that Waldemar Haffkine had effectively cured not just one but two notoriously lethal diseases led Joseph Lister to acclaim Haffkine as "a great savior of mankind."[98,99] The Russo-Franco-English scientist working in India was remarkable not just for these two achievements but also for inspiring an international blend that applied the French technique to develop vaccines (from his Pasteur days) with the pathogenic organisms identified by the German team led by Robert Koch. Nonetheless, memories are notoriously short, and after Haffkine's death in 1930, his contributions faded into relative obscurity.

Returning to the theme of international cooperation, we will now turn to the years surrounding the Second World War. This era proved instrumental in defining new approaches pioneered by multinational scientific findings. It also witnessed the transition of vaccine breakthroughs from the Old World to the international melting pot of the United States.

8
Breathing Easier

With the minor exception of Cotton Mather's pioneering efforts to deploy variolation in the colonies, or the adoption of antitoxins in places such as St. Louis, the United States has not yet figured prominently in our story. While virtually all the research and development of innovative vaccines had occurred primarily in France, Germany, or Britain throughout the 18th and 19th centuries, this would abruptly change immediately after the conclusion of the Great War. This chapter will detail the transition of pioneering vaccine research from the Old World to the New. The list of vaccines and the pioneers responsible for their discovery are well beyond what could be addressed in a single chapter, so we will focus on one example: the role that American-led vaccine research played in preventing death and diseases of the lung. This chapter will focus not merely on the development of the scientific establishment in the United States but upon another American contribution to the subject, the rise of an organized anti-vaccinator movement. As an example of both trends, we will focus upon the development and deployment of a vaccine targeting the age-old disease known as whooping cough. As our story transitions away from France and Germany and towards America, our flight plan momentarily takes us to a country torn between the two main European powers that have featured so prominently in the history of vaccines.

Geography has not been terribly kind to the small nation of Belgium. Wedged between France and Germany, "the Battlefield of Europe" has all too frequently been the focus of its larger neighbors. It was host to major carnage in both the Hundred Years' War in the 14th century and the Thirty Years' War that began in the 17th. A century and a half later, the resurgence of Napoleon following his escape from Elba was brought to a halt by the combination of Wellington and Blucher's stand in the small Belgian town of Waterloo. A century after that, the invasion of Belgium by the Kaiser's army served as a means to bypass French defenses further to the south and led to the British entry into the Great War. Yet another quarter century later, a Nazi invasion of Belgium occurred not once but twice, first during the lightning campaign to take Western Europe in 1940 and again during the Battle of the Bulge in late 1944.

Despite these challenges, geography has imparted some distinct advantages upon tiny Belgium. One example is its role as the birthplace of Jules Jean Baptiste Vincent Bordet.[1] The budding scientist was born in 1870 in the Walloon village of Soignies, a site near the French border that would gain later notoriety as the location where the British Expeditionary Force would first collide with the Germans during the Battle of Mons in 1914. His father was an itinerant schoolteacher from a town near the German border, but the family later settled in Brussels, where the young Bordet completed his primary and secondary education, graduating as a doctor of medicine in 1892. Having gained early distinction for his early work with viruses, Bordet was recognized by the Belgian government in the form of a paid fellowship in 1894 to work at the Pasteur Institute under the auspices of Elie Metchnikoff, who would soon receive the Nobel Prize for his work in discovering the basic tenets of immunology.

During a remarkably productive seven-year stint in Paris, Bordet first discovered the underpinnings of the complement system whereby antibodies can kill bacterial invaders (recall a brief overview of this subject in chapter 6), an accomplishment that would earn Bordet the 1919 Nobel Prize. Towards the end of his time in Paris, Bordet also partnered with another Belgian, Octave Gengou, on a particularly challenging project to identify the cause of whooping cough.[2]

Four centuries earlier and also in Paris, the physician Guillaume de Baillou was the first to document an epidemic of a disease known as pertussis.[3] The symptoms, which overwhelmed the primitive medical establishment of Paris in 1578, were initially indistinguishable from a common cold but quickly progressed to loud fits of violent spams of coughing, accompanied by a characteristic "whooping" sound, which often triggered uncontrollable vomiting. These symptoms could persist for months, earning the disease the moniker "100-day cough."[4, 5, 6] Although this Parisian outbreak is often cited as the first historical description of the disease, new data suggests the disease might have first been detected in Korea as early as 1433 and in Persia in 1484.[7]

Though the disease had been experienced by children for centuries, its cause remained largely unknown until the early years of the 20th century. Repeated attempts had failed to isolate the organism. Indeed, Bordet and Gengou were repeatedly frustrated by the fact that while the symptoms of the disease could persist for months, the organism responsible for the disease could only be isolated during a very narrow window of time (indeed, the duration of the symptoms reflected the injury done early in the infection and the time needed to fully repair the damage). Compounding the problem of isolating the germ responsible for whooping cough, the bacterium pathogen was restricted to the lung and could not be detected in the blood of an infected child. Worse still, even when small amounts of the bacterium could be isolated from the lung or mucous of infected children, the organism that caused such powerful outcomes was itself surprisingly fragile. The bacterium grew agonizingly slowly and quickly died outside the body.[8] The team was finally able to see the pertussis bacterium under a microscope as early as 1900, but the organism died before they could propagate it in the laboratory. Success finally came in 1906 when Bordet and Gengou developed novel ways to isolate and keep the bacterium alive. The breakthrough involved a customized broth that allowed the bacterium to proliferate outside the body.[9] The bacterium was named in honor of the more senior of the pair and is now known as *Bordetella pertussis*.

Having at last isolated the organism, the team then focused on developing a vaccine. Although the fragility of the organism provided a consistent and aggravating impediment, the breakthrough broth they had developed allowed researchers to routinely culture *Bordetella pertussis* in the laboratory. This ignited a version of the Oklahoma Land Rush to stake the ground for a successful vaccine. On October 9, 1909, the first reports of a vaccine developed by Bordet and Gengou emerged from a study conducted in London by the prominent British physician John Freeman.[10] Much excitement accompanied the rumors of success in London, but the effectiveness of this vaccine in preventing whooping cough in children was marginal at best.

By this time, Bordet and Gengou had returned to their native Belgium to lead a branch campus of the Pasteur Institute in Brussels. This seemingly insignificant relocation is rather symbolic in that it coincided with a dispersion of vaccine research beyond the confines of Paris or Berlin. Indeed, just eight days before the publication of the Bordet-Gengou vaccine, John Zahorsky of Washington University in St. Louis was technically the first to report clinical findings of a whooping cough vaccine he had developed. The report appeared in an obscure Midwestern journal.[11]

Though the promise conveyed by studies of the Zahorsky vaccine was also minimal, this report marks a sort of milestone in that it established the Americans as an alternative to the European domination of vaccine research. Lacking a focused center of gravity such as the Pasteur Institute in Paris or the Koch Institute in Berlin, a more dispersed group of laboratories in the United States quickly rose to become the dominant players in vaccine research and development. We will witness this transition to North America as a dominant theme of this chapter.

Staying on the topic of a pertussis vaccine, little progress, but much hope and frustration, was generated throughout most of the first third of the 20th century. A series of scientific and medical studies were conducted with small populations of patients that in many cases were not necessarily representative of the wider population. The first large-scale studies to optimize a pertussis vaccine were conducted by Louis Sauer in 1933, but this vaccine likewise did not convey the necessary efficacy required to fully protect children from the dreaded whooping cough.[12] Consequently and despite the testing and occasional marketing of various pertussis vaccines from 1909 onwards, the incidence of whooping cough from 1909 through the 1940s remained high, with most years witnessing more than 100,000 new cases (and some with many more). Tragically, five thousand to eight thousand children continued to die annually from this dreaded disease throughout this time period.[13]

Three Women and a Baby

The war against pertussis changed dramatically as the result of contributions from three remarkable American scientists, all of whom happened to be women, one of whom has been relegated to an almost forgotten historical footnote.

Pearl Kendrick was a three-year-old living an otherwise unremarkable life as a preacher's daughter in Wheaton, Illinois, when she contracted whooping cough in 1896 but survived the ordeal.[14] Ever a precocious child, she began questioning some of the tenets of her religious upbringing during debates with her father about the volatile subject of evolution (a prescient topic, since the University of Michigan has since named a professorship for her in evolutionary biology). Evolution, along with women's right to vote, were the hot-button issues of the day, and the Kendrick family debates occurred two decades before the far more visible Scopes Monkey Trial of 1925. Kendrick's love of science led her to enroll at nearby Greenville College and later to transfer to far-off Syracuse University, where she obtained a bachelor's degree in zoology in 1914. At a time when a woman's choice of career remained highly restricted, she took the conventional approach of teaching school for three years in upstate New York. Less conformist was her decision to commute down to New York City on the weekends to volunteer as a research assistant. Kendrick had seized upon an opportunity to work with Hans Zinsser, a pioneering typhus researcher, who not only helped discover the bacterium responsible for the disease but later developed a vaccine for its prevention.

Compelled by this experience, Pearl Kendrick committed herself to starting a full-time career as a scientist, first with a job at the New York State Department of Health and two years later with the 1919 offer of a job with the Michigan Department of Health in Grand Rapids. These two job offers were extended to Kendrick not based solely on her past accomplishments and potential but because governmental financial constraints were favorable to hiring less expensive (i.e., underpaid) women like Pearl rather than investing larger sums to attract her male peers.[15]

Pearl soon became established in Lansing as an assistant to the director of the Bureau of Laboratories, Cy Young (not to be confused with the baseball legend of the same name). In part, Kendrick was attracted to the position based on Young's policy of supporting the professional development of the women he hired. Seizing this opportunity, Kendrick rose quickly through the ranks and was assigned by Young to lead a satellite laboratory in Grand Rapids, Michigan. In another example of her multitasking capabilities, Kendrick managed to not only fulfill the service requirements expected of this laboratory but also to make it stand out as one of the nation's most innovative and efficient bacteriological laboratories. All the while, Pearl was earning her doctorate from Johns Hopkins, which was awarded in 1932. In that same fateful year, Pearl Kendrick hired and began nurturing another rising upstart by the name of Grace Eldering.

At roughly the same time Pearl Kendrick began questioning her father about evolution, Grace Eldering was born in 1900 to an immigrant Scottish mother and Dutch father, who left their homelands to settle in the central Montana town of Rancher, which is a ghost town today.[16] At the tender age of five, Grace was diagnosed with a particularly virulent case of whooping cough, which left lifelong and vivid memories of painful coughing and vomiting. This experience motivated Eldering to seek a degree in science at the University of Montana. She remained in the state as a school teacher to earn tuition money for college. Grace continued to teach English and science at Hysham High School but was desirous of launching a career in the medical sciences. She found her chance in an advertisement from the Michigan State Department of Health and received an offer soon thereafter. Grace Eldering relocated to Michigan, first to Lansing and then to Grand Rapids, where the team of Kendrick and Eldering would soon begin to make headlines.

In the early years of the Great Depression, Kendrick and Eldering had developed a diagnostic test to detect *Bordetella pertussis*, and had used this to determine the period in which an infected child was contagious.[18] This advance was extraordinary in that it allowed doctors to know when and how long children should be quarantined to curb the spread of the disease.

The pair had also begun to identify opportunities to progress from autogenous therapies to vaccines that could be manufactured en masse. Given the limited time that *Bordetella* could be isolated and remain alive to be cultured into a vaccine, only small amounts of the bacterium could be obtained. Often, the bacterium from a patient was cultured and then killed to create a personalized vaccine (used only for the donor). This autogenous process was necessary given the inability to produce large amounts of pertussis bacteria, but it was wildly inefficient.[19] Kendrick and Eldering were successful in developing a single vaccine that could be given to many, but their ability to deploy this vaccine was limited by a scarcity of funds arising from the financial impact of the Great Depression. This problem was unexpectedly resolved by a 1936 visit from the First Lady of the United States. Eleanor Roosevelt was passionate about public health and had learned of Kendrick and Eldering's work on improving pertussis vaccine. A later interview with Kendrick detailed the visit of the First Lady to Grand Rapids, revealing that Eleanor "was the only lay person to really understand what we were doing."[20] With help from such a high-profile source, funds were suddenly made available from the federal government. By the end of 1939, the first mass production of what is now known as the "whole cell" pertussis vaccine began to be manufactured in earnest but the production was still insufficient to meet the needs of the entire state of Michigan. The term *whole cell* refers to the use of the entire (or whole) bacterial cell that has been inactivated prior to use (this distinction will be important later in our story).

The duo that was destined for such great fame was in fact a trio. The third member of the team was the oft-forgotten researcher Loney Clinton Gordon.[17] Gordon, fifteen years Grace Eldering's junior, was an African-American woman born on October 8, 1915 in rural Arkansas. Like many poor Southern blacks, her family participated in the Southern Diaspora (also known as the Great Migration), which witnessed the uprooting of a largely rural population to large cities in the Midwest and North. Whereas nine out of ten African-Americans lived south of the Mason-Dixon line in 1910, that fraction declined to just over one half by 1970. Loney's parents settled in Grand Rapids, Michigan. At the age of twenty-four, she earned a degree in home economics and chemistry from Michigan State College (now University). Straight out of college, she landed a job as a dietician at a mental sanitarium in Virginia but soon quit, given the poor accommodations of the broken-down institution and equally dismal treatment from its administrators.

Returning to her native Grand Rapids, Loney applied for dietician positions but was declined, being informed that the white male cooks would not take orders from a black female dietician. A mutual friend informed Loney of an open position under Kendrick at the Michigan Department of Health. Gordon was enthusiastic, persistent, and ambitious, and she was offered the job on the spot to conduct a project to improve the way that *Bordetella pertussis* could be grown in culture. She soon discovered a concoction involving sheep's blood that served the purpose.

With the arrival of Gordon in the early 1940s, Kendrick and Eldering sought to improve the efficiency of manufacturing large amounts of the vaccine. This problem was one of the first where Gordon's contributions proved essential. Soon the Michigan Health Department was able to not only make enough material for their home state but also to export to other states throughout the nation and eventually the entire population of the United States.

The team further improved the vaccine by including an adjuvant. To explain the concept of adjuvants, we go back to 1926. In that year, the same Alexander Glenny who, in parallel with Gaston Ramon, had discovered diphtheria toxoid (and named it as such) also demonstrated the value that adjuvants provided for vaccination.[21] An adjuvant is a chemical that can amplify the effectiveness of a vaccine. The first adjuvant discovered by Glenny was the one used by the Michigan team—aluminum hydroxide. This chemical is the active ingredient in antacids such as Gaviscon or Mylanta, but Glenny demonstrated that alum (a shorthand used by immunologists for aluminum hydroxide) tweaks the macrophages of the immune system. Specifically, the chemical helps the macrophages gather the perceived foreign antigens of the vaccine, to interact productively with the lymphocytes of the immune system (B and T cells) and become concentrated within lymph nodes and other immunological structures. Each of these properties increases the intensity, breadth, and speed of an immune reaction and thereby amplifies the effectiveness of the vaccine.[22] As one example familiar to all who have been immunized, adjuvant-based activation of macrophages (and other immune cells) is responsible for the rapid (sometimes within minutes) redness, swelling, heat, and occasional discomfort associated with the vaccination site.

The inclusion of alum increased the potency of the whooping cough vaccine. The whole cell vaccine began to show promise in the clinic starting in 1943.[23] At the same time, the Michigan team initiated the practice of combining the whole cell pertussis vaccine with the toxoids for diphtheria and tetanus. This triple combination, later known as DTP, thereby provided a means to protect children from three notorious pediatric infections at the same time. The vaccine was widely adopted in the years following the end of the Second World War and largely continued in the same form developed by the three women scientists until the mid-1990s. As we will now see, the pertussis component of the vaccine has been a subject of considerable controversy, which has persisted in one form or another to the present day.

A Shot Heard Round the World

The prevalence and deadliness of pertussis meant that any vaccine was warmly embraced by the medical community. This all changed following a presentation by the English pediatrician John Wilson to the Royal Society of Medicine in 1973. In this talk, Wilson suggested a subset of children immunized with the pertussis vaccine demonstrated a spike in fever that preceded seizures and could later progress to coma, permanent brain damage, and death.[24] Prior to this, any adverse effects of the vaccine were found to be rare and not particularly damaging.[25] Wilson's work, though preliminary, claimed to definitively link the pertussis vaccine with neurological damage.[26]

Wilson's study was focused upon the posh Bloomsbury neighborhood of London. The high-profile, affluent neighborhood of Bloomsbury inspired the likes of E. M. Forster, Virginia Woolf, and John Maynard Keynes and also happens to house much of the British medical and legal elite. The report of children suffering in this high-profile neighborhood unleashed a firestorm of alarmist reports from British tabloids. Wilson fanned the flames through multiple appearances on television, where he adamantly warned parents and physicians not to use the pertussis vaccine.

A year after the publication of the British study, a review by the Japanese government created further consternation. Japan had mandated vaccination with the whole cell pertussis vaccine by the age of three in the months following matriculation in school. Thereafter the rates of pertussis dropped dramatically. Whereas almost twenty thousand children died annually from pertussis in 1947, the number dropped to zero by 1972.[27] During the winter of 1974–75, two high-profile deaths were recorded in infants within twenty-four hours following DTP immunization. The tragedy of these deaths was amplified by a hungry media, and the resultant uproar caused the government to suspend vaccination and form a study group to analyze the problem. After intensive investigation, the committee recommended a resumption of the immunization mandate. However, parents began ignoring the mandate, and the rates of infection and death from pertussis again climbed, exceeding forty deaths per year by the end of the decade. In response, the Japanese government instructed the creation of a new pertussis vaccine that would be required to convey "one-tenth" of the side effects associated with the whole cell vaccine.

Throughout many established nations, outraged parents, particularly those with the financial means and passion to back their words with action, began organizing anti-vaccine crusades and pushed their pediatricians to suspend vaccination. As attention on the subject grew, more and more parents refused to allow their children to be jabbed with the DTP vaccine. Immunization rates plummeted throughout England, Japan, and the rest of the world. Consequently, the rates of pertussis rebounded to more than 100,000 per year in England alone in 1978 and 1979.[28] Despite the opportunities afforded by antibiotic intervention (a luxury not available to physicians before the Second World War), dozens of children again began to succumb to the whooping cough each year.

The firestorm that had engulfed Japan and the United Kingdom worked its way across the oceans on April 19, 1982, when the Washington, D.C., television station WRC-TV broadcast a special report with the title *DTP: Vaccine Roulette.*[29] The documentary was masterminded by local consumer reporter, Lea Thompson. Thompson conveyed a heart-wrenching tale featuring countless scenes of sincerely agonized parents whose children had suffered very real hardships. The sensationalized and widely hyped documentary aired during a period known as "sweeps week." This is a key time when ratings services such as Nielsen determine audience size and thus influence the advertising rates that local television stations can charge. The gamble by WRC-TV to produce *DTP: Vaccine Roulette* paid off, as ratings were very high. Immediately after the report aired, the company's switchboard lit up with parents concerned—some truly panicked—about the safety of their young children. According to one chronicler of the event, Seth Mnookin, the station "provided callers with the phone numbers of other people who had also called" and thereby nucleated a grass-roots movement that quickly gained momentum.[30]

The heartbreak conveyed by the parents in *DTP: Vaccine Roulette* was sincere, but the report on the tragedies was simplistic, unscientific, biased, and inaccurately implicated the DTP vaccine. According to a well-researched account of the subject in *The Panic Virus*, by Seth Mnookin, the documentary was based on incomplete data analysis and highly edited quotes.[31] In a later article, Mnookin cited an official of the American Academy of Pediatrics who indicated that "Thompson asked the same question, repeatedly in slightly different ways, apparently to develop or obtain an answer that fitted with the general tone of the program."[32] As detailed further by Paul Offit in his outstanding book, *Deadly Choices*, Lea Thompson was described as selective in her presentation of details, subjecting unwitting viewers to a cacophony of half-truths and manipulations of incomplete or inaccurate data.[33] Whereas such oversights are frequently identified by the scientific peer-review system that guides medical research, no such filter for objective, evidence-based truth constrains the media.

Despite a lack of medical and scientific credibility, *DTP: Vaccine Roulette* was critically acclaimed, and Lea Thompson received a local Emmy award.[34] As the sensation associated with the documentary began a long chain reaction, its impact was felt well beyond the local Washington, D.C. market. Soon parents across the country were consumed by the same fear that their British counterparts had felt a decade before. In the more litigious United States, lawyers were the big winners as class-action lawsuits began being leveled at most major manufacturers of the DTP vaccine.

The selective filtering of information was not limited to *Vaccine Roulette.* A similar cherry picking of information led to the publication of a best-selling book and the creation of an anti-vaccinator advocacy group that will feature prominently in our next chapter. On the heels of the sensation created by the sensationalistic local television show, Dr. Harris Coulter and Barbara Loe Fisher founded an organization with the impressive-sounding name of the National Vaccine Information Center (NVIC). This nonprofit amplified unsubstantiated reports arising from the rapidly-organizing anti-vaccine, movement.

Dr. Coulter, who passed away in 2009, was not trained in medicine or science; he was a social scientist who advocated homeopathy, an 18th-century form of pseudoscientific alternative medicine based on long-since refuted concepts such as miasma.[35] Likewise, Fisher's draw to the field was anchored in a tragic history with her child, who suffered a convulsion coincident with a DTP immunization in 1980. Although she had not made the association prior to viewing *Vaccine Roulette*, the documentary apparently convinced Fisher to recall precise details that led her to maintain that the DTP vaccine was responsible for permanent brain damage her son had suffered eighteen months before she and Coulter had organized the NVIC.[36] Trained in public relations, Fisher was well placed to organize fearful parents into the emerging NVIC.

In 1985, the pair published *DTP: A Shot in the Dark*, which advocated that vaccination was responsible for a variety of neurological impairments, including chronic encephalopathy, seizures, and a rather obscure malady (at the time) known as autism.[37] Despite Coulter and Fisher's lack of scientific or medical credentials, sales of *A Shot in the Dark* were propelled by carefully selected anecdotes and heart-wrenching descriptions of pediatric neurological damage. The book became a best seller. Many readers embraced the book despite its failure to convey objective scientific or medical information. As the fears swelled, many doctors were caught unawares and were not sufficiently armed with prior training or objective facts to adequately address the complex questions related to the risks and benefits of DTP vaccination. This would have dire implications for public health for years to come.

Increasing public concern raised by the television documentary and book led governmental and public health officials in the United States and United Kingdom to directly confront the questions surrounding the safety of the DTP vaccine. In response to the uproar, on November 14, 1986 President Ronald Reagan signed into law the National Childhood Vaccine Injury Act.[38] One provision of the new law required a comprehensive analysis of the safety of the pertussis vaccine by the Institute of Medicine of the prestigious National Academy of Sciences. A nonpartisan task force comprised of prominent investigators was tasked with a twenty-month assignment to assess all medical and scientific literature and carefully review the relationship between vaccines for pertussis (and rubella, a subject to which we will return in the next chapter). Their deliberations included a series of workshops and public meetings to gather all relevant data and in a transparent manner to refute future allegations of bias.

The conclusions of this comprehensive National Academy of Sciences study were summarized in a 1990 report, which wholly discredited any relationship between the DTP vaccine and autism, meningitis, chronic neurologic damage, spasms, and many other potential side-effects.[39] The first line of the executive summary of the report began with, "Next to clean water, no single intervention has had so profound an effect on reducing mortality from childhood diseases as has the widespread introduction of vaccines." The report declared "insufficient evidence to indicate a causal relation between DPT vaccine and a large number of side effects advocated by vaccine skeptics, including chronic neurologic damage." Nonetheless, the report did link the whole cell pertussis vaccine with rare cases of shock, anaphylaxis, and protracted crying. A parallel report concluded, "There clearly is an increased risk of a convulsion after diphtheria-tetanus-pertussis immunization but no evidence that this produces brain injury or is a forerunner of epilepsy. Studies have also not linked immunization with either sudden infant death syndrome or infantile spasms."[40] A key figure in the investigation summed up the outcome in the title of a manuscript published in the *Journal of the American Medical Association*: "'Pertussis Vaccine Encephalopathy': It Is Time to Recognize It as the Myth That It Is."[41]

Meanwhile in Britain, the government had quickly commissioned a smaller study, which was led by Dr. David Miller of London. Miller organized a questionnaire, known as the National Childhood Encephalopathy Study (NCES), which was distributed to pediatricians between 1976 and 1979 to report any evidence of either "fever" or "other" arising within seventy-two hours after DPT immunization.[42] Compiling the results of the survey, Miller shocked the world in 1982 with the statement that vaccination could cause acute neurological symptoms, most notably a spike in fever. He captured more headlines by speculating that DTP vaccination might be expected to cause permanent damage in 1 of 100,000 children, a staggering risk if this were determined to be accurate in more thorough studies to be conducted in the future.

The reporting of the early findings from Miller's NCES study created a firestorm on all sides of the vaccine debate. On one side, the anti-vaccine community claimed their beliefs had been vindicated. On the other, the design of Miller's study was rigorously scrutinized and found to be fundamentally flawed. For example, the number of participants studied by

NCES was insufficient to make the broad conclusions originally voiced by Miller. Looking further, the design of the questionnaire, as well as the responses received, were likely to have conveyed inaccurate outcomes. For example, cases of viral encephalitis or Reye's Syndrome, two indications wholly unrelated to DPT vaccination, were included in the analysis.[43] These deficiencies, along with public questions about whether vaccines were safe, inspired a flurry of studies all throughout the world, including the American study being led by the National Academy of Sciences. Each study independently concluded there was no link between the DPT vaccine and neurological damage. Indeed, even a 10-year follow-up of the NCES study had discounted the link.[44,45] Nonetheless, the damage had been done

Building upon inconsistent reports from the United States and United Kingdom, the battle over pertussis vaccine began being waged in the courts, where the pecuniary motivations of both sides had the unintended consequence of bringing all the facts into focus. The most thorough assessments were performed in Britain, largely because American cases tended to be settled out of court in the interests of expediency and to avoid protracted, high-profile headlines. In the United Kingdom, a complex series of court cases began in the mid-1980s following lawsuits filed by the anti-vaccine coalition.

One of the most vocal advocates against vaccines arose from a most unusual source. Gordon T. Stewart was born on February 5, 1919 in the west central Lowlands of Scotland. He earned his bachelor and medical degrees at the University of Glasgow. With the outbreak of war, Stewart spent some of his time as a ship's surgeon on the dangerous convoy escort groups. His remaining efforts were spent in the laboratory studying penicillin. Although some people erroneously credited him with codiscovering penicillin with Alexander Fleming, that event occurred in 1928, when Stewart was nine years old. Rather, the naval physician published his first laboratory-based studies of penicillin in 1945, evaluating early clinical investigations of the antibiotic.[46] Two decades later, in 1963, Stewart published early evidence that bacteria could become resistant to methicillin, a derivative of penicillin that was a mainstay of infection control.[47] Indeed, the subject of drug-resistant pathogens has grown in the years since.

As the 1960s were drawing to a close, Stewart's research began tacking towards the fringes of conventional science. One example was his controversial paper entitled "Limitations of the Germ Theory," which was published in the May 18, 1968 issue of the *Lancet.*[48] This paper argued that Koch's postulates of disease were a "gross simplification" as they ignored the complexity of other factors that determine if and how a disease arises. These factors included genetic constitution (*i.e.,* race), behavior, and socioeconomics.

Whether Stewart originally intended it as such, the paper quickly became a manifesto for far radical fringe elements, which eventually came to include Stewart himself. For example, by the early 1980s, Stewart was publically on record stating that AIDS was not due to HIV but was a disease of the gay lifestyle. He further maintained a debunked theory that the deposition of sperm in the rectum triggered a vigorous immune response that caused a later collapse of the immune system, eventually manifesting itself as AIDS.[49, 50] He further antagonized the conventional scientific and medical establishments with statements such as "Every time an avowedly homosexual or bisexual rock or film star dies of the disease, he is elevated to martyr and hero."[51] Perhaps most damaging, Steward excoriated public health officials who were trying to limit the spread of the disease through the use of antiretroviral drugs such as AZT.

Many otherwise uninformed people listened to Stewart and were irreparably harmed or killed by HIV/AIDS. For example, the president of South Africa, Thabo Mbeki, invoked Stewart's strident views in support of the false claim that HIV was not the cause of the AIDS epidemic sweeping his nation. Citing the John Le Carré thriller *The Constant Gardener*, Mbeki announced his belief that AIDS was a conspiracy propagated by the pharmaceutical industry, which was profiting from the sales of antiretroviral drugs targeting an irrelevant virus known as HIV.[52] Gordon Stewart actively supported Mbeki's misguided views, which had been inspired by Stewart's 1968 manuscript, until August 2007, when Stewart unveiled a public letter renouncing Mbeki to the high commissioner of the Republic of South Africa, who was probing Mbeki for his reckless statements and policies.[53] The death toll arising from Mbeki and Stewart's bellicose and irresponsible actions rose quickly. A 2008 report from the Harvard School of Public Health estimated 365,000 South Africans had unnecessarily died because of the Stewart-supported policies of Mbeki based on HIV denial.[54]

Courting Disaster

Pertussis provided another opportunity for Stewart to rail against the widely accepted germ theory of disease. Starting in the late 1970s, Stewart published a series of papers decrying the safety and efficacy of some of the most important medical breakthroughs of the 20th century, including antibiotics and the pertussis vaccine.[55] Based on his increasingly extreme views on these subjects, Gordon had become an obvious choice for a key interview profiled in *Vaccine Roulette*, where he stated, "I believe that the risk of damage from the vaccine is now greater than the risk of damage from the disease."[56] Unsurprisingly, Stewart was also set to be the superstar witness in support of a lawsuit targeting the pertussis vaccine.

Prior to Stewart's involvement in the legal wrangling over the pertussis vaccine, the first trial took place in his native Scotland in 1985. The parents of a nine-year-old child, Richard Bonthrone, sued their doctor, nurse, and health department, claiming their son had suffered a series of seizures, culminating in severe retardation, that began nine days after he received a dose of DTP vaccine.[57] The judge ruled against the parents, citing the testimony of experts that cast significant doubt upon the parents' claims of negligence.

The Scottish trial was but a small skirmish in a rapidly escalating war. The second encounter occurred months later a bit further south in England. This case was brought by the parents of Johnnie Kinnear, who claimed their child had begun to suffer from seizures seven hours after receiving a dose of pertussis vaccine.[58] The litigants further claimed the doctor discounted their ordeal and that the seizures persisted for many months and caused irreparable brain damage. As pointed out by Paul Offit in *Deadly Choices*, the doctor and health department were the primary defendants; the Wellcome Foundation (the vaccine manufacturer) was excused, since the exact manufacturer responsible for the vaccine was unclear.[59] In a surprising turn, Wellcome voluntarily joined the case, risking considerable liability to clear its name, which had been sullied by the Kinnears' vocal claims in the months leading up to the trial. The judge agreed to allow Wellcome to join the case but determined it would not be liable for damages.

As planned, the charismatic Stewart was the star witness for the claimants, citing various anecdotes implicating the toxicity of the pertussis vaccine.[60] As his testimony dragged on, it became ever more dramatic, and he undermined himself by citing extravagant claims about the toxicity of the vaccine that were far outside the bounds of even the most aggressive anti-vaccinator. For example, Gordon Stewart expounded upon the results of one study implicating the vaccine in brain damage of sensitized children, only to be embarrassed in cross-examination with the revelation that the study he cited had been conducted with rats, not people.[61] All this was for naught, as subsequent statements from the parents uncovered considerable discrepancies in their testimony about the time period between vaccination and the onset of seizures and brain damage. Specifically, the mother initially testified that seizures began seven hours after her son received the vaccine. Subsequent questioning revealed the symptoms had begun not at that time but five months later. Based on such inconsistencies, the judge dismissed the case.

The third and final attempt by the anti-vaccine lobby to derail the pertussis vaccine in the United Kingdom occurred in February 1988, in a class-action suit centered around a seventeen-year-old English girl by the name of Susan Loveday. At first glance, the case of *Loveday v. Renton and Wellcome Foundation* seemed like a proverbial slam dunk.[62] Loveday's parents testified that the girl had suffered from an unusually high fever and persistent crying almost immediately after receiving her first dose of DPT vaccine. The child's second exposure a year later triggered an even more vigorous response, and a third immunization was linked with severe and irreversible developmental retardation. Again, the Wellcome Foundation volunteered to join the case in hopes of clearing their name in the face of what they believed to be spurious claims.

The judge, Murray Stuart-Smith, expressed sympathy with the parents of the more than two hundred different children named in the class-action lawsuit. As evidence of his concern, the judge reviewed not only the specific case of the Loveday child but also those of the other claimants. This was a fortunate outcome for the anti-vaccinators, since deeper investigation had utterly refuted the Loveday's claims. Even a forlorn Gordon Stewart admitted, "She was not vaccine-damaged. She was damaged before."[63]

The case for and against DPT as the cause of permanent neurological damage was rolled out in the trial and supported by many epidemiological and pathological studies that had been conducted over the years. The anti-vaccinators placed particular emphasis upon the publication of

Miller's highly cited and damning findings from the small Bloomsbury case.[64] The pro-vaccine defense cited many other clinical trials that failed to reproduce Miller's findings. Specifically, the outcomes reported with the small Bloomsbury cohort could not be reproduced by larger and more comprehensive analyses conducted in Sweden, Denmark, the United States, or even in England.[65]

After an objective and comprehensive analysis of the data, the judge issued a thoughtful verdict of more than 100,000 words (roughly the size of this book), which absolved the vaccine of any guilt in causing permanent neurological damage.[66] The judge was particularly critical of John Wilson, revealing that of the fifty children Wilson had claimed were harmed by the vaccine, twenty-two had not even received the pertussis vaccine. He further suggested Wilson's desire to make his point caused him to include these individuals to heavily bias the outcome.

The verdict also revealed that David Miller's epidemiological study, hurriedly conducted with a small group of London physicians amidst an increasing anti-vaccine frenzy over the DTP vaccine, was fundamentally flawed and influenced by popular pressure. The judge concluded the study was hastily designed, executed, and analyzed in a desire to satiate a nervous public. Stuart-Smith unequivocally stated, "I think it can be said that this demonstrates a conscious over-anxiety to appease what I may call the vaccine-damage lobby." Given the detailed judgement, the case was forever put to rest in the United Kingdom.

Not so in the more litigious United States, where emotional jurors, rather than objective and dispassionate judges, rendered the verdicts. Starting with the awarding of $15 million to the parents of an infant who had suffered encephalopathy after a DTP vaccination, the floodgates broke loose. The amount of damages awarded rose steadily from $25 million in 1981 to more than $3 billion in 1985.[67]

Because of these high-profile court cases and the media attention they attracted, public health officials became increasingly anxious over two parallel trends. First, parents were hesitant or outright defiant in not allowing their children to be immunized with the DPT vaccine. Compounding this, many traditional vaccine manufacturers were balking at continued sales, and thus liability, from the vaccine. As the vaccine manufacturers were increasingly inundated with individual and class-action lawsuits, the costs of liability insurance for their companies skyrocketed. Such concerns led many companies to remove vaccines from their portfolio of products. According to *Deadly Choices*, seven companies manufactured DPT in 1960, but the number dropped to three by 1982.[68] The production of other vaccines was equally threatened.

Recognizing the public health dangers arising from both trends, in 1986 President Ronald Reagan signed into law a piece of legislation known as the National Childhood Vaccine Injury Act.[69] We briefly touched upon this new law, which in part tasked the National Academy of Sciences with conducting a full evaluation and report on the safety and efficacy of pertussis and rubella vaccines (a subject to which we will return in the next chapter). In addition, the act essentially a priori convicted the vaccine makers by creating a compensation mechanism for families claiming damages. It also created a reporting system for vaccine reactions and toxicities and mandated better education of parents about the risks and benefits of immunization. Nonetheless, the vaccine industry was assuaged by language that limited their liability by establishing a no-fault system to compensate past and future victims of toxicities caused by legally mandated vaccination.

While the National Academy report and lawsuit absolved the whole cell pertussis vaccine of most of the toxicity associated with its use since the early 20th century, the vaccine was not without its risks, as demonstrated by the presence of some side effects. The opportunity to create a new market thus incentivized scientists across the globe, including a multinational collaboration between the United States and Japanese National Institutes of Health. This group reasoned that the broad targeting of the pertussis bacterium might unintentionally trigger unintended ferocity that was responsible for the rare inflammatory toxicities associated with whole cell vaccine.

A new approach was devised to identify the components of the bacterium that elicited the greatest protection against disease and then focus vaccine development to target a small number of proteins. These molecules are broadly known as hemagglutinins, and a race began to develop vaccines against these key parts of the virus. A Japanese team developed and tested a hemagglutinin vaccine starting in 1981.[70] A pivotal study in animals and people revealed that an acellular vaccine (one that does not encompass the entire bacterium) consisting of different "components" was safe and effective.[71] Specifically, the team reported that the vaccine had achieved the desired level of eliminating more than 90 percent of the toxicity associated with the whole cell vaccine. While this vaccine did prove effective, it was rushed into the Japanese market before long-term data in humans could be obtained.

Within months after the pivotal Japanese report of the new vaccine, an American team of scientists and physicians had traveled to Japan to review the findings. By this time, the acellular pertussis vaccine had been administered to more than twenty million Japanese. In a 1987 report to the *Journal of the American Medical Association*, the American scientists concurred with their Japanese colleagues that the vaccine was at least as effective in conveying protection in the short term (weeks or months) following immunization.[72] More importantly, given the lawsuits and payouts plaguing the American vaccine manufacturers, the safety parameters of the new acellular pertussis vaccine were quite promising. They concluded their analysis of the new DTaP (reflecting diphtheria and tetanus toxioids combined with aP, referring to acellular pertussis). The report suggested the vaccine be adopted in the United States pending the results of ongoing studies in Sweden and the United States. These results were similarly encouraging, and the product was hastily launched in the United States in 1992 to much acclaim and relief from concerned parents.

The story of the pertussis vaccine might very well have ended happily here had nagging concerns about its durability not come to the fore. The specific worry was that the same limited antigenic diversity that favored greater safety of the acellular vaccine might compromise its ability to confer lasting protection. Indeed, the foundational Japanese, Swedish, and American studies were still quite fresh when the vaccine was launched in 1992. According to a 2012 article in *Slate*, an expert at the Centers for Disease Control and Prevention, Dr. Tom Clark, suggested fundamental flaws in the duration and perhaps sensitivity of defining protection of the new acellular disease meant "the new vaccine doesn't actually work as long as the old one."[73] These worries were abated somewhat with a 2005 recommendation to provide eleven- to twelve-year-old children with a booster shot to reinvigorate the immune system against pertussis.

Even this measure proved ineffective, as evidenced by a series of epidemics. On June 23, 2010, the California Department of Public Health declared an epidemic emergency in what was the worst outbreak of pertussis in the United States since 1947. Altogether, 1,144 confirmed or suspected cases were identified, 51 children were hospitalized, and at least 10 died.[74] A disturbing report in the Journal of the Pediatric Infectious Diseases Society revealed at least nine of ten of the sickened had been immunized at least once and many had received the booster vaccine within the past three years, suggesting the DTaP vaccine that had been used had failed to confer lasting protection.[75]

Two years later, the disease returned yet again, hitting almost randomly in the United States, including California, Wisconsin, Vermont, and Washington State. Again, the epidemiological follow-up revealed the longer the time from immunization, the greater the likelihood and severity of disease. Worse still, the additional booster shots appeared to confer little if any benefit. Perhaps most telling, those children who had received the whole cell vaccine had remained protected, whereas their contemporaries immunized with the acellular vaccine did not.[76] Such findings revealed that the hasty switch from the whole cell to acellular pertussis vaccine might have presaged a new susceptibility to an old killer.

Perversely, the higher the education, the more the parents tend to embrace the idea that vaccines cause more harm than good. While the associations conveyed by *A Shot in the Dark* and *Vaccine Roulette* had been thoroughly discredited in many scientific reports, these ideas have remained in the back of the minds of many parents for years.

Perhaps strangest of all, rate of vaccination has decreased the most—by 4 percent in 2009 alone—among families with health insurance as compared with the uninsured.[77] These individuals can afford the vaccine but elect not to immunize their children (and, in some cases, to subvert local or state law). According to a 2010 interview by CNN with Jason Glanz, an epidemiologist with the Kaiser Permanente Institute for Health Research, "A subset of the population, typically well-educated, white and in the upper-middle class, have grown skeptical of immunizations."[78]

While it would be unfair to attribute the terrifying return of pertussis simply to the actions of a small number of educated parents who refuse to immunize their children, the growing problem can more accurately be traced back to the anti-vaccinator campaign. Specifically, sensitivity about the potential side effects of the whole cell pertussis vaccine caused the biomedical community to embrace a less effective variant. The motivations of the anti-vaccinators were pure: enhancing the safety of our most precious resource: children. However, the outcome ironically created the conditions that instead endangered the lives of far more children. These choices were driven primarily by pressures exerted by *Vaccine Roulette* and *A Shot in the Dark*, which were unknowingly complicit in the deaths of future children who were not immunized or were immunized with the less effective acellular vaccine.

The ongoing experience with pertussis reveals vaccination can be a necessary trade-off between two terrible outcomes. On one hand, the whole cell pertussis vaccine can convey occasional and exceedingly rare catastrophic events. One of these terrible examples might even have been responsible for the experience of Barbara Loe Fisher's child (though this can never be conclusively determined, given the time elapsed from whatever occurred that tragic day and the time she attributed the outcome to the DPT vaccine). On the other hand, experience has taught us that a failure to vaccinate against pertussis will inevitably kill vastly larger numbers of children. Seeking a middle ground, a combination of scientific and public health experts conspired in the most constructive sense of the word to develop the safer acellular vaccine that, in the short term, seemed just as effective as the whole cell vaccine. As often happens in life and medicine, the test of time revealed the new vaccine to be less protective. In the early years of the 21st century, we are nudging back to the catastrophic losses experienced at the beginning of the 20th.

Such realization has caused many to advocate for the creation of a new pertussis vaccine at least as safe as the acellular vaccine and with increased potency and durability of protection. In the meantime, the biomedical community and its regulators would be wise to consider the reintroduction of whole cell vaccines, which carry the risk of undesired toxicities in a few rare but tragic instances, but which have been demonstrated to confer greater protection for a larger number of children.

Sadly, the damage done by the anti-vaccinator movement was not limited to pertussis. Indeed, the power and damage exerted by organizations such as the NVIC were yet to be fully realized. This would all change with the campaign against the measles, mumps, and rubella vaccine as the cause of autism, a subject to which we will now turn.

9

Three Little Letters

This part of our story will be replete with three-letter acronyms, many of which have been unnecessarily mired in controversy. The first three-letter acronym is one well known to all parents: MMR. The acronym designates a single vaccine that prevents three of the greatest mass murderers of children: measles, mumps, and rubella. Despite the extraordinary public good that this vaccine has and continues to deliver, these three little letters have been the source of some of the most dangerous misinformation, both blatant and unintended, that increasingly threatens the lives of billions around the world.

Measles is a highly contagious infection first described in the 10th century by the Persian doctor Abubakr-e Mohammad ibn Zakariyya al-Razi (or Rhazes, as he is known in Europe).[1] In the darkest days of the Middle Ages, Islamic scholars dominated virtually all sciences and promulgated knowledge that had been largely lost in Europe following the fragmentation of the Roman Empire. A published work of Rhazes, translated as *The*

Book of Smallpox and Measles, not only conserved knowledge that had been discovered over the generations but advanced our understanding of many maladies, including a bevy of infectious diseases.[2] For example, he was the first to accurately discriminate smallpox from measles, a distinction based in part on subtle differences in symptoms and epidemiology; whereas smallpox could strike people of any age, measles tended to be restricted to more youthful victims.[3]

Rhazes's description of measles is rather remarkable because he might have been one of the first to not only record the disease for posterity, but may have been one of the very first people ever to witness the disease. The reason for this is that the history of measles is surprisingly short in humans. Detailed genetic analyses have recently revealed that measles jumped from cattle to humans at almost the exact moment Rhazes was recording it.[4] Despite the relatively recent jump of measles to man around the 10th century, its parentage had been a legendary source of plague for millennia. It seems that what we today call measles is but a few mutations separated from a disease known as rinderpest. Rinderpest is a plague that consumes cattle, antelopes, giraffes, and other hoofed animals. Historic fascination with the disease is based in part on the fact that death is almost certain following infection. Because of its rapid spread, rinderpest could quickly decimate food sources for entire civilizations. For example, rinderpest virus is believed to be one of the "ten plagues" inflicted upon the Egyptian pharaoh as described in the book of Exodus.[5,6] In addition, an outbreak of the disease in Napoleonic Europe added famine to the list of miseries suffered during that bloody period.

Diseases that are new to humans tend to be the most dangerous, as they have not yet evolved to form a mutually beneficial long-term relationship with their prey. One way of viewing this is that the pathogen seeks to keep its quarry (you and me) alive long enough to ensure its propagation to future generations of victims. On one hand, the microbe will undergo evolution such that those variants that improve their ability to spread to others will win out over those that might be deadlier and kill prior to being passed on. At the same time, genetic variation in people may distinguish those individuals who are more susceptible to the disease than others. Consequently, the evolution of measles and humans has not yet reached a point where the disease has reached a truce with humanity to ensure the survival of both species.

As evidence of its destructive power, measles has been responsible for the deaths of at least 200 million people over the past century and a half alone.[7] As we saw at the beginning of the book, the introduction of smallpox to the New World was devastating to its native inhabitants, and measles soon conveyed the coup de grace to entire societies. One illustration is the experience suffered by the native peoples of Cuba in 1529. Shortly after the arrival of Columbus, the islands of the Caribbean began to be decimated by the spread of smallpox. This began in earnest in Cuba in 1518 and is estimated to have killed as many as one third of the population of that island and those surrounding it.[8] Just as the population was beginning to recover, measles visited the island. In 1529, the disease is estimated to have killed two thirds of the remaining native population.[9] Similar outbreaks occurred throughout the world, and the devastation of measles extended beyond the Caribbean into both American continents, Hawaii, the South Pacific, and isolated peoples in the North Atlantic, to name but a few.[10, 11, 12]

The symptoms of measles include a very high fever accompanied by a cough, runny nose, and red, swollen eyes. Within a few days, an infected individual will develop a series of small, white spots that resemble "grains of bright, white salt set on a wet background" on the inside of the cheek towards the back of the mouth.[13] This rather unique symptom was first described in 1896 by the New York physician Henry Koplik and is more than esoteric, because the presence of Koplik spots is an early predictor of a greatly increased contagiousness that will follow over the next few days.[14] Consequently, the identification of Koplik spots in an infected child provided an early means to isolate the child and thus limit the spread of the disease to siblings and playmates.

A day or two after the emergence of Koplik spots, a series of red bumps will start to emerge on the skin, often beginning on the face behind the ears and near the hairline. These bumps grow in number and eventually merge into a flattened and itchy rash that can coat the body's surface from head to toe. This type of rash is also associated with other diseases (including Ebola and other hemorrhagic fever viruses) and provoked the name for scarlet fever. One feature distinguishing measles from these equally horrific maladies is that the rash often changes color from red to brown. For those lucky enough to have a less severe infection or some immunity (albeit not enough to prevent the disease), the brown rash will fade, and all that remains is a bad memory of the terrible itching. The most likely survivors are those infected while young, who tend to fare better. For older children and adults, many severe complications greatly increase the morbidity and mortality caused by measles.

As evidenced by the Koplik spots, the rash is not restricted to the outer skin surfaces, as the infection can spread to internal tissues in the lining of the mouth, lung, eye, ear, and other body surfaces. An infected patient may begin to cough up blood and experience chest pains as the heart and lung recoil from the viral assault. The resulting damage can trigger complications such as pneumonia, diarrhea, and corneal damage. As the virus wends its way through the body, the liver, spinal cord, and brain can become involved. Complications can include permanent neurological damage and blindness, particularly in adults. As indicated above, these symptoms can progress to death directly caused by the virus, or they can render the body severely susceptible to secondary infections (usually bacterial in origin) that finish the job.

Despite considerable human experience with measles since Rhazes's reports, both in terms of its public health devastation and symptoms, the discovery of the measles virus did not occur until the middle of the 20th century. The virus arose from a most interesting source.

Enders Game

Virtually every photograph publically available of John Enders portrays a dignified man, almost invariably in a tweed jacket, who looks every bit an Ivy League patrician. Although stereotypes are frequently dangerous, they are surprisingly accurate in this case. John Franklin Enders was born on February 10, 1897, in West Hartford, Connecticut.[15] John's father, John Ostrom Enders, was a wealthy Hartford banker, and his grandfather had founded the Aetna Life Insurance Company.

Although better known today as the center of the insurance business (including Aetna, Travelers, and, of course, The Hartford), Hartford was also a mecca for the intelligentsia in turn-of-the-century America. The local newspaper, the *Connecticut* (later *Hartford*) *Courant* is the longest-serving newspaper in the United States, while the number of publishing houses calling Hartford home at the beginning of the 20th century eclipsed even nearby New York City. Indeed, the list of prominent authors, who lived here included Harriet Beecher Stowe and Samuel Clemens (Mark Twain), both of whose finances were managed by Enders' father. Clemens made a strong impression on John during the writer's frequent visits to the Enders home in the early years of the 20th century. Given his pedigree and interest in erudition, John Enders, like his brother after him and many other members of his blue-blooded family, attended Yale University. Whereas his brother Ostrom joined the family business after graduating from Yale, John's experience took a different route.

John Enders enrolled at Yale University in the autumn of 1915, a halcyon time in America despite the fact Europe had already lost millions to a year-long war. John was soon captured by the new fad of flying aircraft. In the handful of years after controlled flight had first been demonstrated by Wilbur and Orville Wright on the dunes of Kitty Hawk beach on December 17, 1903, access to flying machines had remained quite rare—limited to an affluent few who could afford such diversions. As detailed in Marc Wortman's outstanding 2007 book, *The Millionaire's Unit*, Yale hosted an elite group of flyers, many of whom volunteered to fly in France once the United States entered the First World War in the spring of 1917.[16] Although Enders did not see battle, he did join the fledgling Army Air Corps as an ensign and played an equally important role as an early military flight instructor in Pensacola, Florida. The dangers associated with aviation, particularly in those early days, clearly left an impression on Enders, who tended to avoid flying as a means of transportation for the rest of his life thereafter.[17]

At war's end, Enders finished his studies at Yale. Upon graduation in 1920, he returned to Hartford and tried for a year to make his name in the real estate business, but his heart was simply not in it.[18] Instead, he enrolled at Harvard as an English major, where he focused upon Celtic and early English literature. As Enders was completing his master's degree, he happened to share a boardinghouse in Brookline, Massachusetts, with an Australian by the name of Hugh Kingsley Ward.[19] Ward was the youngest of eight children from an aristocratic Sydney family (his father had served as a scout in the Maori wars and later became editor of the *Sydney Mail* and the *Daily Telegraph*). After graduating from the University of Sydney, he was awarded a Rhodes Scholarship to Oxford. As one sign of his assimilation into English society, he took up the sport of rowing, participating in the 1912 Olympic Games. Conveniently, the games were held in London that year, and while this might seem to have presented a certain advantage for a member of the Australian team already acclimated to the time and location, his Australian teammates denigrated Ward as a foreigner inserting himself into an established team and not being "an Australian rower."[20] Ward was therefore largely excluded by the team from most events in the 1912 London Olympics.

After completion of the games, Ward finished degrees at Oxford in anthropology and public health in 1913. Ward was enamored with the emerging field of microbiology, but his studies were interrupted by the outbreak of the Great War. He served in France as part of the Royal Army Medical Corps Special Reserve. Despite being wounded and gassed multiple times, Ward survived the war and was embraced upon his return to Australia as a hero, receiving the Military Cross with two bars. After the war, he returned to Oxford and continued his work on microbiology.[21] Ward desired to perform research in America and was awarded a one-year Rockefeller Foundation fellowship to work with Hans Zinsser at Harvard. Zinsser was an American bacteriologist who had discovered the pathogen responsible for typhus.

As we have already seen, the Harvard roommate of the Aussie Ward was the Connecticut Yankee John Enders. As Enders was completing his master's thesis on Middle English and was preparing to embark on a PhD on the subject, the two struck up a lasting friendship. Enders would frequently accompany Ward to his laboratory, where the latter would often be engaged in long experiments.

This initial exposure to bacteriology fascinated the American, and Enders was soon infected by the rapidly spreading field. Enders was particularly intrigued by Zinsser and his work. Enders switched majors in 1927, and three years later, he successfully defended a doctoral dissertation on tuberculin and endotoxins.[22] Although a Yale man by family tradition and early training, Enders accepted an offer to remain in Cambridge, and he would remain at Harvard for the rest of his professional career (though he did ultimately donate his papers to his *alma mater*).

As a junior professor, Enders's first breakthrough occurred in experiments performed just as the Second World War was taking hold in Europe. Enders and two colleagues, Alto E. Feller and Thomas H. Weller, grew vaccinia virus in tissues derived from chicken embryos.[23] This may sound rather banal, but the study was a landmark that introduced the world to opportunities arising from the new science of cell culture. The fundamental breakthrough came from Enders's findings that viruses could be propagated in the laboratory using eggs or even small dishes of cells rather than relying upon more laborious, expensive, and slower techniques that required producing viruses in chickens, cows, monkeys, or other animals. The field of cell culture would soon change the scientific and medical worlds, but a series of tragic events slowed its adoption for a time.

The slide towards war drained both the resources and personnel needed to follow up on the work. In addition, Enders's boss and inspiration, Hans Zinsser, was slowly dying of leukemia, to which he would succumb in 1940. This loss was not just a personal blow for Enders; he also inherited the responsibility and burden of running Zinsser's laboratory studies of typhus, as well as Zinsser's entire department at Harvard.[24] Worse still, Enders's wife of sixteen years, Sarah, died from acute myocarditis in 1943, a tragedy that led Enders to surrender his responsibilities as department head to a colleague.

Freed from the administrative burden and perhaps needing to put his mind elsewhere to overcome his grief, Enders focused on his work. In 1946, he was recruited by Drs. Charles Janeway (whom we met in an earlier chapter) and Sydney Farber (for whom the Harvard Cancer Center is named) to Children's Hospital, also in Boston. Around this time, Enders began to study mumps, a disease that captures the second letter of MMR and whose viral pathogen had been discovered in 1934 by the Vanderbilt University investigator Dr. Ernest Goodpasture.[25]

Although mumps is not one of the deadlier viruses known (roughly one in ten thousand infected people dies), it is one of the most infectious, being easily spread among individuals in confined settings such as nurseries or military barracks. Mumps could temporarily incapacitate a group of soldiers and thus had become a high priority for the United States military. Using an approach similar to the one Enders used with vaccinia virus, the NIH scientist Karl Habel demonstrated that mumps virus could be cultured in chicken embryo cells (specifically, fibroblasts) isolated from eggs.[26]

Inaugurating his new laboratory at Children's Hospital, Enders appointed Thomas Weller and his new pediatric resident, Dr. Frederick Chapman Robbins, to optimize his own approaches using chicken embryo cells to create a vaccine. Starting in 1948, Robbins began the process of propagating mumps virus. Weakening the virus through repetitive passages through eggs, he soon adapted an attenuated form of the mumps virus, a subject to which we will soon return after a short interlude.

This important work with mumps was a mere footnote relative to Enders's impact on polio. Over a remarkably short time, the team of Weller, Robbins, and Enders likewise developed the ability to culture poliovirus in eggs—itself a key achievement that would be needed to mass-produce future vaccines—and to use this system to evaluate the ability of antisera and vaccines to block polio. These accomplishments would earn each member of the team the 1954 Nobel Prize for Physiology or Medicine.

Having made key contributions to the development of both mumps and polio vaccines, Enders turned his sights upon two forms of measles in 1954. This work briefly led to a side project on rubella, also known as German measles (and the third letter depicted in the acronym MMR). The virus responsible for rubella and the discovery of its vaccine has recently been detailed at length in the outstanding book *The Vaccine Race.*[27] While rubella is relatively innocuous in children, viral infection in pregnant women can be devastating. Infection of the mother can spread to the fetus, where rubella causes massive damage. Consequently, the pregnancy often suffers from spontaneous abortions, or, for those fetuses that survive, debilitating injury.

The fame and fate of rubella were strongly influenced by a tragic incident involving a Hollywood superstar, Gene Tierney, whose child was severely injured by an infection with rubella *in utero.*[28] A stunning actress best known for her starring role in the 1944 film *Laura,* Tierney married Oleg Cassini, the famous fashion designer, in 1941. The couple became pregnant in 1943, but Tierney was exposed to German measles during a performance at the Hollywood Canteen. She gave birth to a severely disabled daughter, Antoinette Daria Cassini, who suffered from severe neurological symptoms, including profound mental deficiencies, blinding cataracts, and deafness. Tierney entered a long depression that shortened her career and broke up her marriage. She then fell in love with John Fitzgerald Kennedy (the future president) and had a series of high-profile and soon-to-be-divorced beaus, including Charles Feldman (who was divorcing Jean Howard), Prince Aly Khan (as he was divorcing Rita Hayworth), and a Texas oil baron (as he was divorcing Hedy Lamarr).

Despite this string of high-profile affairs, Tierney is perhaps best remembered in popular culture for her connection with German measles. Years after giving birth to Antoinette Daria Cassini, Tierney was approached by a woman who told her that she had snuck away from quarantine and had gone to the Hollywood Canteen. According to an article in the *New Yorker*, "Everyone told me I shouldn't go," the starstruck woman told Tierney years later at a tennis match, not realizing what she was responsible for, "but I just had to go. You were my favorite."[29]

This experience later served as a plot vehicle for one of Agatha Christie's most famous novels, *The Mirror Crack'd From Side to Side*, which became a best seller and spawned multiple adaptations for the large and small screens.[30] In part driven by the publicity generated by the highly publicized experience of Gene Tierney, many resources were dedicated to propel a race for a vaccine. Enders's primary contribution was to provide the key techniques utilized by Dr. Leonard Hayflick—the primary subject of *The Vaccine Race*—to propagate the virus in cultured cells.[31]

Although rubella (German measles) and rubeola (measles) share considerable similarities in their names and symptoms, the viruses are quite distinct. Another key contribution of Enders was the definitive discovery of the measles virus. Based on the cell culture techniques developed in his laboratory, a new team consisting of Thomas C. Peebles and Kevin McCarthy was able to propagate a virus isolated from an eleven-year-old boy from the Fay Boarding School in Southborough (Massachusetts), facilitating its identification in 1954.[32] Indeed, the viral isolate from young David Edmonston (who is still alive and well as of press time) would go on to provide the material that would be developed into a vaccine to prevent measles.[33] The aforementioned www.scienceheroes.com website estimates that John Enders has contributed to the saving of more than 120,000,000 lives as of early 2017 (twice as many as Gaston Ramon and three times more than Paul Ehrlich or Shibasaburo Kitasato).[34]

Shrunken heads & Big Impacts

Joining John Enders in the pantheon of vaccine innovators is a figure born into tragedy but who would survive to save countless numbers of children. Maurice Hilleman was born on August 30, 1919 in Custer County, a cattle ranch and dirt farming region of almost four thousand square miles in Wyoming that has been populated by no more than five thousand houses holding no more than thirteen thousand people at any time for more than a century.[35] The county received its name and *raison d'etre* in the days following the disastrous Battle of the Little Bighorn, which occurred about 130 miles to the southwest. Led by a flamboyant, controversial, and reckless General George Armstrong Custer, the 7th Cavalry Regiment was utterly decimated by the Sioux chief Sitting Bull in June 1876. A panicked federal government threw up a series of forts throughout the Montana Territory to help contain the victorious Native Americans. One of these forts, really a mere outpost, was known as the Tongue River Cantonment. Given its proximity to the Little Bighorn battlefield, the garrison grew for a time, and the site was formally named for one of Custer's fallen officers, Myles Keough (whose horse was the sole American survivor of the battle).[36] In the early days after its creation, Fort Keough was commanded by General Nelson Appleton Miles, a Civil War veteran who had gained a reputation as an Indian hunter (controversially claiming the solo capture of the famous Nez Perce chieftain Chief Joseph). Miles would later capture Puerto Rico from the Spanish in a war for primacy in the Caribbean and become a popular author based on his many experiences.[37] An eponymous civilian village near the fort served as the birthplace of Maurice Hilleman.

In the middle of the Spanish flu outbreak (a subject to which we will return in the final chapter), the very Christian Anna Hillemann (née Uelsmann) performed one of her last acts on earth, giving birth to Maurice and a stillborn twin sister, named Maureen.[38, 39] Within a few hours of birthing, Anna began to suffer from seizures, symptoms of eclampsia, and would be dead within hours. Just prior to her death, she instructed her husband about the future disposition of her six older sons and daughter, relating that the newborn Maurice was to be raised by Maurice's uncle. According to Paul Offitt's outstanding account of Hilleman's life, *Vaccinated*, Maurice was raised in the rough-and-tumble environment of the Montana farmlands, almost succumbing to a particularly virulent case of diphtheria as a child.[40]

Perhaps reflecting the wild and primitive natural beauty surrounding him, Maurice grew into a rebellious and independent thinker. As a child, he was raised under the strict fundamentalist view of the Missouri Synod Lutheran Church, a conservative German-language denomination that rejected humanism and strongly advocated creationism. For this reason, a childhood confrontation with church elders is particularly instructive to gain a view of Maurice's developing personality. It seems that the young Maurice was caught in the act of reading a library book during a Sunday sermon.[41] This was not just any book but Darwin's *Origin of Species*, and an outraged minister attempted to grab the book away from the child. Maurice fought back, indicating the library book was checked out under his name and that the confiscation of the book, as public property of the library, would be a violation of the law. The minister reeled back from this response, and the satisfied Hilleman would never again conform to the teachings of the Synod. Such rebelliousness continued to characterize Hilleman throughout his career, during which he was known as a brilliant but brutal taskmaster. He proudly and prominently displayed a cabinet full of shrunken heads, one symbolically placed for each employee he had fired.[42]

Perhaps motivated by a combination of self-imposed brilliance, strong work ethic, and a fear of becoming a shrunken head in someone else's cabinet, Hilleman became the most prolific discoverer of new vaccines who has ever lived. Hilleman's specialty was the efficient creation of attenuated forms of viruses that could be used as vaccines. As an employee at E.R. Squibb and later as head of vaccinology at Merck & Company, Hilleman and his team developed a series of vaccines that continue to save the lives of millions each year, including those for measles, mumps, rubella (all three components of MMR), as well as Japanese encephalitis, hepatitis B, and chickenpox viruses (the latter of which was later reformulated and approved as a shingles vaccine months after Hilleman's death).[43, 44]

Hilleman was a workaholic. He frequently combined his personal and professional lives, perhaps never more than when his five-year-old daughter, Jeryl Lynn, awoke with a sore throat in the early hours of a spring morning in March 1963.[45] As an expert virologist, Maurice immediately realized that her sore throat was accompanied by a swelling neck, the telltale sign of the mumps. Knowing that nothing could be done for her other than supportive care, and despite being a single parent (his wife had died of breast cancer just four months before), Hilleman immediately drove to his laboratory and returned with some sterile equipment. He swabbed Jeryl Lynn's throat and immediately returned to work, seeing the material scraped from his daughter's upper esophagus as an opportunity to create an attenuated version of the mumps vaccine.[46] Indeed, the Jeryl Lynn strain of mumps remains a staple of the vaccine armamentum around the world. Independent of this early contribution, Jeryl Lynn herself has also contributed to public health, albeit from a different angle. After completing a BA at Brown and an MBA at the Wharton School of Business, Jeryl Lynn began a career at Merck and later became the chief financial officer at a variety of successful biotechnology companies.[47]

The source of Hilleman's samples for a measles vaccine was not quite as close to home; they came from John Enders. When we last left Enders, he had developed a vaccine from a sample obtained from young David Edmonston. This vaccine was a good first step but tended to convey undesirable side effects such as rashes and fever. Upon receiving a sample of the attenuated version from Enders, Hilleman applied his well-worn techniques and further weakened the virus while retaining its ability to elicit protective immunity. Indeed, the resultant Moraten strain of the virus is an acronym for "more attenuated Enders" and still serves as the basis for the vaccines used today.

A very similar story surrounds the rubella vaccine. Hilleman further refined an existing rubella virus vaccine and introduced a safer version in 1969 (although the rubella component was later replaced with another virus developed by Stanley Plotkin that demonstrated even greater safety).[48, 49] Two years later, Hilleman and Merck combined these three safer versions of the measles, mumps, and rubella vaccines into a single injection that was appropriately named MMR. The emphasis, bordering on obsession, that Hilleman devoted to increasing the safety of the measles, mumps, and rubella vaccine was ironic, given how his creations were misrepresented in the years to come.

Maurice Hilleman would undoubtedly be even more appalled to know that nemeses such as measles, mumps, and rubella are again on the rise. To understand this, we again turn our attention to the anti-vaccine movement, which seized upon the MMR vaccine following its failure to turn the public against DPT. This struggle may seem familiar as it involves some familiar faces and arguments.

Fake News

The *Lancet* is one of the most revered medical journals in the world. This medical periodical was founded in 1823 by Thomas Wakley, an English surgeon who frequently railed against medical incompetence and privilege.[50] Wakley joined an impressive group of politicians, physicians, and attorneys to create a publication intended to decry the frauds and charlatans of the medical community. As evidence of success, the journal became subject to multiple libel suits in the years following its founding (though most were successfully defended). Over time, its reputation grew, and the impact of the *Lancet* ranks second only to that of the *New England Journal of Medicine*. Given this rich history and adherence to fact, it is truly ironic that the *Lancet* disseminated what may be the most notorious and arguably most dangerous fraud in medical history. Despite overwhelming evidence to the contrary and very public stripping of his professional credentials, the mastermind behind this deception has continued to profit from the propagation even as the damage toll rises. Tragically, thousands and perhaps millions of children throughout the world are increasingly at risk of death or disability from diseases that are thoroughly avoidable. It is to this fraudster that we now turn.

Andrew Wakefield was born on September 3, 1957, the son of two upper-middle-class physicians, one of whom specialized in neurology.[51] As a member of the gentry growing up in the affluent town of Bath, Wakefield attended the best schools that would accept him. He obtained his medical degree in 1981 from the St. Mary's Hospital (the same institution that has housed reputable scientists such as Sir Alexander Fleming, a discoverer of penicillin, and Lord Moran, Winston Churchill's personal physician).

Wakefield began practicing medicine at the tender age of twenty-six. Seeking to make a name for himself, he became interested in research, first coming to the widespread attention of the English medical community in 1993. While employed with the Bowel Disease Study Group at the Royal Free Hospital School of Medicine in London, Wakefield wrote a report claiming a link between measles virus infection and Crohn's disease.[52] Using samples from a small number of patients (a total of twelve), Wakefield claimed that persistent measles virus infection contributed to the inflammation associated with Crohn's disease, an autoimmune disorder. Despite the small number of samples involved in the study, which should have raised questions about its reproducibility, the scientific and medical community embraced this work. This outcome does not reflect the design, execution, or interpretation of Wakefield's study as much as it reflects the fact that the community investigating Crohn's and other inflammatory bowel diseases was desperate for a breakthrough and grasped at the study as a potential inroad to future improvements. As often happens in science, Wakefield's hypothesis that persistent measles virus infection causes Crohn's disease was completely debunked by 2001 and many precious resources and time were devoted to disproving it.[53]

Such an outcome itself is not necessarily an indictment, as peer review and reproducibility are the hallmarks of the scientific method and commonplace in medical research. However, what followed flies in the face of good methodology. Wakefield's true damage to public health would arise in 1998 as a follow-up to his discredited study.

On February 28, 1998, Wakefield published a bombshell report in the *Lancet.*[54] This study reported the findings of a study of twelve children and claimed that regressive development disorders were linked with chronic entercolitis and, in particular, with MMR vaccination. This study, like his previous report on Crohn's disease, was based on a small population of twelve children. His study linked impaired neurological development with MMR vaccination in eight of twelve children evaluated. Translated into a more user-friendly headline that was broadcast throughout the world: Autism is caused by a chronic infection of the gut in general and with immunization with the MMR vaccine in particular. In just under 3,000 words (2,857 to be precise), Wakefield made a claim whose negative reverberations and implications for public health were propelled by numerous television interviews and amplified by the full mobilization of the public affairs resources of his employer, the Royal Free Hospital of Hampstead, which was bathing in the glow of publicity generated by one of its young stars.

Given rising concern and sensitivity surrounding the subject of autism, a disease that had been quickly rising in prominence and coming to the attention of many for the first time in the months prior to Wakefield's report, the negative publicity of Wakefield's small study expanded like a nuclear chain reaction. To understand how and why this claim is both patently false and actively harmful, it is necessary to introduce the concepts behind these bold assertions.

Misunderstandings, Hopes, and Frauds

The disease captured by the word *autism* today is one of the most emotionally charged diagnoses in the English language. The meaning of the word itself has changed over time in parallel with our understanding of the disorder. As originally described in 1919 by the Swiss psychiatrist Eugen Bleuler, and remaining as the first definition used by the *Oxford English Dictionary* (the OED), autism is "a condition or state of mind characterized by patterns of thought which are detached from reality and logic, formerly sometimes regarded as a manifestation of schizophrenia or other psychiatric illness."[55] Though now recognized to be a flawed interpretation, Bleuler reasoned that the disease was characterized by an "autistic withdrawal of the patient to his fantasies against which any influence from outside becomes an intolerable disturbance."[56]

Over the next few decades, our understanding and description of autism have been subject to many refinements, most famously by the Viennese child psychiatrist Hans Asperger, who lives on eponymously in the naming of one form of the disorder. The current definition of autism, again citing the OED, is "A neurodevelopmental condition of variable severity with lifelong effects which can be recognized from early childhood, chiefly characterized by difficulties with social interaction and communication, and by restricted or repetitive patterns of thought and behavior."[57]

As our understanding of autism has increased, the disorder is now understood to represent a wide spectrum of symptoms, ranging from relatively mild impairments of social interaction and communication skills to truly incapacitating disruptions in neurodevelopment. Likewise, behavioral traits can range widely from repetitive activities to the extreme, savant-like recollection as stereotyped by Dustin Hoffman in the 1988 Hollywood hit *Rain Man*.

In the years following the premiere of *Rain Man*, reports of the disease skyrocketed. Whereas fewer than one case of autism per thousand children had been registered in 1997, this number ballooned fivefold over the following decade.[58] Like many aspects of this baffling disease, the reasons for the increased incidence are unclear but many explanations have been invoked.

With many medical conditions, increased public awareness can cause an upsurge in reporting as more patients, parents, and physicians appreciate that disparate symptoms may be part of a larger disease or spectrum. Indeed, a 2015 report in the *British Medical Journal* assessed children in Sweden who had been diagnosed with the disease between 1993 and 2002.[59] The report concluded that while the prevalence of the disease had not changed, its diagnosis had increased dramatically. Stated another way, the likelihood of getting the disease had not changed, only its diagnosis had. On one hand, this is a positive finding and suggests that the disease rates have been stable and that our ability to detect autism is improving, which can translate into providing patient care. The flip side is that increasing diagnosis rates have increased the level of anxiety about the disease manifold. Importantly, this Swedish study indicated nothing about causation, which will occupy much of the remainder of our story.

With many neurological and psychiatric impairments, assessing causation is challenging, if not impossible, a fact that will dominate the remainder of our story. However, Wakefield's 1998 *Lancet* report offered a simple and much-needed hope to desperate parents of those affected by the disease. The problem, according to Wakefield, was the MMR vaccine.

There was only one problem with Wakefield's conclusion: his data were fabricated. Worse still, especially for the parents duped into making decisions about their children's health and well-being, the motivations behind the paper were far more dubious than might have been imagined.

Scientists tend to be a rather quiet and cautious lot, trusting in the process of the scientific method (discoveries verified or refuted by peer review and additional studies). Compounded by the technical nature and specialized language of different scientific fields, researchers can sometimes come across to the general public as uncommunicative or uncommitted. Fortunately, this is not so with journalists, and our story now turns to an aggressive and successful investigative reporter whose beat included the pharmaceutical and public health fields.

Left and Right

Brian Deer is an investigative reporter who graduated with a degree in philosophy from the University of Warwick in Coventry, England. This location is unintentionally symbolic, as Coventry was the site where Leofric, Earl of Mercia, founded a monastery in 1043.[60] While Leofric is not a household name, his wife is: Lady Godiva. According to legend, Godiva rode nude through the streets of Coventry as a protest (actually more of a dare) directed at Leofric to relieve the tax burden on the people of Coventry.

Although the lore has it that most of the townspeople averted their gaze to avoid despoiling Godiva, a tailor by the name of Tom snuck a look; and in doing so, gave birth to the popular notion of a *Peeping* Tom. As conveyed by a decline in vaccination in the first two decades of the 21st century, a failure to see the truth is far more dangerous than shielding your eyes. This fact was brought to the world's attention by a graduate of a university located in Godiva's Coventry.

Following Brian Deer's graduation, he sought a career as a writer.[61] The University of Warwick has a reputation as a progressive campus, and Brian's early writing career reflected his immersion within a liberal campus culture. After joining a collective and publishing a few pieces in radical left-wing magazines, Brian learned about opportunities to publish freelance articles in a health care journal. This provided an income and allowed him to continue working for liberal causes. For example, an article arguing for the rights of homosexuals within the British National Health Service, published in 1979, was progressive in its political views and would be thrust to the forefront with the emergence of the HIV/AIDS crisis months later.[62] Over time, Deer's reputation for hard-driving opinion pieces, with emphasis upon public health issues, led to a job offer at the *Times* (London).

The *Times* was founded in 1785 and is one of the most popular dailies in the United Kingdom, with a print circulation of almost a half million in 2016 (all the more remarkable given the demise of the print media over the past two decades). The center-right *Times* formed an unexpected home for a left-leaning investigative journalist, who focused his articles on social issues, including but not limited to poverty, child abuse, and homelessness. Although these articles were generally well received, Deer realized his most widely read articles touched upon revelations of medical scams, and his reporting progressively became more investigative than opinion-based.

Among Brian Deer's early breakthrough investigations was the revelation that an eminent British scientist by the name of Michael Briggs had fabricated research pertaining to the safety of contraceptive pills. Briggs had made a name for himself by publishing an enormous amount of research, as well as writing books with his wife, about the safety of contraceptives.[63] As Deer exposed in an article for the *Sunday Times* on September 28, 1986, the work was entirely fabricated.[64]

A few years later, Deer broke a story on an antibiotic known as Bactrim or Septra. The story graced the front page of the *Sunday Times* on February 27, 1994 and exposed the fact that the drug had been implicated in more than a hundred deaths in England alone.[65] Later that year, Deer revealed that the will of Henry Wellcome, founder of the British pharmaceutical powerhouse Burroughs Wellcome, essentially allowed a charitable foundation to retain control over a pharmaceutical empire and its profits (the same foundation and company sued in the DTP trials).[66] Although Deer has similarly investigated many other aspects of public health, his work on vaccine safety has arguably had greatest impact, saving the lives of thousands, if not millions, worldwide.

As you may recall from our vignette about Gordon Stewart's skepticism toward vaccines, the DTP vaccine was the subject of a series of lawsuits throughout the world, but the British trials were among the most widely covered and deeply litigated. Upon learning that Judge Murray Stuart-Smith had summarily dismissed the claims put forth in *Loveday v. Renton and Wellcome Foundation*, Deer, as a knee-jerk reaction and perhaps reflecting some remaining concerns with the Wellcome Foundation, "wrote a piece snootily dismissing the judge's finding, substituting my own opinion."[67] The piece appeared in the April 3, 1988 edition of the *Sunday Times*, and Deer began receiving messages of adoration from anti-vaccinators.[68]

Shortly thereafter, Deer had a fateful encounter with Margaret Best, who had won £2.75 million in damages against a DPT vaccine manufacturer. An intensive interview revealed inconsistences suggesting that Margaret wasn't telling the truth about her child's experience in suffering from neurological damage following immunization.[69] Thereafter, Deer began questioning his assumptions about vaccines. Diving deeper, Deer assessed all of the data about the DPT vaccine that he could, consulted with experts and, after a ten-year investigation, concluded the vaccine did not cause brain damage (thus aligning himself with Stuart-Smith's verdict). His work led to an award-winning expose in the November 1, 1998 edition of the *Sunday Times.*[70] This piece revealed scams being perpetrated by attorneys who preyed upon families that were convinced that DTP jabs had caused their children to suffer brain damage.

Deer's understanding of vaccines increased further following his 1999 exposure of a bogus HIV vaccine, AidsVax, being developed and bankrolled by the World Bank, the World Health Organization, and the American Centers for Disease Control and Prevention.[71] This investigation not only revealed that the highly touted vaccine was ineffective but also suggested it might accelerate or worsen an already deadly disease. Having had his fill of reporting on immune-modulating frauds, Deer was "sick to death of vaccines" and looking to stories on anything except that subject.[72] Little did he realize that his work on vaccines had barely begun.

In late 2003, Brian Deer began hearing rumors about Andrew Wakefield, the high-born physician who published the blockbuster findings linking the MMR vaccine to autism.[73] Similar to how the story with DPT unfolded, Deer interviewed one mother who claimed there was a link between MMR vaccine and autism, but again he did not find her story convincing. He then began conducting what he referred to as a "forensic interview" with the mother about the circumstances that linked MMR to autism. Deer quickly concluded her recollections "contradicted Wakefield's paper." Based on his past experiences with DPT and the instincts that distinguish the most gifted investigative reporters, what Deer believed would be a "three-week work" ballooned into a detailed investigation that exposed a "100% scam" by Wakefield that Deer describes as a greed-motivated "heist."[74]

To summarize a long and detailed series of careful analyses conducted over more than a half decade, Deer first revealed in the February 22, 2004 edition of the *Sunday Times* that Wakefield had in fact been paid to fabricate the findings linking MMR to autism.[75] Specifically, English attorney Richard Barr had financed Wakefield's studies and put him in contact with the twelve children profiled in the *Lancet* study. These families were either already part of Barr's brewing lawsuit or were being recruited for that purpose. While the study was conducted at a London hospital, none of the children were from the area. Indeed, one of the twelve was flown in from the United States specifically to be included in the cohort under investigation. Given the prevalence of autism, such extraordinary measures were certainly unnecessary at a minimum.

Rather than adhering to the scientific method of testing hypotheses, Wakefield was on a paid mission to prove a link that simply did not exist. Therefore, it was necessary to bend the data towards his needs. For example, Wakefield's data consisted of "associations" made by the parents of autistic children (many being the same parents who were actively seeking remuneration) that vaccination was linked with the onset of autism. At a minimum, this is poor science and itself would raise questions regardless of the source of funds.

That Wakefield was accepting money from an attorney was not necessarily an issue. Indeed, many scientists serve as expert witnesses for both civil and criminal trials around the world. However, the fact that Wakefield had failed to disclose that he was paid by the attorney behind many MMR lawsuits was an early sign of a larger problem. This was not a simple oversight. At a March 1998 meeting of the British Medical Research Council, Wakefield was queried as to the source of his data. He not only failed to report the link with the subjects of the preexisting lawsuit but also claimed "no conflict of interest."[76] These assertions were later proven as false by Deer, who revealed that by that time, Wakefield himself had already personally received more than £50,000 for his work in the lawsuit.[77]

Roughly a year after Deer began researching the supposed link between MMR vaccination and autism, his work was the subject of a November 18, 2004 documentary, *MMR—What They Didn't Tell You*, broadcast on Britain's Channel 4 television network.[78] The documentary included new investigative research and revealed not only that Wakefield had created near panic for the parents but also that he had filed for patents and intended to launch his own measles, mumps, and rubella vaccine products as an alternative to conventional vaccines. It seems that Wakefield was not opposed to vaccines but rather was simply opposed to those vaccines for which he did not own the intellectual property.

Deer and his team further revealed discrepancies between the findings reported by Wakefield and the outcomes reported by other laboratories—including Wakefield's. The documentary was received by the scientific and public health communities with considerable relief, since Wakefield's spurious report had triggered both parental anxiety and decreased immunization rates, particularly among high-income and well-educated Britons. Critical review was also positive with the *British Guardian*, who gleefully proclaimed, "Deer went for Wakefield like a bull pup with a taste for trousers."[79]

Perhaps unsurprisingly, given this public outing of his methods and motivations and his proven association with attorneys such as Richard Barr, Wakefield sued Channel 4, the producers of the documentary, and Brian Deer himself. This lawsuit was later found to convey as much merit as Wakefield's *Lancet* study and was dismissed by the judge, who cited that "the Claimant wishes to use the existence of the libel proceedings for public relations purposes, and to deter other critics, while at the same time isolating himself from the downside' of such litigation."[80] The judge further ruled that this was a tactic repeatedly employed by Wakefield and that the now-disgraced scientist had also sought to prevent the public health system from fighting against his disproven ideas.

The coup de grace delivered by Deer against Wakefield was reported in a February 8, 2009 article in the *Sunday Times*. Wakefield had not been satisfied with simply biasing his findings by selectively using children referred by Richard Barr. Wakefield had actively faked the findings, altering the data to support his thesis. The data had been so heavily modified that they'd allowed a small-scale study of twelve children to have outsized implications.[81] A further tragedy revealed by Deer's investigative reporting was that the twelve children in question had been subjected to a grueling and unethical series of medical procedures, including lumbar punctures, general anesthesia, and intrusive bowel imaging.

A detailed investigation by the *British Medical Journal* concluded that Wakefield's study was a fraud.[82] Upon learning of the practices employed by Wakefield, the coauthors on his paper disowned the work and any connections with Wakefield. The *Lancet* conducted its own investigation, starting in July 2007.[83] In a twist of irony, given the passion of the publication's founder to combat medical incompetence and privilege, the journal's leadership agonized about the public humiliation of acknowledging its insufficient review of the original Wakefield paper. However, the journal publically retracted the paper, an exceedingly rare event in scientific publishing circles. Likewise, Wakefield was stripped of his medical license, with the board of ethics citing multiple lapses. These actions are the scientific equivalent to the public stripping of honor and medals suffered by Albert Dreyfus, who was a far more sympathetic and innocent character.

Although publically outed as a money-grubbing fraudster, Wakefield shifted the focus of his entrepreneurial zeal from hocking his own vaccines to writing books. After many losses in court, the now-unemployed charlatan began playing the role of the martyr, portraying himself as the helpless victim of a rapacious Brian Deer. During a January 5, 2011 interview of

Andrew Wakefield by CNN's Anderson Cooper about the forced retraction and the outcomes of the British Medical Journal investigation, which has just been published, Wakefield turned the interview into a plug for his own book. The interview was watched in disbelief by Seth Mnookin, himself a writer for the *New Yorker*, *Washington Post*, *New York Times*, and many other prominent periodicals, who at the time was finalizing his outstanding book about the vaccine-denier movement, *The Panic Virus*.[84, 85] CNN had lined up Mnookin to appear with Wakefield, but the disgraced English doctor refused to appear with Mnookin, a long-time critic of Wakefield. Mnookin was interviewed separately by Cooper that night. Rather than plugging his own upcoming book, Mnookin carefully addressed each of Cooper's questions, revealing the extent of Wakefield's fraud and the risks to public health from avoiding vaccination. As one example of the danger of false correlations, Mnookin compared the rise of autism diagnosis with a corresponding rise in microwave popcorn sales in the same period (for the record, there is no connection—which was Mnookin's point). All of this was detailed in Mnookin's blog the next morning.[86]

Wakefield was still not done. In 2016, he wrote and directed a documentary, *Vaxxed*, which was entered into the lineup of the high-profile Tribeca film festival. This film was described by reviewer Eric Kohn as "a self-aggrandizing approach" by Wakefied "designed to trick you" and was given a grade of D.[87] Kohn related his view that in his biased work, Wakefield had boldly maintained the pharmaceutical industry was masterminding a conspiracy to profit from products that cause autism. Kohn's review further described the film as "Shakespearian hubris" meant to propagate falsehoods, writing that *Vaxxed* was "a tale of sound and fury signaling nothing but its own homegrown idiocy."

Learning that *Vaxxed* had been scheduled to appear at Tribeca, legitimate documentary filmmakers revolted, demanding its exclusion. Likewise, the medical community reacted with outrage, as typified by a statement from pediatrician Dr. Mary Anne Jackson, who stated, "Unless the Tribeca Film Festival plans to definitively unmask Andrew Wakefield, it will be yet another disheartening chapter where a scientific fraud continues to occupy a spotlight."[88] The next day, Robert De Niro, who founded the festival and is the parent of an autistic child, reversed himself by withdrawing the film from the festival in response to the uproar.[89]

Still Wakefield persisted in his desire to profit from the charade. Realizing additional monies could be made in high-priced speaking engagements, Wakefield, stripped of his medical license, moved to Texas and hit the lecture circuit. He began giving talks to groups of receptive audiences, for a price. His ideas were particularly popular among wealthy Americans, and his devotees included Hollywood celebrities with autistic children, including Alicia Silverstone, Charlie Sheen, and Jenny McCarthy.[90]

Notable anti-vaccinators from the political sphere included Robert Kennedy, Jr. and Donald Trump. Despite Trump's busy schedule on the campaign trail, he made time to meet with Andrew Wakefield in August 2016 and invite him to one of Trump's presidential balls, where Wakefield "called for the need for an overhaul of the Centers for Disease Control and Prevention."[91] Indeed, in the days before Trump's inauguration, Kennedy, a long-standing and outspoken critic of vaccines, announced from the lobby of Trump Tower he had accepted a position in the Trump administration to chair a vaccine safety panel. This statement was walked back a bit by a Trump spokesperson, who conveyed that such an offer had not yet been finalized.[92] In response, the senior fellow for global health at the Council for Foreign Relations and Pulitzer Prize–winning journalist Laurie Garrett responded with a scathing response, contemplating the millions of lives that could be at stake were such a reckless action to proceed.[93] Sadly, long before Garrett's impassioned plea, the body count had begun to climb.

Recurring Nightmares

The March 24, 2011 edition of the *Star Tribune*, the largest daily newspaper in Minnesota, reported the third visit by Andrew Wakefield to Minnesota since December. Wakefield's message of skepticism about vaccination, the report went on, had been particularly embraced by a displaced Somali community in Minneapolis.[94]

It may come as a surprise to many, but the chilly climes of Minneapolis host "the largest Somali community in North America, perhaps in the world outside of East Africa," reported CNN in February 2017.[95] A 2014 article in the *Journal of the American Board of Family Medicine* reported that "the Somali community is more likely to believe that MMR vaccine causes autism and are more likely to refuse the MMR vaccine than non-Somali parents."[96] For example, the rates of vaccination among Somalis in Minnesota dropped from 92 percent in 2004 to 42 percent in 2014. Consequently, this population has suffered disproportionately from outbreaks of measles. Periodic outbreaks had begun in 2011, coincidental with Wakefield's visits. Perhaps sensing legal liability, he was reported by the *Washington Post* to have stated, "The Somalis had decided themselves that they were particularly concerned [about vaccine safety and] I was responding to that."[97] He further claimed no responsibility for the 2011 outbreak or a more vigorous measles outbreak that began in 2017 and racked up more than four dozen confirmed cases in the span of two months.[98]

Although the anti-vaccinator response is particularly high among Somali Minnesotans, it reflected a disturbing trend that has persisted since Wakefield's fraudulent publication linking MMR vaccine with autism. According to a 2008 publication in the medical journal *Pediatrics*, the rates of MMR immunization declined in the United States following Wakefield's discredited publication in the *Lancet*.[99] Although the overall trend did improve as news of the fraud was revealed by Brian Deer and communicated via mass media and word of mouth, the myth has persisted in isolated communities, such as the Somali population. However, the anti-vaccinator movement is not unique to Somali or Minnesota. At least twenty of the fifty American states allow their children to be exempted from vaccination based on philosophical or personal belief. A study of California communities revealed that the largest rate of personal belief exemptions from mandatory vaccination prevailed in white, affluent neighborhoods.[100]

As such, it may come as no surprise that the highest rates of measles, mumps, and rubella in the United States are occurring among unvaccinated people in California, as well as among groups that have historically rejected technology, such as Amish communities. A prominent example is the rise of measles, mumps, and rubella in American universities. As conveyed in the introduction to this book, vaccine-preventable infectious diseases are breaking out in universities across the United States. One of the most troubling aspects of this recent trend is the increasing frequency with which immunized students are themselves the victims of their classmates' failure to be immunized. For this, we need to take a moment and review the concept of herd immunity.

Arthur William Hedrich was born on July 7, 1888 in Chicago, where he would serve as a public health officer. Interested in the spread of epidemics, he continued working while seeking an advanced degree under Lowell Reed at Johns Hopkins University. Although only two years older than Hedrich, Reed had become a superstar in the field of public health by blending cutting-edge mathematics with epidemiology. In 1928, he unveiled the Reed-Frost model, which was developed in collaboration with a more senior Hopkins investigator, Wade Hampton Frost.[101] This equation provided the foundations of modern epidemiology as it predicted how infectious diseases spread in a community over time. In that same year, Arthur Hedrich defended his doctoral thesis, titled "Epidemic Studies: The Monthly Variation of Measles Susceptibility in Baltimore Maryland from 1901 to 1928."[102] Rather than return to Chicago, Hedrich remained at Johns Hopkins to continue his research.

Expanding upon this first study in Baltimore, and combining these findings with data from Boston, Hedrich revealed that measles infections tended to peter out when something like two thirds of the population had already become infected.[103] This might not seem like a terribly exciting finding, but it saved the lives of millions and formed the foundation of modern mass-vaccination campaigns. For each infectious agent, there is a particular fraction of the population that, if rendered resistant to the disease because of immunization or generated immunity from a prior infection, will convey protection upon the larger community. This key fraction confers what is known as "herd immunity."

A simple way of thinking about herd immunity is that even if an infected individual comes into the community, he or she is unlikely to convey the disease to susceptible individuals since "the herd" of protected people will squelch the pathogen before it can spread. Such an outcome is a statistical probability and changes with alterations in frequency and distance of travel (particularly big problems in an age of common air travel). Consequently, we now know that the fraction of the population needed to ensure protection from a highly contagious disease like measles is often greater than 95 percent.[104] (There remains considerable debate as to the ideal fraction of people who need to be immunized to protect the larger population.) What we do know is that if the level of protection falls below this critical level, then the entire population becomes susceptible to the disease and epidemics will invariably follow.

Compounding the problem of herd immunity, a vaccination received during childhood conveys protection for a finite amount of time. Taken together, this means that a population of densely packed individuals, such as those in a large city or on a college campus, may be more susceptible to infectious diseases conveyed by the unvaccinated. This is exactly the scenario unfolding at college campuses around the United States. The good news is that the outbreaks have been relatively tame thus far, due to a combination of luck and diligent public health care services found at most major universities. However, the increasing frequency of these epidemics could suggest that our luck may be running out. As one example, we briefly turn to a mumps outbreak in Missouri.

An assessment of the incidence of mumps in the United States conveys the extraordinary benefits of vaccination.[105] Whereas 5.5 per 100,000 Americans reported the disease in 1987, the number plummeted to fewer than 0.1 by the early 2000s. What one can presume to be the "Wakefield effect" then kicked in, characterized by a spike in mumps incidence, reaching an incidence of 2.5 in 2006. Mumps tends to wax and wane naturally, and the 2006 spike was followed by two years of relative calm before a return of the disease in 2008–11, followed by another brief lull and an increase starting in 2013. In 2014, mumps came to prominence in the media when it took its toll on the National Hockey League, infecting both players and referees. In that same year, mumps returned to school, with high-profile outbreaks at America's largest college campus, Ohio State University.[106] From Columbus, the disease spread to the west, overtaking the University of Illinois in 2014 and the University of Missouri starting in 2016.[107, 108] In the latter case, the case incidence continued to increase, even between academic years, exceeding more than three hundred cases in 2016 at the flagship campus in Columbia and spreading to Southeast Missouri State and other campuses soon thereafter.

Due to the inadequate vaccination rates and the relatively low durability of protection conveyed by the current mumps vaccine (which you may recall is a weaker form of an earlier vaccine), mumps appears to be a disease that will return to the lexicon of horrors facing the American population for the foreseeable future. This is not welcome news, for, as we will see, measles, mumps, rubella, and other vaccine-preventable diseases pose a distraction for public health workers preparing for even greater threats arising at an unprecedented rate.

Paraphrasing the subtitle of Stanley Kubrick's 1964 political satire *Dr. Strangelove*, we will need to stop worrying about vaccines and instead embrace them with even more vigor and love in the coming years. Hence, we will close our story with examples of why we need vaccines now more than ever and why vaccines might, ironically, provide an opportunity to decrease the incidence of autism for future generations of children.

10

When Future Shocks Become Current Affairs

The name Entebbe periodically enters the news. A small town of fewer than seventy thousand on the shores of Lake Victoria in southern Uganda, the town first came to international prominence in 1952 as the setting for the airport where Princess Elizabeth II and her husband, Prince Philip, ended an African safari upon learning of the untimely death of her father, King George VI.

The 1970s were times of upheaval and revolution, particularly in terms of Mideast and post-colonial African peace. A one-sided Israeli victory in the Six Day War increased the fervor of Palestinian separatists, encouraging splits even within the most extreme terrorist groups such as the Popular Front for the Liberation of Palestine (PFLP). A splinter group, the PFLP External Operations Group (PFLP-EOG) was founded by Wadie Haddad, a Palestinian nationalist whose expertise was in the hijacking of commercial airliners.[1] Haddad had masterminded many hijackings, most notoriously, the Dawson's Field hijackings of September 1970, where five airliners were hijacked and diverted to a remote desert airstrip in Zarka, Jordan. Upon landing, the passengers were separated between Jewish (and some high-profile Americans) and others, with the latter captives being released almost immediately. The remaining hostages were dispersed on the planes and sites throughout Jordan, thereby demonstrating the impotence of the Jordanian government to reign in the Palestinian terrorists that had become embedded throughout the Hashemite Kingdom. Because of increasing threats of military retaliation sprinkled with diplomacy, the hostages were all eventually released (as were high-profile PFLP members serving jail time in Britain), though the five airliners were destroyed with explosives in spectacular images filmed and viewed throughout the world. Because of this terrorist event, the Jordanian government cracked down on Palestinian extremists in a series of raids memorialized as Black September, causing Waddad to seek sanctuary and patrons elsewhere. Waddad soon found an ally on the shores of Lake Victoria.

Idi Amin, at six foot four inches, was an imposing figure, not least because his large frame combined with erratic behavior to make him brutal, merciless, and surprisingly charismatic.[2] After rising to power in a military coup, Amin inaugurated a campaign of violence that earned him the moniker "The Butcher of Uganda." In June 1976, Amin sympathized and colluded with Haddad to support the hijacking of Air France flight 139, an Airbus A300 commercial airliner carrying 246 passengers, mostly Israelis. While flying over Greek airspace, the plane was hijacked by two PFLP-EOG and two German Revolutionäre Zellen terrorists and diverted first to Benghazi, Libya, and then to Entebbe airport.[3] This final destination placed the passengers and crew under control of Amin, who held all under guard in the terminal. The dictator then began negotiations for the release of other terrorists imprisoned in Israel and Europe (though on the behalf of the terrorists, not the hostages). As occurred at Dawson's field, the Jewish hostages were separated from the non-Jewish, with the latter being released. With threats the Jewish hostages would be executed if terrorists' demands for the release of Palestinian inmates were not met, the IDF mobilized, and, on the fly, developed an intricate plan for rescuing the hostages despite the fact that Uganda was over 2,500 miles away. In a daring assault, code-named Operation Thunderbolt and led by Yonatan Netanyahu (Yoni, brother of future Prime Minister Benjamin Netanyahu),

IDF forces stealthily traversed halfway across a continent by air, rescued all hostages (except four who had been murdered by the terrorists), neutralized the terrorists, and immobilized the entire Ugandan air force with the loss of only one Israeli, Lt. Col. Netanyahu, who was struck down by a sniper lurking at the edge of the jungle surrounding the airport.

The neighborhood surrounding the famous airport is quite heavily wooded and dotted with forests teeming with verdant hills, which afford an occasional glimpse of a powerful mountain gorilla. Such greenery is due to a tropical climate, with temperatures in Entebbe rarely dropping below 60°F at night or exceeding 80°F during the day. Steady temperatures that vary day to day by less than 3°F, combined with abundant rainfall exceeding 60 inches per year, provide the perfect climate for mosquitoes.[4] Not just any mosquitoes but *Aedes aegypti* mosquitoes, a species that has bred and continues to breed terror far worse and more expansive than the PFLP-EOG or Idi Amin himself.

The cause of the suffering is not from the *Aedes* mosquito itself but from the viruses that teem within. The apes living in this jungle, not to mention their many human neighbors surrounding the region, have provided an opportunity for the mosquitoes, their passenger viruses, and hominids to coevolve, which has created a witches' brew of killer pathogens. Because of this threat, Entebbe has again entered the modern lexicon, this time as home to a microbial terrorist named for where it was discovered—a small, twenty-five-acre park just to the north of the airport known as the Ziika forest.

Zika is one member of a family of killers known as flaviviruses (*flavi* being a Greek term for 'yellow'). Indeed, the best-known flavivirus is the cause of a notorious disease known as yellow fever, named for the pallor that its victims frequently exhibit prior to a painful death.[5] The disease is endemic in Africa, and the Ziika forest is host to both the virus and the insect that spreads the virus, *Aedes aegypti*. Consequently, the Uganda Virus Research Institute (UVRI) was founded in Entebbe's Ziika forest in 1936 by investigators at the Rockefeller Foundation. Eleven years later, a caged rhesus money at the UVRI began displaying a fever, and the virus was isolated and identified as the modern-day Zika virus.[6] Like a bad movie script, it was not until 1952 that Zika was known to have spread to humans. Just as this knowledge was being processed, investigators found themselves in the midst of a growing pandemic. In the years following the end of the Second World War, the virus had already been spreading rapidly to the west, where it had already gained a foothold in the coastal regions and small nations of west Africa. This region was particularly susceptible, as the public health infrastructures were stretched, broken, or never created. Additionally, this was a time when these newly independent or aspirational countries were throwing off the yoke of colonial rule, thereby severing ties that might have helped quash the emerging virus.

The Zika virus made a fateful trip, either by plane or ship, to Malaysia in the 1960s, where it percolated for decades. Much more recently, the virus crossed the Pacific Ocean to South America and then even more quickly north to Central and North America.[7] The earliest signs of a problem in the Western Hemisphere were revealed in Brazil, coincident with that country's hosting the 2014 FIFA World Cup. This fact has triggered persistent speculation that a participant or fan, possibly there for an international rowing event (which included athletes and fans from many Pacific island nations), might have brought Zika to South America. Such causation was particularly troubling, as the rise of Zika coincided with Brazil's preparations to host the 2016 Summer Olympic Games.

Although Zika infection generally manifests itself with flulike symptoms, its most notorious effect is the potential to infect the nervous system of growing fetuses, thereby imparting a horrific and debilitating condition known as microcephaly (which literally translates into 'small heads'). As such, the 2015–16 outbreak was the source of dramatic headlines throughout the world, accompanied by heartbreaking photographs of babies born with abnormally small heads and profound nervous system defects.

These concerns raised apocalyptic thoughts that the virus might spread further, causing some athletes and fans, particularly women of childbearing years, to cancel their plans to attend the 2016 Olympics. However, these acute concerns quickly abated in the months following the conclusion of the Olympics as the rate of virus spread slowed. Nonetheless, the rate of Zika virus spread has remained at a dangerous level for those concerned about the disease over the long term.

Yellow Fever

While Zika clearly remains a prominent threat to public health, its impact has not yet come close to equaling two of its relatives. Historically, yellow fever virus has killed millions, particularly following its spread to North America as part of the Columbian Exchange in the late 15th century.[8] In a tragic irony, one group of repressed peoples likely created a tragedy for another. Specifically, yellow fever is conjectured to have been imported into the New World by enslaved Africans. This hypothesis is based on epidemiological evidence that the disease was most prominently associated with 17th-century outbreaks in the slave-holding Caribbean Islands of Barbados and Guadalupe.

The slave-holding Caribbean island of Hispaniola was the likely source of a yellow fever pandemic in the United States.[9] On September 20, 1697, the first true world war was concluded with the Treaty of Ryswick. A conflict known as the Nine Years' War or the War of the Grand Alliance pitted the Sun King, Louis XIV of France, against a coalition of major European powers.[10] The war was fought throughout Europe and in Asia (India), the Caribbean (Hispaniola), and the North American colonies (where it was known as King William's War). Despite being outnumbered, the French nearly carried the day. In the resulting treaty at William of Orange's palace of Ryswick, various concessions were awarded, including French hegemony of part of Hispaniola, which was renamed Saint Domingue and is known today as Haiti. In 1791, the slaves and free people of color rose in rebellion against their French overlords, and large number of French refugees began an exodus by sail from Saint Domingue to coastal towns throughout the United States.[11] Many settled in Philadelphia, which was serving as a temporary capital of the newly formed United States of America as a new capital was being constructed on the Potomac River.

The abrupt surge in immigration was accompanied by equally unexpected guests in the form of mosquitoes infected with yellow fever.[12, 13] The disease spread throughout the City of Brotherly Love, creating the greatest levels of devastation in the environs near the Delaware River, which provided ample breeding grounds for mosquitoes. Fear grew as the epidemic killed one in ten Philadelphians, and the city was evacuated by President George Washington and his newly formed government. The spread threatened the new republic in the days following the start of Washington's second term and was worsened by the dispersion of people (and mosquitoes) to other major American towns, most of which were similarly positioned on or near major bodies of water (and thus harboring the mosquitoes that often accompanied the refugees).

This is not the only time Philadelphia enters the story of yellow fever, as this city later contributed to the eradication of the disease in the United States. In 1855, Philadelphia served as the temporary home for Cuban-born Carlos Juan Finlay.[14] Finlay's father was a physician working for the Venezuelan freedom fighter Simon Bolivar. Juan sought to emulate his father by becoming a doctor as well.[15] Infectious diseases almost cut down Finlay's aspirations on two different occasions, when his travels to France were stymied by local eruptions of cholera (1844) and typhoid fever (1848). Both outbreaks forced Carlos Finlay to return to his native Cuba. However, a third attempt went according to plan, and Finlay completed training as a physician at Jefferson Medical College in Philadelphia, where he demonstrated particular aptitude and interest in infectious diseases. Finlay trained under John Kearsley Mitchell, an early advocate of the germ theory. This was fortunate for our story, since the "miasma" theory of disease still held sway at the time and Mitchell was bucking a belief that was still predominant in most medical schools of the day.

Finally gaining an opportunity to train in France, Finlay moved from Philadelphia to Paris to study under Armand Trousseau, an internal medicine specialist who seemed to have a particular penchant for Caribbean trainees. Specifically, Trousseau had also served as preceptor for the nationalist Ramon Emeterio Betances, the "Father of the Puerto Rican nation."[16, 17] Upon Finlay's return to Cuba, he specialized in ophthalmology but remained devoted to understanding how and why yellow fever spread throughout the population.

In 1879, Finlay was tasked by the Cuban government with identifying the means by which yellow fever spread throughout the population. Contemporary research into the disease revealed that the infectious agent was within the bloodstream, which caused Finlay to postulate, "It occurred to me that to inoculate yellow fever it would be necessary to pick out the inoculable material from within the blood vessels of a yellow fever patient and to carry it likewise into the interior of a blood vessel of a person who was to be inoculated. All of which conditions the mosquito satisfied most admirably through its bite."[18] In 1881, Finlay appeared at the 5th International Sanity Conference in Washington, D.C., where he presented evidence that the disease was primarily spread by mosquitoes. To Finlay's amazement and considerable disappointment, his ideas were largely ignored, though not entirely.

In the same year Finlay was investigating how yellow fever spreads in Cuba, a young assistant surgeon by the name of Walter Reed was in his fourth year in the US Army Medical Corps, serving as a physician in Ft. Apache, where his wife Emilie gave birth to their first daughter (also named Emilie).[19, 20] Over the next decade and a half, Reed shuffled between various Indian reservations in the western United States. His Native American patients included none other than the great Apache leader Geronimo. In 1893, Reed returned east, joining the faculty of the George Washington University School of Medicine. He was stationed at the new Army Medical School (whose medical institute would later be adorned with his name). Whereas the popular assumption had been that yellow fever was spread through contaminated drinking water, Reed contrasted the habits of those with and without the disease. A major insight occurred as Reed realized infected soldiers tended to spend time walking in the swampy areas of Washington, whereas uninfected soldiers tended to avoid those areas.[21, 22]

Expanding upon this finding with yellow fever, Reed traveled to Finlay's native Cuba in the days following the 1898 Spanish-American War to study typhoid fever, a scourge of the troops serving in the newly conquered territory. This first Cuban trip was a rousing success as Reed revealed the disease was spread by flies that contaminated food and water with human waste materials.[23]

Two years later, Reed returned to Cuba, this time tasked with finding the cause of the pervasiveness of yellow fever in the country. The disease was a particularly high priority for the US military, given the fact that incidence of the disease had delayed and fundamentally threatened the completion of a strategic canal being carved through nearby Panama (outbreaks had frustrated and then ended construction of a 48-mile canal due to extraordinary mortality on construction teams). Based on Finlay's breakthrough ideas, Reed worked with another army officer, William Crawford Gorgas, to implement a campaign of layered defenses against mosquitoes by purging sources of stagnant water, purifying public sources of water, and promoting the use of insecticides and insect netting. Though Walter Reed appropriately credited the Cuban physician with the breakthrough discovery, history has tended to forget this exchange and instead attributes the finding primarily, if not exclusively, to Walter Reed.[24]

By targeting mosquitoes rather than the virus itself, yellow fever was largely eradicated from many tropical climates throughout the Western Hemisphere, including Panama. The public health miracle facilitated the final construction of the canal and more fully opened the United States as a blue-water power on both the Atlantic and Pacific Oceans. Likewise, the incidence of yellow fever and other mosquito-borne diseases began to disappear in the United States and throughout the world following the widespread dissemination of Finlay and Reed's breakthrough.

The disease of yellow fever had been largely eliminated, but not entirely. The eradication of the disease could be achieved through public health measures in some affluent countries, such as the United States. However, the eradication or durable suppression of mosquitoes has proven particularly challenging in countries and regions with fewer resources and more challenging climates (such as Entebbe, which serves as the perfect breeding ground for mosquitoes). In these locations, the disease could only be eliminated by the production and widespread adoption of a safe and effective vaccine.

The disease that arose in Africa was fittingly conquered by an African. Max Theiler was born in Pretoria in 1899. He completed training as a physician in South Africa and received additional training in tropical medicine in London prior to moving to Harvard University for a time. He later left Cambridge to settle down at the Rockefeller Institute, which had a strong interest in yellow fever and would create the Uganda Virus Research Institute in 1930.

In 1918, the Rockefeller Institute dedicated itself to identifying and eliminating the source of yellow fever. However, the program suffered from a rather inauspicious start.[25] A Rockefeller bacteriologist, Hideyo Noguchi, had traveled to Ecuador to obtain samples from infected patients. After many days and weeks of analyzing the samples under a microscope, Noguchi reported that the disease was caused by a type of spiral-shaped bacterium, appropriately named a spirochete. Based on this finding, Rockefeller Institute researchers developed an antiserum and vaccine against the bacterium and tested the vaccine on thousands of volunteers.[26] Unexpectedly, the vaccine proved ineffective. Moreover, the evidence linking yellow fever with the spirochete could not be duplicated by his colleagues. The conclusive evidence refuting the bacterial origin of yellow fever was delivered by another Rockefeller scientist, Max Theiler.

In 1927, Rockefeller moved its yellow fever operations to Africa, and Noguchi joined the team seeking the cause of the disease, maintaining until the end that the spirochete was responsible despite negative results from studies of more than one thousand different animals. Sadly, Noguchi's end came soon, as he perished from yellow fever in May 1928. Just as the fever was claiming this high-profile victim, a study was being conducted that revealed the disease was caused by a filterable virus. The evidence implicating the virus had been collected in Dakar, capital of the French colony of Senegal. The sample had been taken from a patient, Francois Mayali, who had been infected with a relatively mild case of the disease. This sample was subjected to the Chamberland filter technique, which isolated a transmissible virus and not a spirochete (though spirochetes were later found to contribute to a variety of other diseases that occur in the same patients that have yellow fever).

Using the same clinical specimen obtained from Mayali, Theiler began to further attenuate what was already known to be a rather weak pathogen. This was accomplished by passaging the newly discovered virus repeatedly through mice (much as Pasteur had done with rabies virus to create his vaccine decades before).[27,28] The passaging did indeed weaken the virus, yielding a vaccine that retained the ability to elicit a protective immune response without causing disease. The vaccine was soon deployed throughout Africa and other regions where mosquito control had not been sufficient to eliminate the disease. In appreciation for his contributions, Theiler was awarded the only Nobel Prize ever awarded solely for the discovery of a new vaccine.[29]

Double Jeopardy

Despite the success in controlling yellow fever all those years ago, one of its closest flavivirus cousins is the fastest-growing infectious disease in the world and is certain to grab future headlines in a hotter, wetter, and more densely packed world. Dengue is another mosquito-borne flavivirus with a penchant for urban and semi-urban areas in both tropical and subtropical climates.[30] The name of the virus is derived from the Swahili word *Ka-dinga pepo*, which roughly translates into "disease caused by the devil that causes terrible pain."[31] This appellation accurately reflects the symptomology of dengue, which has also been referred to as "break-bone fever," reflecting the extraordinary arthralgia (joint pain) that is a primary symptom of infection. This pain can cause permanent damage to the joints, giving the disease the additional moniker of Dandy fever, based on the fact that previously afflicted patients were said to acquire a stance and walking style comparable to a wealthy fop.[32, 33]

A dark joke about dengue fever is that the disease can strike only twice. The first time a person becomes infected, they want to die. The second time, they do.

Deconstructing the statement, dengue infection causes a series of symptoms collectively known as a viral hemorrhagic fever. In the days following initial exposure to the contagion (generally arising from the bite of an infected mosquito), the patient may suffer flulike symptoms, which include high fever, general malaise, and the beginnings of muscle and joint pain. Over time, the pain greatly increases and is accompanied by loss of appetite, nausea, and vomiting.

What distinguishes hemorrhagic fevers from simple fevers is that the former triggers a powerful immune response. This might, at first glance, seem like a positive thing, since, as we have seen, the job of the immune system is to eliminate foreign invaders like dengue virus. However, the type of response triggered by dengue virus is so acute and exaggerated that it tends towards utter destructiveness. Much as we have seen with bacterial endotoxins, a vigorous activation of the immune system can trigger the production (or, using the scientific term, "dumping") of massive amounts of cytokines, chemicals designed to alert the body to the presence of an infectious invader. One job of these cytokines is to "loosen" the blood vessels to allow immune cells to gain access to the tissues. This is a beneficial reaction when, for example, an infection is localized to a specific site. When a person cuts him- or herself and thereby exposes the body to external pathogens, the loosening of the blood vessel lining near a site of infection will allow immune cells to enter and accumulate, thus speeding the clearance of the infection.

The problem with dengue and other viral hemorrhagic fevers is that the cytokines are conveyed en masse into the bloodstream and lymph systems throughout the body. The results tend to be rapid and lethal. For example, a sudden rush of cells out of the circulation and into tissues can cause rapid swelling (and much of the pain associated with the disease). Despite the fact that the blood mostly remains inside the body, the rapid blood loss can cause a form of shock like that seen with gunshot or accident victims. In extreme cases, the loosening of the blood vessels can be sufficiently extreme that blood begins to seep out of the body altogether. Consequently, an infected patient can begin bleeding from the eyes, ears, mouth, rectum, genitals, or other sensitive parts of the body that are rich with tiny blood vessels near the body's surfaces. These responses are identical to those made famous by Richard Preston in *The Hot Zone*, an outstanding chronicle of early studies of Ebola virus infection (although the accounts overdramatized the symptoms a bit, since most patients do not bleed out or live long enough to experience liquefaction of vital organs).[34]

Returning to dengue, assuming the patient survives this first encounter with dengue virus, the body's immune system eventually reverts to its more beneficent role of developing protective antibodies and T cells, and the virus is eliminated. However, any relief experienced by the patient is sadly temporary because a second exposure to dengue virus (even years later) will cause the same set of symptoms, only magnified manifold because the immune system (and the cytokine storm) has been trained to recognize the pathogen and thus will convey an even more vigorous response that in turn amplifies the symptoms. The result of a second infection is invariably death.

Our understanding of how dengue fever infection has impacted people over time has been growing. Ancient accounts of an arthritis-causing infection have been unearthed in ancient Chinese texts and 17th-century Egyptian reports and have been assumed to be early written descriptions of dengue infection.[35] However, it was not until a 1780 outbreak in Philadelphia that the disease received its first conclusive documentation. A report by the American physician Benjamin Rush (a Forrest Gump–like character, who befriended many of America's Founding Fathers and thereby influenced an extraordinary number of events in the revolutionary and postrevolutionary United States) described a "bilious remitting fever" whose "more general name among all classes of people was the break-bone fever."[36]

Despite these early accounts of dengue fever, its incidence remained low, with occasional high-profile outbreaks. For example, the queen of Spain was afflicted in 1801 following a period of protracted rain (which increased the local mosquito population) but survived.[37] However, all this began to change in the second half of the 20th century as the global incidence began to skyrocket and has continued to do so at an alarming rate.[38] Further, the geographic span of the outbreak has comparably expanded, with infections being recorded on all continents (excluding Antarctica), in part as a result of climate change, which continues to increase the range of its mosquito vectors. As of press time, no vaccine or antiviral treatment beyond supportive care for its symptoms has been developed to counter this rapidly growing scourge. While dengue is unquestionably growing in incidence (as evidenced by precise DNA and other epidemiological assessments), recent years have witnessed considerable confusion, as many other diseases mirror the symptomology of dengue.

Mkomaindo Hospital is a small structure located at the intersection of the A19 and B5 roads in the town of Masasi in the southern part of Tanzania, near its border with Mozambique. Even today, the hundred-bed hospital is staffed by no more than twenty clinical officers (CO), a title certifying two years of training. It serves 300,000 regional patients. Despite the hospital supporting a local CO training college, the turnover rate is very high, since most trainees quickly leave Mkomaindo for the lures of the capital, Dar-es-Salaam, and its higher pay and amenities.

The current two-story building lacks running water and stable electricity and yet is a marked improvement over its immediate predecessor, a rickety structure torn down in 1952. The hospital is the only source of medical support for the regional population. Its construction was overseen by two remarkable women doctors. Dr. Frances Taylor had come to Masasi in 1922 and would remain as the only physician in the area for most of the four decades of her missionary service. At the time the new hospital was being built, Dr. Taylor had the relative luxury of working with another English-trained missionary-physician by the name of Marion C. Robinson.

Starting in late 1951 and extending into the early months of 1952, the fortune of having two staff physicians was nonetheless stretched to the limit by an outbreak of a new disease. Although most of the infected survived, the disease was characterized by high fever and joint pain, which caused patients to stoop over as a consequence of debilitating arthritis triggered by the infection. After the worst of the crisis had passed, Dr. Robinson realized that this disease was unlike any other she had treated, including dengue fever, and reached out to a colleague at the Yellow Fever Research Institute in Entebbe.

Dr. William Hepburn Russell Lumsden was a Scottish physician who had trained in Glasgow, receiving his medical degree in 1938 and specializing in tropical medicine.[39] This choice of careers proved fortunate for his homeland, since Britain was soon inundated with a world war that saw Dr. Lumsden serving as a commanding officer in charge of malaria field units in the European, Middle Eastern, North African, and Asian theaters of war. He served in Palestine, Transjordan, North Africa, Italy, India, and Sicily. Indeed, in this last location, Lumsden's career and life were nearly ended when the truck he was riding in struck a land mine and was blown off the road. Months after the declaration of peace, Lumsden accepted a position in Entebbe, where he would remain for the next nine years.

On a fateful day in 1952, Lumsden learned about the situation in Masasi and agreed to work with Dr. Robinson to investigate its causes. They worked diligently over the coming years and, in 1955, introduced the world to their new discovery.[40,41] The epidemic in Tanganyika was not dengue but rather a new virus they named *chikungunya*, which is the local Makonde language word for 'that which bends up.' Like the etiology of dengue, this name reflects a primary symptom of the disease, which reflects the joint pain and dysfunction suffered by the afflicted.

As we saw with Zika, the whole of continental Africa was ablaze in revolutionary change and the unshackling of colonial bonds. These realities directly impacted Marion Robinson and the fate of chikungunya. In the months after publication of her seminal scientific articles with William Lumsden, Marion fell in love with an Anglican priest by the name of Mark Way. At the same time Robinson was battling a chikungunya outbreak, Way was being appointed bishop of the Anglican Masasi Diocese.[42] However, Bishop Way's battles had yet to begin, as he was soon assailed by his clergy with overt and subtle claims of micromanagement and racism. The local clergy revolted and Way, who was preparing a trip to the United Kingdom in late 1959 to marry Marion Robinson in their homeland, made the absence permanent with his resignation and the couple's return to England.

The appointment of a new bishop who was known for his strong anti-apartheid convictions was more to the liking of the local clergy.[43] The new bishop, Ernest Urban Trevor Huddleston, was widely praised, but his appointment created its own crisis, as his authoritarian tendencies led to the loss of half the trained physicians in Masasi. This condition would deteriorate further in 1963, as evidenced by an open letter submitted by Huddleston to the entire Anglican community, pleading for a doctor to replace Frances

Taylor, who herself was set to retire later that year. He further related that all attempts to replace Robinson over the past few years had failed (though not mentioning her by name), and he feared (rightfully so) that the same challenges would confront the looming retirement of Taylor.

These symptoms of a larger transition in African society and infrastructure would help propel the chikungunya virus well beyond its Tanganyika home. The virus was being experienced in India at the same time Bishop Huddleston was composing his open letter and would soon spread throughout the subcontinent and the entire Indian Ocean basin. From there, the virus expanded globally, being experienced throughout the world but too often being confused with dengue fever.[44] This was more than a minor nuisance, as it confounded epidemiological understanding of both diseases and has led the medical community to appreciate we are simultaneously facing not one but two particularly dangerous pandemics in parallel.

The dangers from hemorrhagic fever viruses are not limited to dengue and chikungunya, as this classification and these horrific symptoms are shared by a frighteningly large and growing number of hemorrhagic fever viruses, which include many names that are increasingly familiar to physicians throughout the world. These viruses include Ebola, Marburg (which we met earlier), Lassa, Nipah, Hendra, Kyasanur Forest, Alkhurma, LCM, Omsk, Chapare, Lujo, and Sabi. Hemorrhagic fevers (each caused by a different pathogen) are associated with Venezuela, Argentina, Bolivia, Crimean-Congo, Kaysanur Forest, and the Rift Valley of east Africa, to name but a few. The simple reality is that the list of viral hemorrhagic fevers is a growing problem and is certain to worsen in the coming months and years.

Despite their rapid proliferation, modern medical science is almost entirely unprepared for this growth. At best, the most wealthy public health systems can try to address the symptoms of these diseases, but targeting their causation can only be accomplished by a type of medicine that has lost the attention of the pharmaceutical industry and is worryingly shunned by the general public: vaccines.

Hopes and Opportunities

Vaccines provide a very real hope for addressing dengue, chikungunya, and all the other viral hemorrhagic fevers. Although each pathogen provides its own set of scientific challenges, the experiences gained with yellow fever, not to mention the myriad vaccines successfully developed to date, demonstrate that it is feasible. Vaccines have saved millions (perhaps even billions of lives) and generate many billions of dollars in revenue each year for their manufacturers. Despite such impressive numbers and the fact that we know more about the breadth and destructive potential of the many lethal microorganisms in our environment, the field of vaccine production has largely stalled. Given ongoing trends in the business of vaccines, it might not continue in the future.

I initiated a project in early 2011 to document the sources of innovation in the pharmaceutical and biotechnology industries. As described in greater detail in *A Prescription for Change*, the rationale for doing so was to assess both the science and business of making new medicines and to see how both have changed over time.[45] In parallel with our work to track the sources of all new medicines, we began to track the same information for vaccines. These studies revealed a total of 135 different passive (i.e., antibody) or active (i.e., vaccines) innovations. These breakthroughs provided much-needed immune-based therapies approved for use in the United States by the FDA or its predecessors.

Our initial reaction to assessing the data was that 135 is a remarkably small number, particularly when one considers that many of the new products were simply improvements upon earlier versions of vaccines. Specifically, the 135 novel products altogether were able to prevent disease caused by a mere 28 different microbial pathogens.

Looking further, our studies revealed the number of companies devoted to vaccines expanded starting in the 1940s, reaching a peak of twelve different companies in 1970. Thereafter, a combination of industry consolidation, declining revenue potential, and increasing concerns about liability (arising from the experiences with DPT and MMR) cut in half the number of companies that develop new vaccines. Despite varied incentives to encourage new vaccine development to address rising problems such as Zika or dengue virus, the number of organizations focused on vaccines remains at a perilously low level.

Reflecting a larger trend of waning research and development by the pharmaceutical industry, more of the early research and development of new vaccines is being performed by relatively small upstart companies with names such as BioVex, Protein Sciences, and Acambis. Although start-up companies can bring new ideas and energy to a field, the relative reluctance of large, experienced companies to more fully engage in vaccine research could unnecessarily limit our ability to develop much-needed vaccines. All of these changes are occurring amidst an ever-increasing array of pathogens encountered as the planet grows warmer and more densely populated and when intercontinental travel can spread a disease at an unprecedented rate.

Consistent with these concerning trends, a comprehensive analysis revealed that the number of different pathogens that can be targeted with vaccines has not changed since the early 1990s. Extending this further, almost all the growth in developing vaccines to prevent infectious disease since the 1950s has been focused upon viral diseases. A 2014 study by investigators at Brown University identified twelve thousand different outbreaks that affected forty-four million people worldwide from 1980 to 2013.[46] To put this into perspective, the list of the world's fastest growing infectious diseases discovered since 1980 includes but is not limited to: Lyme disease (discovered in 1982); Escherichia coli O157:H7 (1982) HIV/AIDS (1983), Rotavirus (type B in 1984 and type C in 1986), hepatitis c virus (1989), Hantavirus (1993), Hendra virus (1994), Vancomycin-resistant Staphylococcal aureus (1996), Nipah virus (1999), Human metapneumovirus (2001), and SARS (2003).[47, 48] This list does not include other pathogens such as dengue virus, chikungunya, or pandemic influenza virus, which had been known to medicine years before and still do not yet have a protective vaccine.

A few years ago, I was dragged into the murky world of bioterrorism (and the development of countermeasures to protect the public) after serving under a boss who himself was a former head of the Defense Advanced Research Projects Agency (DARPA). This secretive agency is the research and development arm of the Defense Department and responsible for the introduction of Global Positioning Satellite (GPS) technology, the internet, and autonomous vehicles (as well as Area 51 and its antecedents). At the time, the US government was still reeling from the 2001 anthrax attacks on the media and Congress (attacks that we now know were launched by a military colleague developing a vaccine to anthrax). In a knee-jerk response, the government invested heavily in a variety of programs such as the 2004 Project BioShield Act, which allocated billions of dollars to support the development of countermeasures to protect against terrorist attacks.

Amidst all the uproar about man-made threats, my old boss stated calmly, "Nature is the worst terrorist of them all." This accurate statement reflects the fact that the emergence of new infectious diseases does not require malevolent intent but can arise from a combination of natural and unnatural circumstances. The natural contributors include spontaneous mutations that allow a virus that had been restricted to animals (e.g., rinderpest), to take a new interest in humans (where it became measles) or for a chance encounter between a monkey and a human to facilitate the spread of HIV, launching one of the most deadly plagues our species has suffered. Likewise, influenza virus tends to undergo a natural shift every few decades that gives rise to a new form that had not been encountered for at least a few generations. Unlike the seasonal forms of influenza that are moderately deadly (killing thirty to fifty thousand Americans per year), these periodic and completely natural pandemic influenzas can kill tens or hundreds of millions and perhaps, given our globalized society, billions.

Compounding these natural problems, human activity is changing both the rate and types of infections that one can expect in the future. For example, the warming of the planet increases the range of tropical infectious organisms and the vectors (*e.g.,* mosquitoes, ticks, and bats) that harbor them. As population rates increase in underdeveloped nations, so does encroachment upon the natural reservoirs of these vectors, rendering it more likely that humans will encounter pathogens that have long existed but have been blissfully unaware that humans might provide their next meal. Compounding this further, urbanization, combined with greater availability and speed of travel, means that an infection in one place can be broadcast widely before the public health system becomes aware of its existence. No example of these mutually reinforcing dangers is more representative of this change than antibiotic-resistant bacteria. It is with this example that we will conclude our story, adding an ironic twist.

Vaccines and Autism: A Different View

The growth of drug-resistant infections has been the source of growing concern for many decades, and the potential for a "post-antibiotic world" is rapidly becoming a reality. As we have already discussed, resistance to these miracle drugs known as antibiotics accompanied their introduction in the 1940s. This resistance is a natural outcome of the overuse of antibiotics combined with the propensity of bacteria to swap plasmids that encode for resistance mechanisms (recall toxin-antitoxin from an earlier chapter). The consequences of these actions frequent the news cycles with reports of "superbugs," flesh-eating bacteria, and drug-resistant forms of many different diseases. Prominent among the latter is a wave of drug-resistant forms of tuberculosis. These threaten the return of a disease once known as "consumption"—one of the world's most deadly diseases—whose victims have included Eleanor Roosevelt, Frederic Chopin, and George Orwell.

The final victim is symbolic of the Orwellian twist to our story. The same structure that gave rise to antibiotics undercut the deployment of vaccines that might have prevented some of the drug resistance we face today. Continuing with the example of tuberculosis, we encounter the character of Leon Charles Albert Calmette, a French physician who learned bacteriology from Emile Roux and started a branch of the Pasteur Institute in Saigon (then a part of French Indochina).[49] Upon his return to France, Calmette began studying what we now call *Mycobacterium tuberculosis*, a bacterium discovered by Robert Koch that was responsible for the disease of "consumption" (a.k.a. tuberculosis). Specifically, Calmette and his colleague, Camile Guerin, began to grow tuberculosis bacteria on potato slices soaked in glycerin and bile. The bacteria that could survive under these conditions were weakened by repeating the process for more than a dozen years (from 1908 until 1921).[50] The attenuated *Bacille Calmette Guerin* (or BCG) strain of bacteria had a bit of a rough start when early testing caused the deaths of seventy-two infants in Lubeck, Germany, in 1930.[51] In the subsequent follow-up, it was revealed that the BCG strain had become contaminated with a more virulent natural form of tuberculosis during its manufacturing and the more aggressive contaminant caused the disaster. This inauspicious start unsurprisingly rendered many physicians and parents wary about the BCG vaccine. Ultimately, the need for the vaccine was eroded by the introduction of antibiotics in the years during and immediately after the conclusion of the Second World War.

The widespread availability and low cost of antibiotics decreased the attractiveness of many vaccines. In our analysis of innovative vaccines, we found the number of different bacterial pathogens that could be targeted with vaccines grew rapidly from 1900 until 1940 and barely changed thereafter despite an ever-increasing discovery of new bacteria that cause various diseases. A pharmaceutical industry that had relied upon the development and deployment of new vaccine products instead embraced new antibiotic wonder drugs, and with an average of three new antibiotics being introduced each year from the 1950s through the 1980s, the concept of resistance seemed no more likely than the sky falling.

As the years passed, circumstances changed. A glut of newly introduced antibiotics and increasing generic competition arising from the inexpensive marketing of older medicines fractured the market. New drugs could not command high prices, and the rate of new antibiotic introduction plummeted. Indeed, our work at Washington University revealed new antibiotic approvals dropped by more than 90 percent from the 1990s onwards. At the same time, the well-worn drugs of the past had become so overused that drug-resistant bacteria underwent selection in a manner that would have made Darwin proud. The result, as a casual glance at the headlines will reveal, is that most antibiotics will have limited effectiveness in the future. Compounding the problem, the expertise and personnel that developed earlier generations of antibiotics have largely atrophied. Even if an economic model could be made to support the development of new antibiotics, the pharmaceutical industry is ever less able to seize upon the opportunity.

Returning to tuberculosis, a suitable, albeit imperfect vaccine exists. However, the low prevalence of tuberculosis during the antibiotic era has generally meant that public health officials do not routinely advise parents to immunize their children with BCG. As tuberculosis continues a relentless drive towards antibiotic resistance, the use of BCG might have to be reconsidered. Even better, a new generation of tuberculosis vaccines that improves upon the safety and efficacy of a vaccine that was developed almost a century ago using potato slices could undoubtedly be achieved.

In a truly ironic twist, vaccines might also provide an opportunity to decrease the incidence of autism and developmental disorders. The idea is that vaccines tend to be scalpels, whereas antibiotics are blunt objects. Extending the analogy, a tuberculosis vaccine such as BCG should selectively eliminate a disease-causing organism with no beneficial value to humans, whereas penicillin depletes both the bad actors and the beneficial organisms of the human microbiome.

In a series of scientific reports and a recent book targeting the general population, the director of the Human Microbiome Program at New York University, Martin Blaser, conveyed a theory that one outcome of the overuse of antibiotics is a disruption of the normal microbial component of the human body.[52, 53] Using the bacterium *Heliobacter pylori* as an example, Blaser proposes that antibiotics have altered the microbiome in such a manner that, to paraphrase the title of one chapter of his 2014 book, *Missing Microbes*, it has rendered many people "fatter and taller" than might be expected or be considered healthy.[54] Indeed, antibiotics have been used for decades to fatten and increase the overall size of livestock prior to slaughter (which has also contributed to the development of drug-resistant bacteria). Likewise, Blaster postulated that an altered microbiome might have links with increasing rates of autism and other developmental disorders.

As we have already seen, the link of any causation to autism is complex. Oversimplifications have been abused repeatedly by desperate parents or charlatans advocating a link with vaccines. Thus, any claims as to causation must stand up to rigorous scrutiny that often takes years or decades of intensive, peer-reviewed work. Nonetheless, early reports suggest that the microbiomes of autism patients differ from those of "healthy" counterparts. Furthermore, a fecal transplant, the process of transferring the healthy gut bacteria from a non-disease patient to a disease-bearing individual, may show promise in ameliorating some of the symptoms of autism. It is essential to point out that this early study is small, akin to the size that Wakefield used to support his bogus claims, but hopefully not biased. Additional investigation will be needed to assess whether the results can be reproduced and expanded.

Were such findings to hold, utilizing vaccine-based scalpels to selectively eliminate disease-causing bacteria and viruses may convey considerable advantages over chemical treatment such as antibiotics. Indeed, one advantage has already been established with the introduction of the first pneumococcal vaccines in the 1980s. *Streptococcus pneumoniae* was first isolated in 1881 by Louis Pasteur and, in parallel, by George Miller Sternberg (the commanding officer of a young Walter Reed). The disease was a leading killer across the world, and the first, crude vaccines were developed in 1909. Greatly improved versions of the vaccine were introduced in the 1980s, and an unexpected consequence was a 47 percent plunge in the overall use of antibiotics in children.[55] In a world facing increasing rates of infection from drug-resistant bacteria such as *Clostridium difficile*, *Pseudomonas aeurugi-nosa*, and others, which are currently treated with the blunt objects known as antibiotics, new generations of safe and effective vaccines could convey benefits well beyond their intended applications by decreasing the overall need for antibiotics.

It will not be easy. Throughout this book, we have witnessed that virtually every introduction of a new vaccine has been countered by skepticism, if not outright hostility. From the time before the first vaccines (since variolation does not qualify as such), vaccines have encountered resistance, often from unexpected sources. The fears have ranged from the reasonable opposition that variolation can itself be deadly to the irrational fears that a cowpox-based vaccine might cause the recipient to sprout horns and begin to graze in a meadow. No one might have predicted the extreme responses to DPT and MMR precisely because the motivations behind the opposition were based on fabricated data or ulterior motives.

Other unexpected reactions have included charges that the vaccine created to eradicate cervical cancer, a sexually transmitted disease, was instead intended to promote promiscuity. Perhaps the most damning response is an utter lack of response by many otherwise intelligent individuals, who elect not to receive an influenza vaccination because it is inconvenient for their schedule or because they don't care for needles (even in the face of an intranasal option).

Thus, the rosy scenario painted above, of a future where vaccines act as scalpels to selectively eliminate certain obnoxious pathogens, must face certain realities. First, any new vaccines must and indeed should face considerable scrutiny to ensure they are safe. Based on past experiences, some of which have been reviewed herein, vaccine manufacturers have been responsible for self-inflicted wounds, which have lingered long in individual and institutional memories. Given immunity generally lasts for years, this may require extensive and expensive needs for assessing safety.

Furthermore, the vaccines must be sufficiently efficacious, and, under extreme circumstances, certain trade-offs must be considered. We have already experienced this with the mumps vaccine. Older versions of the mumps vaccine were associated with decreased duration of efficacy, so that individuals receiving this vaccine could become sensitive to infection with mumps within a few short years after vaccination. This risk is particularly high in the current environment, where the "herd immunity" of the population is crumbling because of the misplaced efforts of anti-vaccinator extremists.

The growth of a small but dangerous anti-vaccine resistance is instructive. Much greater care must be taken to train physicians, especially pediatricians, about the importance of proper vaccination and how to discuss the fears, both rational and otherwise, that they will inevitably confront with understandably anxious parents, who might have been exposed to inaccurate propaganda from the vocal anti-vaccine lobby. Inaccurate rumors must be dispelled, and accurate information conveyed in its stead. Much care must be taken to respect the sensitivities of their patients (and parents) while conveying that the danger to children, and indeed the much wider community, is fundamentally threatened by a failure of even one parent to have his or her child immunized.

A more difficult challenge faces the public health community, lawmakers, and regulators in their attempts to restart a sputtering vaccine industry. It is crucial for lawmakers, regulators, and indeed the entire general public to place themselves in the shoes of a vaccine manufacturer, if only via a simple thought experiment. Imagine that you are an executive at a large pharmaceutical company and you have a choice of where to invest limited research and development dollars. You could choose a designer medicine for a sick population, such as those with cancer, which will garner a price tag of five or six digits. This drug would take, on average, ten to fifteen years to develop and cost about a billion dollars.

Alternatively, you could develop a vaccine for an infectious disease. Whereas the cancer drug is being administered to sick patients in need of a cure, the vaccine is administered to healthy people, many of whom will never encounter the infectious disease you seek to prevent. The product you seek to develop is a commodity, intended to be administered to a mass population rather than to a subset of needy patients. Your pricing structure must reflect this. This fact precludes a five- or six-digit price tag. Moreover, your pricing assumption may not include factors beyond your control. For example, any perceived side effect arising at or near the time your vaccine is administered, may be attributed to your product. Despite the fact you have spent enormous resources and time to assure that the vaccine is safe, this may all be neutralized by a string of coincidences, or, worse still, the active intervention of a bad actor such as Andrew Wakefield. These perceptions (or realities) will substantially decrease your market size and may jeopardize the sustainability of the product and even your company.

By spending just a few moments contemplating such decisions, it is easy to understand why drug manufacturers view vaccine products as less attractive than conventional medicines. Compounding the problem, the drug development community itself is shrinking rapidly, as are the resources devoted to research and development, largely because of industry consolidation meant to restrain a persistent decrease in efficiency.

What can be done? In a word: incentives. If vaccine manufacturers can be assured stable markets through advanced market commitments, they are more likely to address unmet medical needs. The stability of the market could be assured if regulators and legislators required immunization at the national level. Such actions are a commonsense approach to ensure that herd immunity sustainably protects the population. However, true to the cliché, common sense is increasingly rare as it pertains to current trends in vaccine mandates. Rather than insisting that individuals be immunized against diseases that threaten the population, a more libertarian approach has held sway in which individuals can exempt themselves or simply ignore the advice of physicians and public health boards. Indeed, the individual appointed in 2017 to lead the public health infrastructure in the United States advocates that the federal government reverse its long-standing advocacy of vaccines and instead return such responsibilities to individual states.[56] Such an approach could be disastrous in terms of herd immunity. As we have seen in the example of mumps, it would threaten not just those individuals who make the unwise decision not to immunize but the entire community as well. Such trends must be reversed, both to protect the population and to ensure that manufacturers remain confident enough of the market potential to ensure competition that favors further improvements in the safety, efficacy, and breadth of products that prevent the myriad disease-causing organisms that we face.

For many emerging diseases, as well as those that are reemerging due to antibiotic-resistant bacteria, additional incentives may be necessary to entice innovators to develop novel vaccines. In the case of the West African Ebola virus outbreak of 2014, quick responses by the scientific and biopharmaceutical communities facilitated the rapid identification and testing of vaccines meant to arrest a fast-growing epidemic before it could spread even further. On one hand, much of the work was initiated after the proverbial horse had left the barn. On the other hand, the Ebola crisis demonstrated the value of incentivizing experimental therapies and vaccines in a just-in-time manner (that was quite literally true in this case).

The regulators and developers of future vaccines for emerging and reemerging diseases could learn much from the 2014 Ebola crisis. First, incentives could be placed upon the development of certain vaccines. The incentives might be passive, such as ensuring market exclusivity for a longer period, or more active, such as providing tax breaks for a fraction of the research and development costs for vaccines that address particularly important needs. Such incentives do work, as evidenced by the extraordinary growth in drugs meant to target low-incidence diseases in the years following congressional passage of the 1983 Orphan Drug Act.[57] Such incentives might be made more powerful by increasing cooperation between vaccine manufacturers and the epidemiologic community. For example, the Centers for Disease Control and Prevention could help prioritize the pathogens for which vaccine treatments are developed. The incentives provided by governmental agencies could include provisions such as those in the Orphan Drug Act but might also include additional incentives as conveyed by social impact bonds.

In July 1988, an economist by the name of Ronnie Horesh was asked to present at a conference of the New Zealand Branch of the Australian Agricultural Economics Society held at Lincoln College in Canterbury, New Zealand.[58] Horesh's expertise was social and environmental policy, but the impact of his presentation would comparably apply to many other fields. What Horesh proposed on this winter day was that the public sector should establish a contract that is payable to a recipient that meets certain predefined outcomes. The beauty of the program was that the payer would not have to expend funds until the payer had obtained their desirable goal. This concept, known as a "social impact bond" or "pay for success bond" was the subject of discussion for a few decades thereafter. It was finally tested in March 2010 when British justice secretary Jack Straw announced an impact bond meant to reward innovators who could decrease inmate recidivism at Peterborough Prison.[59] Within six months, a group of investors had posted £5 million pounds in an effort that could return £8 million, an attractive profit in a time when bank interest rates hovered at and occasionally below zero percent.[60]

In the years since the United Kingdom's bold embrace of this new project, the number and breadth of social impact bond projects have expanded to include multiple private, volunteer, and public-sector organizations. We propose herein that the framework of a social impact bond could be one means of addressing the growing need to incentivize new vaccines. Specifically, programs might be developed to build a pipeline of different projects that identify and reward organizations and individuals that meet predetermined goals in the development of much-needed vaccines. Had such a program been announced in the UK a few centuries earlier, the farmer Benjamin Jesty might have been rewarded with more than a painting and a letter of thanks for his development and testing of the first smallpox vaccine.

Clearly, governmental incentives and mobilization will need to be deployed for urgent and critical needs such as a regional outbreak of Ebola or global risk of pandemic influenza. Though lacking the speed of such public-sector deployments, the bandwidth of most governmental agencies (e.g., CDC and NIH) is limited and often must be reserved for meeting urgent and important needs. However, new mechanisms such as social impact bonds could be deployed to address other important needs such as the discovery and testing of improved mumps vaccines. Because the rate of scientific innovation in the discovery and understanding of infectious diseases has never been higher and is expanding to an exponential degree, comparable innovation and entrepreneurship in financial instruments and other incentives are needed if we are to stem the inevitable plagues that constantly loom over humanity.

We end our story where it began. Despite all the old and new infectious diseases that are being constantly unearthed, our greatest susceptibility may again reside in one of our oldest nemeses. You may recall from chapter 2 that the United States and Soviet Union each agreed to store a single vial of smallpox material, one in Atlanta, the other in Moscow, which could be used as a source for the future development of a vaccine. However, much speculation suggests that the smallpox virus may be far more widely dispersed.

In 1992, a Kazakh colonel of the former Soviet (now Russian) army requested asylum in the United States. This might seem an odd request, since the Soviet Union had ceased to exist a few years before and was not deemed a significant threat. However, American officials were quick to grant the Kazakh, Kanatjan Alibekov (now known as Ken Alibek), the protection he requested, since Alibek arrived with a story that even Ian Fleming might not have been able to concoct on his most creative days.

Alibek told his CIA handlers that he was an expert in military medicine who had risen through the ranks to lead an offensive bioweapons program for the Soviets and their Russian successors. Specifically, he had been tasked with creating both microorganisms and toxins that could be used as tactical or strategic bioweapons. Alibek then went on to detail descriptions of a bioweapons program that both utilized the wild-type versions of many deadly pathogens and fabricated novel variants to maximize their ability to kill and spread disease. Among the disease-causing pathogens that were weaponized during Alibek's reign were smallpox and variants meant to overcome Western safeguards. Alibek quickly discarded his Soviet upbringing and embraced capitalism, penning a memoir of his exploits as a bioweaponeer and founding a company meant to counter the obnoxious agents he developed as a Communist.[61]

The stories conveyed by Alibek were terrifying and, to be frank, may not be entirely accurate. Questions about his motives and the accuracy of his statements remain. However, the potential and perhaps likelihood that the Soviets and Russians have maintained not just a biological weapons program but one that has utilized natural and engineered smallpox is truly terrifying. Since its presumed eradication, smallpox immunization is no longer performed, which renders the world more susceptible than it has been for decades.

Even if Alibek's claims are revealed to be less than accurate, the risk of a resurgence of smallpox is not inconceivable. For example, global warming has begun melting areas that have for millennia been frozen. As a result, graves of smallpox victims, especially those in the permafrost, may be warming to the point where the frozen virus within could regain the potential to rise from the grave. Worse still, knowledge of the DNA sequence could be exploited to facilitate the laboratory creation of a Frankenstein-like virus that could wreak havoc on an unsuspecting world.

Regardless of its source, natural or otherwise, the possibility that smallpox might reemerge is a source of considerable concern for the United States and its Western allies. Compounding this are continuing reports that smallpox exists within arsenals developed by North Korea and Russia. The alarm triggered by such concerns has caused the US government to quietly advance smallpox countermeasures, which include the deployment of new vaccines, as evidenced by the 2007 FDA approval of ACAM2000, a vaccine that is managed by the Centers for Disease Control and Prevention and the Department of Homeland Security.[62] The approval of ACAM2000 three decades after the last natural smallpox infection and the fact that it has been included in stockpiles maintained by the federal government lend credibility to the stories propagated by Alibek and others. These facts should serve as a reminder that we face dangers from infectious organisms, both old and new, which can only be contained or prevented using safe and effective vaccines. At a minimum, such realizations must force us to recognize that vaccines have saved countless lives and that we will continue to rely upon these miraculous substances to ensure that we avoid future public health calamities.

Epilogue

Both the title and my goals for the book have sought to reflect the countervailing forces of hope and fear that have surrounded the subject of vaccines throughout their history. From the standpoint of public health, hope should—and ultimately will—dominate the more negative emotion, because modern science has conveyed practical and affordable means to prevent (and in rare cases utterly eliminate) scourges that have long plagued society, chief among them smallpox. Other successes include the management of myriad diseases that have historically killed or maimed innumerable of our youngest and most vulnerable population, as well as the potential to eliminate polio within the coming years. Similarly, vaccines hold the potential to finally manage pandemic scourges such as bubonic plague and influenza, the latter tending to cut down double- and even triple-digit millions of people worldwide on a recurring basis every few decades. In the United States alone, 30,000–40,000 people die each year from influenza in a "good year," when vaccine uptake is high, and the vaccine is properly matched with the viruses circulating in the American population. Far worse outcomes arise periodically due to vaccine mismatches with circulating virus strains and, even more rarely, when new influenza viruses arise.

As we have seen, each vaccine success has consistently been met with skepticism and outright hostility. Such fears began with the ostracism of Benjamin Jesty and the broadly-held, early nineteenth century view that subjects receiving a cowpox-based vaccine would sprout horns and ruminate in the field. Jesty's contemporaries can be forgiven their trespasses because they fundamentally lacked an understanding of the disease, not to mention the concept of vaccination. Indeed, most if not all of Jesty's contemporaries, even the most highly trained scientists of his day, would likely have raised an eyebrow—if not actively tried to dissuade him—from the bold actions that led him to protect his family from smallpox using his wife's dirty knitting needles.

Modern-day vaccine deniers no longer have the convenience of claiming ignorance. The facts exist, and it is worth repeating a sentiment expressed by the late Senator Daniel Patrick Moynihan: "Everyone is entitled to his own opinion, but not his own facts." The facts show that vaccines save the lives of countless millions worldwide each year. Denial of this fact exceeds the boundaries of common sense and would entail malpractice if a decision not to vaccinate were made by a physician. However, it is not. Such decisions to avoid vaccines are made by parents—well-intentioned, perhaps, but exceedingly selfish.

In justifying this venal act, many parents cite the mantra of "herd immunity." This statement reflects the idea that if more than a certain fraction of a population is immunized, then the selfish few will be protected as well. The facts are that the fraction of the population needed to invoke herd immunity varies from pathogen to pathogen and tends to be very high, often requiring effective protection of more than 99.9 percent of the population. This goal is quite daunting, given additional challenges faced by parents, who mght lack access to health care or the resources to get their children immunized, not to mention the number of children who are not able to be immunized because they suffer from a preexisting malady such as cystic fibrosis or who are undergoing chemotherapy for childhood cancer. Furthermore, certain cultures, such as the Somalis in Minneapolis, have long-standing traditions of vaccine resistance. Adding vaccine deniers to the problem only increases the inevitably that the threshold required for herd immunity will not be achieved. Consequently, we are seeing increased incidence of preventable disease and death, spiking with tragedies such as an ongoing tragedy that is cutting through unvaccinated children in Minnesota.

Compounding the problem, an anti-vaccine movement has already done lasting damage in terms of diluting the efficacy of the pertussis, or whooping cough, vaccine. An overreaction to the side effects of the vaccine, first in Japan and then in the United Kingdom, compelled an efficacious product to be replaced by one we know to be inefficient, often requiring boosts every two years. This weakened form of the pertussis vaccine has therefore rendered millennials and future generations susceptible to the return of a dangerous old enemy. These same children (and a growing number of adults) are beginning to feel the consequences. At the same time, the rates of measles, mumps, and rubella are spiking on college campuses as the children of the anti-vaccinators matriculate into the nation's universities.

What is to be done?

My own fear for the future of vaccines is swamped by the hope that ignorance, rather than malign intentions, is behind the anti-vaccine movement. Ignorance can be overcome, but this requires diligence. In a population captured by the cult of celebrity, it is far too easy for high-profile vaccine deniers such as Jenny McCarthy and Charlie Sheen to convey their messages. Indeed, the highest-profile celebrity of them all tweeted on March 27, 2014, "If I were President I would push for proper vaccinations but would not allow one time massive shots that a small child cannot take—AUTISM." This individual is clearly ignorant of the fabricated link of vaccines to autism but currently resides in 1600 Pennsylvania Avenue, where he oversees the National Institutes for Health and Centers for Disease Control and Prevention, among other responsibilities.

As a scientist, I believe ignorance is best countered by facts. The facts need to be delivered consistently, accurately and by trusted individuals. With this in mind, public health officials should consider using every possible venue—from physicians to clergy, teachers, and even hairdressers—as a source of information. These populations could be trained in the facts behind immunization, the real benefits and potential risks (and yes, there is some risk, albeit exceedingly rare). In the same way that the much-needed #MeToo movement was initiated by a celebrity scandal but sustained by a groundswell of brave survivors coming forward and conveying their stories, a sustained movement to convey the dangers of a resurgence in measles, mumps, rubella, and other diseases is needed. From #MeToo we learned that sexually motivated abuse is everywhere in our society, affecting women and girls rich and poor, and of all ethnicities. Pathogens are just as indiscriminate, and it is only by understanding how vaccines are vital for the health of our communities that can we stop the resurgence of the deadly diseases that plagued past generations.

It is critical to remember key decisions about immunization are generally made by young adults, who tend not to have as much worldly experience and are arguably more susceptible to the cult of celebrity. Therefore, a counter-denial program should target venues such as MTV, social events, and, most importantly for today's youth, social media. Social media outlets should be included in a campaign to educate teens and twenty-somethings about the importance of vaccination BEFORE they become pregnant. In the same way that special vitamins are prescribed for prenatal care, information about vaccines needs to be conveyed to this same demographic.

Overcoming this ignorance should be a comparatively easy task, as the primary population from which the anti-vaccinators are generally drawn is not from the inner city or rural countryside but rather comprises educated and comparatively wealthy individuals. While composing this book, I have been asked in social occasions about what I am working on and have watched a small fraction of people recoil in horror at the prospect of promoting vaccines. However, it is comforting that a continuation of the conversation generally reveals that a typical anti-vaccinator is unaware, for example, that Andrew Wakefield's data was fraudulent and that the rest of the anti-vaccine movement is built almost entirely on a house of cards. In these conversations, I happily convey the search terms they can use to check this out for themselves. I frequently use the "white box truck" example that introduced this book to address the inevitable response that "so and so's child was diagnosed with autism after they were vaccinated" as conclusive proof.

Only by actively engaging, rather than excluding, anti-vaccinators can we hope to convey the actual facts and overcome the ignorance. Indeed, this was the motivation for the book you are reading. It was heartening that the inclusion of a high-profile anti-vaccinator in a televised 2018 New Year celebration triggered a wave of social media responses that related how this same person had not hesitated to have a deadly bacterial neurotoxin injected into her face. I am confident this is a reflection of a growing countermovement that will not only ensure the saving of many more lives but will also dispel the ignorance inspired fundamentally by the same fears that led Benjamin Jesty's neighbors to be convinced his family would soon sprout horns and run wild through the streets.

ENDNOTES

Introduction

1. M. V. Narayanan, K. K. Shimozaki, "Mumps Outbreak Grows to 5, Cases Suspected at Yale," *The Harvard Crimson*, http://www.thecrimson.com/article/2016/12/1/mumps-outbreak-fall-five-yale-cases/, 2016.
2. A. Hviid, S. Rubin, K. Mühlemann, "Mumps," *The Lancet* 371(9616) (2008) 932–944.
3. M. McKee, S. L. Greer, D. Stuckler, "What will Donald Trump's presidency mean for health? A scorecard," *The Lancet* 389(10070) (2017) 748–754.
4. L. Garratt, "Donald Trump and the anti-vaxxer conspiracy theorists," *Foreign Policy*, http://foreignpolicy.com/2017/01/11/donald-trump-and-the-anti-vaxxer-conspiracy-theorists/, January 11, 2017.
5. C. Davenport, "The White Box Truck That Wasn't," *Washington Post*, https://www.washingtonpost.com/archive/local/2002/11/18/the-white-box-truck-that-wasnt/bd5639ab-fc06-4312-99a2-12efe7ca6ddf/?utm_term=.000914c871f2, November, 18, 2002.
6. A. Cannon, *23 Days of Terror: The Compelling True Story of the Hunt and Capture of the Beltway Snipers*, (New York: Simon and Schuster, 2010).
7. A. M. Arvin, "Varicella-zoster virus," *Clinical Microbiology Reviews* 9(3) (1996) 361–381.

Endnotes

Chapter 1: Pox Romana

1. W. L. MacDonald, *The Architecture of the Roman Empire: An Introductory Study*, (New Haven: Yale University Press, 1982).
2. W. Scheidel, "Roman population size: the logic of the debate" in *People, Land Politics: Demographic Developments and the Transformation of Roman Italy, 300 B.C. to 14 A.D.* ed. L. De Ligt and S. Northwood (Leiden: Brill, 2007).
3. E. Gibbon, *The history of the decline and fall of the Roman Empire*, ed. J.J. Tourneisen (London: T. Cadell, Strand, 1789).
4. U. Wilcken, *Alexander the Great*, (New York: W.W. Norton & Company, 1967).
5. P. Wheatley, "The Diadochi, or Successors to Alexander" in *Alexander the Great: A New History* ed. Waldemar Heckel (New York: John Wiley & Sons, 2009) 53–68.
6. M. Scott, *From Democrats to Kings: The Downfall of Athens to the Epic Rise of Alexander the Great*, (London: Icon Books Ltd, 2010).
7. U. Wilcken, *Alexander the Great*, (New York: WW Norton & Company, 1967).
8. J. C. Moore, *The History of the Small Pox*, (London: Longman, 1815).
9. D. Oldach, R. Richard, E. Borza, R. Benitez, "A mysterious death," *The New England Journal of Medicine* 338(24) (1998) 1764–9.
10. B. A. Cunha, "The death of Alexander the Great: malaria or typhoid fever?," *Infectious Disease Clinics* 18(1) (2004) 53–63.
11. P. Wheatley, "The Diadochi, or Successors to Alexander" in *Alexander the Great: A New History* ed. Waldemar Heckel (New York: John Wiley & Sons, 2009) 53–68.
12. R. A. Gabriel, *Hannibal: The Military Biography of Rome's Greatest Enemy*, (Lincoln, NE: Potomac Books, Inc., 2011).
13. Ibid.
14. M. A. Speidel. *The Encyclopedia of Ancient History* (New York: John Wiley & Sons, 2012).
15. J. F. Gilliam, "The plague under Marcus Aurelius," *The American Journal of Philology* 82(3) (1961) 225–251.
16. K. Harper, "Pandemics and passages to late antiquity: rethinking the plague of c. 249–270 described by Cyprian," *Journal of Roman Archaeology* 28 (2015) 223–260.
17. R. J. Littman, M. L. Littman, "Galen and the Antonine plague," *The American Journal of Philology* 94(3) (1973) 243–255.
18. R. McLaughlin, *Rome and the Distant East: Trade Routes to the Ancient Lands of Arabia, India and China*, (London: Bloomsbury Publishing, 2010).
19. R. J. Littman, M. L. Littman, "Galen and the Antonine plague," *The American Journal of Philology* 94(3) (1973) 243–255.
20. B. G. Niebuhr, M. Isler, *Niebuhr's Lectures on Roman History*, (Los Angeles: HardPress Publishing, 2012).
21. M. Aurelius, C. Gill, *Meditations*, (Oxford: Oxford University Press, 2013).
22. E. Gibbon, *The History of the Decline and Fall of the Roman Empire*, ed. J.J. Tourneisen (London: T. Cadell, Strand, 1789).
23. Y. Li, D. S. Carroll, S. N. Gardner, M. C. Walsh, E. A. Vitalis, I. K. Damon, "On the origin of smallpox: correlating variola phylogenics with historical

smallpox records," *Proceedings of the National Academy of Sciences* 104(40) (2007) 15787–15792.

24. D. R. Hopkins, "Ramses V: earliest know victim?," *World Health* (May, 1980), 220.
25. R. P. Duncan-Jones, "The impact of the Antonine plague," *Journal of Roman Archaeology*, 9 (1996) 108–136.
26. F. P. Retief, L. Cilliers, "The epidemic of Athens, 430–426 B.C.," *South African Medical Journal* 88(1) (1998) 50–53.
27. R. J. Littman, M. Littman, "The Athenian plague: smallpox," *Transactions and Proceedings of the American Philological Association*, 100 (1969), pp. 261–275.
28. M. J. Papagrigorakis, C. Yapijakis, P. N. Synodinos, E. Baziotopoulou-Valavani, "DNA examination of ancient dental pulp incriminates typhoid fever as a probable cause of the Plague of Athens," *International Journal of Infectious Diseases* 10(3) (2006) 206–214.
29. G. Hardin, "The tragedy of the commons," *Science* 162(3859) (1968) 1243–1248.
30. M. Menotti-Raymond, S. J. O'Brien, "Dating the genetic bottleneck of the African cheetah," *Proceedings of the National Academy of Sciences* 90(8) (1993) 3172–3176.
31. D. M. Hopkins, *The Bering Land Bridge*, (Redwood City, CA: Stanford University Press, 1967).
32. M. W. Pedersen, A. Ruter, C. Schweger, H. Friebe, R. A. Staff, K. K. Kjeldsen, M. L. Mendoza, A. B. Beaudoin, C. Zutter, N. K. Larsen, "Postglacial viability and colonization in North America's ice-free corridor," *Nature* 537(7618) (2016) 45–49.
33. A. Curry, "Coming to America," *Nature* 485(7396) (2012) 30.
34. S. R. Holen, T. A. Deméré, D. C. Fisher, R. Fullagar, J. B. Paces, G. T. Jefferson, J. M. Beeton, R. A. Cerutti, A. N. Rountrey, L. Vescera, "A 130,000-year-old archaeological site in southern California, USA," *Nature* 544(7651) (2017) 479–483.
35. P. Skoglund, S. Mallick, M. C. Bortolini, N. Chennagiri, T. Hünemeier, M. L. Petzl-Erler, F. M. Salzano, N. Patterson, D. Reich, "Genetic evidence for two founding populations of the Americas," *Nature* 525(7567) (2015) 104–108.
36. C. C. Mann, *1491: New Revelations of the Americas Before Columbus*, (New York: Knopf, 2005).
37. J. D. Daniels, "The Indian Population of North America in 1492," *The William and Mary Quarterly* 49(2) (1992) 298–320.
38. S. F. Cook, W. W. Borah, Essays in Population History: Mexico and the Caribbean, (Berkeley, CA: University of California Press, 1971).
39. T. E. Emerson, R. B. Lewis, *Cahokia and the Hinterlands: Middle Mississippian Cultures of the Midwest*, (Champaign, IL: University of Illinois Press, 1999).
40. A. W. Crosby, *The Columbian Exchange: Biological and Cultural Consequences of 1492*, (Westport, CT: Greenwood Publishing Group, 2003).
41. F. Fenner, D. Henderson, I. Arita, Z. Jezek, I. Ladnyi, *Smallpox and its Eradication* (Geneva: World Health Organization, 1988).
42. A. W. Crosby, *The Columbian Exchange: Biological and Cultural Consequences of 1492*, (Westport, CT: Greenwood Publishing Group, 2003).
43. J. Needham, "China and the Origins of Immunology" (Lecture, Centre of Asian Studies, University of Hong Kong, Hong Kong, November 9, 1980).

44. J. B. Tucker, *Scourge: The Once and Future Threat of Smallpox* (New York: Grove Press, 2002).
45. L. M. W. Montagu, *The Complete Letters of Lady Mary Wortley Montagu: 1708–1720*, ed. Robert Halsband (Oxford, UK: Clarendon Press, 1965).
46. S. L. Plotkin, S. A. Plotkin, "A short history of vaccination," *Vaccines* 5 (2004) 1–16.
47. G. Miller, "Putting Lady Mary in her place: a discussion of historical causation," *Bulletin of the History of Medicine* 55(1) (1981) 2.
48. A. M. Behbehani, "The smallpox story: life and death of an old disease," *Microbiological Reviews*, 47(4) (1983) 455.
49. K. Silverman, *The Life and Times of Cotton Mather* (New York: Harper Collins, 1984).
50. T. H. Brown, "The African connection: Cotton Mather and the Boston smallpox epidemic of 1721–1722," *Journal of the American Medical Association*, 260(15) (1988) 2247–2249.
51. E. W. Herbert, "Smallpox inoculation in Africa," *The Journal of African History* 16(04) (1975) 539–559.
52. N. Sublette, C. Sublette, *American Slave Coast: A History of the Slave-Breeding Industry* (Chicago: Chicago Review Press, 2015).
53. E. W. Herbert, "Smallpox inoculation in Africa," *The Journal of African History* 16(04) (1975) 539–559.
54. A. Boylston, "The origins of inoculation," *Journal of the Royal Society of Medicine* 105(7) (2012) 309–313.
55. G. L. Kittredge, *Some Lost Works of Cotton Mather* (Cambridge, MA: John Wilson and Son, 1912).
56. L. H. Toledo-Pereyra, Zabdiel Boylston. "First American surgeon of the English colonies in North America," *Journal of Investigative Surgery*, 19(1) (2006) 5–10.
57. T. H. Brown, "The African connection: Cotton Mather and the Boston smallpox epidemic of 1721–1722," *Journal of the American Medical Association* 260(15) (1988) 2247–2249.
58. G. L. Kittredge, *Some Lost Works of Cotton Mather* (Cambridge, MA: John Wilson and Son, 1912).
59. T. H. Brown, "The African connection: Cotton Mather and the Boston smallpox epidemic of 1721–1722," *Journal of the American Medical Association* 260(15) (1988) 2247–2249.
60. Ibid.

Chapter 2: Vaccination & Eradication

1. B. Schaefer, "Campaign 2016's Theme: A Pox On Both Your Houses." *The Blaze*, March 7, 2016, Web. February 17, 2018.
2. E. Jenner, "Observations on the natural history of the cuckoo. By Mr. Edward Jenner. In a letter to John Hunter, Esq. FRS," *Philosophical Transactions of the Royal Society of London* 78 (1788) 219–237.
3. P. M. Dunn, "Dr Edward Jenner (1749–1823) of Berkeley, and vaccination against smallpox," *Archives of Disease in Childhood*. 74(1) (1996) F77–F78.
4. S. Riedel, "Edward Jenner and the history of smallpox and vaccination," *Proceedings of the Baylor University Medical Center*, 18(1) (2005) 21–25.

5. E. Jenner, *An Inquiry into the Causes and Effects of the Variolae Vaccinae, a Disease Discovered in Some of the Western Counties of England, Particularly Gloucestershire, and Known by the Name of the Cow Pox* (London: D.M. Shury, 1801).
6. "America Invents Act of 2011," Pub. L. No. 112–29, 125 Stat. 284 *through* 125 Stat. 341 (2011).
7. R. Jesty, G. Williams, "Who invented vaccination?" *Malta Medical Journal* 23(02) (2011) 29.
8. Thucydides, *History of the Peloponnesian War*, ed. M.I. Finley (London: Penguin Classics, 1972).
9. D. Van Zwanenberg, "The Suttons and the business of inoculation," *Medical History* 22(01) (1978) 71–82.
10. D. R. Flower, *Bioinformatics for Vaccinology* (London: John Wiley & Sons, 2008).
11. G. Peachey, "John Fewster, an unpublished chapter in the history of vaccination," *Annals in the History of Medicine,* 1 (1929) 229–40.
12. B. Knollenberg, "General Amherst and germ warfare," *The Mississippi Valley Historical Review* 41(3) (1954) 489–494.
13. R. Jesty, G. Williams, "Who invented vaccination?" *Malta Medical Journal* 23(02) (2011) 29.
14. L. Thurston, G. Williams, "An examination of John Fewster's role in the discovery of smallpox vaccination," *The Journal of the Royal College of Physicians of Edinburgh*, 45(2) (2014), 173–9.
15. D. A. Kronick, "Medical publishing societies in eighteenth-century Britain," *Bulletin of the Medical Library Association* 82(3) (1994) 277.
16. J. Hammarsten, W. Tattersall, J. Hammarsten, "Who discovered smallpox vaccination? Edward Jenner or Benjamin Jesty?" *Transactions of the American Clinical and Climatological Association* 90 (1979) 44.
17. Wikipedia, Pilt Carin Ersdotter. https://en.wikipedia.org/wiki/Pilt_Carin_Ersdotter, 2017).
18. J. T. Davies, L. Janes, A. Downie, "Cowpox infection in farmworkers," *The Lancet* 232(6018) (1938) 1534–1538.
19. R. Jesty, G. Williams, "Who invented vaccination?" *Malta Medical Journal* 23(02) (2011) 29.
20. J. T. Davies, L. Janes, A. Downie, "Cowpox infection in farmworkers," *The Lancet* 232(6018) (1938) 1534–1538.
21. R. Jesty, G. Williams, "Who invented vaccination?" *Malta Medical Journal* 23(02) (2011) 29.
22. J. F. Hammarsten, W. Tattersall, J. E. Hammarsten, "Who discovered smallpox vaccination? Edward Jenner or Benjamin Jesty?" *Transactions of the American Clinical and Climatological Association* 90 (1979) 44.
23. P. J. Pead, "Benjamin Jesty: new light in the dawn of vaccination," *The Lancet* 362(9401) (2003) 2104–2109.
24. Ibid.
25. Ibid.
26. P. Hyland, *Purbeck: The Ingrained Island,* (Dove Cote, UK: Dovecote Press, 1978).
27. R. Jesty, G. Williams, "Who invented vaccination?" *Malta Medical Journal* 23(02) (2011) 29.

28. P. J. Pead, "Benjamin Jesty: new light in the dawn of vaccination," *The Lancet* 362(9401) (2003) 2104–2109.
29. P. C. Plett, "Peter Plett and other discoverers of cowpox vaccination before Edward Jenner," *Sudhoffs Archive*, 90(2) (2006), 219–32.
30. L. Thurston, G. Williams, "An examination of John Fewster's role in the discovery of smallpox vaccination," *The Journal of the Royal College of Physicians of Edinburgh*, 45(2) (2014), 173–9.
31. G. Williams, *Angel of Death: The Story of Smallpox* (New York: Springer, 2010).
32. E. Jenner, *An Inquiry into the Causes and Effects of the Variolae Vaccinae, a Disease Discovered in Some of the Western Counties of England, Particularly Gloucestershire, and Known by the Name of the Cow Pox* (London: D.M. Shury, 1801).
33. P. J. Pead, "Benjamin Jesty: new light in the dawn of vaccination," *The Lancet* 362(9401) (2003) 2104–2109.
34. Ibid.
35. R. Southey, C. C. Southey, *The Life of the Rev. Andrew Bell. Prebendary of Westminster, and Master of Sherburn Hospital, Durham. Comprising the History of the Rise and Progress of the System of Mutual Tuition*, (London: J. Murray, 1844).
36. J. F. Hammarsten, W. Tattersall, J. E. Hammarsten, "Who discovered smallpox vaccination? Edward Jenner or Benjamin Jesty?" *Transactions of the American Clinical and Climatological Association* 90 (1979) 44.
37. P. J. Pead, "Benjamin Jesty: new light in the dawn of vaccination," *The Lancet* 362(9401) (2003) 2104–2109.
38. J. E. McCallum, *Military Medicine: From Ancient Times to the 21st Century*, (Santa Barbara, CA: ABC–CLIO, 2008).
39. A. Aly, "Smallpox," *New England Journal of Medicine* 335(12) (1996) 900–902.
40. J. E. McCallum, *Military Medicine: From Ancient Times to the 21st Century*, (Santa Barbara, CA: ABC–CLIO, 2008).
41. J. A. Nixon, "British Prisoners Released by Napoleon at Jenner's Request," *Proceedings of the Royal Society of Medicine*, 32(8) (1939) 877–83.
42. "Jenner & Napoleon," *Nature* 144 (1939) 278.
43. J. Baron, *The Life of Edward Jenner: With Illustrations of His Doctrines, and Selections from His Correspondence*, (London: Henry Colburn, 1838).
44. D. R. Hopkins, *The Greatest Killer: Smallpox in History* (Chicago: University of Chicago Press, 2002).
45. J. D. Rolleston, "The Smallpox Pandemic of 1870–1874," *Proceedings of the Royal Society of Medicine* 27(2) (1933) 177–92.
46. T. Jefferson, To Dr. Edward Jenner, Monticello, May 14, 1806.
47. M. P. Ravenel. "How the President, Thomas Jefferson, and Doctor Benjamin Waterhouse established vaccination as a public health procedure," *American Journal of Public Health*, 27(11) (1936) 1183–4.
48. J. E. McCallum, *Military Medicine: From Ancient Times to the 21st Century*, (Santa Barbara, CA: ABC–CLIO, 2008).
49. J. Voss, N. E. Gratton "American Academy of Arts and Sciences." In *Encyclopedia of Education* ed. J. W. Guthrie (New York: Macmillan Reference, 2003).
50. B. Waterhouse, "Variolae Vaccinae," *The Columbian Sentinel*, Boston, March 12, 1799.

51. H. Bloch, "Benjamin Waterhouse (1754–1846): the nation's first vaccinator," *American Journal of Diseases of Children* 127(2) (1974) 226–229.
52. B. S. Leavell, "Thomas Jefferson and smallpox vaccination," *Transactions of the American Clinical and Climatological Association* 88 (1977) 119.
53. J. B. Blake, "Benjamin Waterhouse and the introduction of vaccination," *Reviews of Infectious Diseases* 9(5) (1987) 1044–1052.
54. M. P. Ravenel. "How the President, Thomas Jefferson, and Doctor Benjamin Waterhouse established vaccination as a public health procedure," *American Journal of Public Health*, 27(11) (1936) 1183–4.
55. E. A. Underwood, "Edward Jenner, Benjamin Waterhouse, and the Introduction of Vaccination into the United States," *Nature* 163 (1949) 823–828.
56. B. S. Leavell, "Thomas Jefferson and smallpox vaccination," *Transactions of the American Clinical and Climatological Association* 88 (1977) 119.
57. Ibid.
58. M. P. Ravenel. "How the President, Thomas Jefferson, and Doctor Benjamin Waterhouse established vaccination as a public health procedure," *American Journal of Public Health*, 27(11) (1936) 1183–4.
59. B. S. Leavell, "Thomas Jefferson and smallpox vaccination," *Transactions of the American Clinical and Climatological Association* 88 (1977) 119.
60. Monticello.org, "Invisible Heroes: Battling Smallpox." (2017).
61. K. B. Patterson, T. Runge, "Smallpox and the Native American," *American Journal of the Medical Sciences* 323(4) (2002) 216–222.
62. Anonymous, "Emery Called Incompetent; Brooklyn Health Commmissioner Accused by Dr. Barney," *New York Times*, September 14, 1894. Page 9. Print.
63. J. K. Colgrove, "Between persuasion and compulsion: smallpox control in Brooklyn and New York, 1894–1902," *Bulletin of the History of Medicine* 78(2) (2004) 349–378.
64. F. Fenner, D. Henderson, I. Arita, Z. Jezek, I. Ladnyi, *Smallpox and its Eradication* (Geneva: World Health Organization, 1988).
65. J. Gillray, The Cow-Pock-or-the Wonderful Effects of the New Inoculation! (1802) Etching. British Museum, London.
66. M. R. Leverson, "Biographical Memoir of Mr. Wm. Tebb," *The Homoeopathic Physician: A Monthly Journal of Medical Science* 19 (1899) 407–417.
67. W. Tebb, *Compulsory Vaccination in England*, (London: E.W. Allen, 1884).
68. D.-L. Ross, "Leicester and the anti-vaccination movement, 1853–1889," *Transactions of the Leicester Archaeological and Historical Society* 43 (1967) 35–44.
69. A. Allen, *Vaccine: The Controversial Story of Medicine's Greatest Lifesaver*, (New York: WW Norton & Company, 2007).
70. E. B. Glenn, *Bryn Athyn Cathedral: The Building of a Church*, (Bryn Athyn, PA: Bryn Athyn Church of New Jerusalem, 1971).
71. G. Williams, *Angel of Death: The Story of Smallpox* (New York: Springer, 2010).
72. R. D. Johnston, *The radical middle class: Populist democracy and the question of capitalism in progressive era Portland, Oregon*, (Princeton, NJ: Princeton University Press, 2003).
73. American Medical Association Bureau of Investigation, "The Propaganda for Reform," *Journal of the American Medical Association* 79(1) (1922) 395–398.

74. U. Sinclair, *The Jungle* (New York: Doubleday, Jabber & Company, 1906).
75. American Medical Association Bureau of Investigation, "The Propaganda for Reform," *Journal of the American Medical Association* 79(1) (1922) 395–398.
76. A. P. Greeley, *The Food and Drugs Act, June 30, 1906: A Study with Text of the Act, Annotated, the Rules and Regulations for the Enforcement of the Act, Food Inspection, Decisions and Official Food Standards*, (New York: J. Byrne, 1907).
77. American Medical Association Bureau of Investigation, "The Propaganda for Reform," *Journal of the American Medical Association* 79(1) (1922) 395–398.
78. D. R. Hopkins, *The Greatest Killer: Smallpox in History* (Chicago: University of Chicago Press, 2002).
79. J. Colgrove, R. Bayer, "Manifold restraints: liberty, public health, and the legacy of *Jacobson v Massachusetts*," *American Journal of Public Health* 95(4) (2005) 571–576.
80. I. Weinstein, "An outbreak of smallpox in New York City," *American Journal of Public Health and the Nations Health* 37(11) (1947) 1376–1384.
81. J. Oppenheimer, "The Panic of 1947." *The Daily Beast,* September 19, 2009. Web. February 4, 2018.
82. C. Franco-Paredes, L. Lammoglia, J. I. Santos-Preciado, "The Spanish royal philanthropic expedition to bring smallpox vaccination to the New World and Asia in the 19th century," *Clinical Infectious Diseases* 41(9) (2005) 1285–1289.
83. J. J. Esposito, S. A. Sammons, A. M. Frace, J. D. Osborne, M. Olsen-Rasmussen, M. Zhang, D. Govil, I. K. Damon, R. Kline, M. Laker, Y. Li, G. L. Smith, H. Meyer, J. W. Leduc, R. M. Wohlhueter, "Genome sequence diversity and clues to the evolution of variola (smallpox) virus," *Science* 313(5788) (2006) 807–12.
84. J. Rhodes, *The End of Plagues: The Global Battle Against Infectious Disease* (London: Macmillan,2013).
85. D. Henderson, *Smallpox: The Death of a Disease. The Inside Story of Eradicating a Worldwide Killer* (Amherst, NY: Prometheus Books, 2009).
86. Editorial Board: "D. A. Henderson and the triumph of science," *St Louis Post-Dispatch*, St Louis, MO, August 23, 2016. Web. February 14, 2018.
87. Obituaries, "Donald Henderson, epidemiologist who helped to eradicate smallpox," *The Telegraph*, London, August 21, 2016. Web. February 14, 2018.
88. F. Fenner, D. Henderson, I. Arita, Z. Jezek, I. Ladnyi, *Smallpox and its Eradication* (Geneva: World Health Organization, 1988).
89. Ibid.
90. Ibid.
91. A. C. Madrigal, "The Last Smallpox Patient on Earth," *The Atlantic*, New York, December 9, 2013. Web. February 14, 2018.
92. J. Donnelly, "Polio: A Fight in a Lawless Land," *Boston Globe*, Boston, February 27, 2006. Web. February 14, 2018.
93. "The End of Smallpox," *Rx for Survival, Public Broadcasting Service.* WGBH, Boston. http://www.pbs.org/wgbh/rxforsurvival/series/diseases/smallpox.html (2005).
94. A. C. Madrigal, "The Last Smallpox Patient on Earth," *The Atlantic*, New York, December 9, 2013. Web. February 14, 2018.
95. Ibid.
96. S. Kotar, J. Gessler, *Smallpox: A History* (Jefferson, NC: McFarland, 2013).

97. M. Lockley, "The smallpox death that locked down Birmingham could have been avoided," *Birmingham Mail*, Birmingham, England, May 15, 2016. Web. February 14, 2018.
98. R. Shooter, *Report of the Investigation into the Cause of the 1978 Birmingham Smallpox Occurrence* (London: Her Majesty's Stationery Office, 1980).
99. T. H. Flewett, "The clinical and laboratory diagnosis of variola minor (alastrim)," *The British Journal of Clinical Practice* 24(9) (1970) 397–402.
100. A. Geedes, "Alasdair Geddes—Emeritus Professor of Infection in the School of Medicine, University of Birmingham, UK." Interview by Pam Das, *The Lancet*. 4(1) (2004) 54–7.
101. R. Shooter, *Report of the Investigation into the Cause of the 1978 Birmingham Smallpox Occurrence* (London: Her Majesty's Stationery Office, 1980).
102. Ibid.
103. M. Lockley, "The smallpox death that locked down Birmingham could have been avoided," *Birmingham Mail*, Birmingham, England, May 15, 2016. Web. February 14, 2018.
104. R. Shooter, *Report of the Investigation into the Cause of the 1978 Birmingham Smallpox Occurrence* (London: Her Majesty's Stationery Office, 1980).
105. Ibid.
106. B. W. Mahy, J. W. Almond, K. I. Berns, R. M. Chanock, D. K. Lvov, R. K. Pettersson, H. G. Schatzmeyer, F. Fenner, "The remaining stocks of smallpox virus should be destroyed," *Science* 262(5137) (1993) 1223–1225.

Chapter 3: Becoming Defensive

1. J. D. Haller, "Guy de Chauliac and his Chirurgia Magna," *Surgery* 55 (1964) 337.
2. G. de Chauliac, *Inventarium sive Chirurgia Magna* (Leiden: E. J. Brill, 1997).
3. D. A. Watters, "Guy de Chauliac: pre-eminent surgeon of the Middle Ages," *ANZ Journal of Surgery* 83(10) (2013) 730–734.
4. Ibid.
5. P. Prioreschi, *A History of Medicine* (Lewiston NY: Edwin Mellen, 2003).
6. J. R. Strayer, *The Reign of Philip the Fair* (Princeton, NJ: Princeton University Press, 1980.
7. J. Burnes, *Sketch of the History of Knights Templars* (London: Blackwood, 1840).
8. R. L. Poole, *Wycliffe and Movements for Reform*, (London: Longmans, Green, and Company, 1889).
9. W. J. Reardon, *The Deaths of the Popes: Comprehensive Accounts, Including Funerals, Burial Places and Epitaphs*, (Jefferson, NC: McFarland, 2004).
10. J. D. Haller, "Guy de Chauliac and his Chirurgia Magna," *Surgery* 55 (1964) 337.
11. J. Enselme, "Commentaries on the great plague of 1348 in Avignon," *La Revue Lyonnaise de Medecine* 17(18) (1969) 697–710.
12. D. A. Watters, "Guy de Chauliac: pre-eminent surgeon of the Middle Ages," *ANZ Journal of Surgery* 83(10) (2013) 730–734.
13. G. de Chauliac, *Inventarium sive Chirurgia Magna* (Leiden: E. J. Brill, 1997).
14. F. Adams, *The Genuine Works of Hippocrates*, (London: Sydenham Society, 1849).
15. S. A. Eming, T. Krieg, J. M. Davidson, "Inflammation in wound repair: molecular and cellular mechanisms," *Journal of Investigative Dermatology* 127(3) (2007) 514–525.

16. M. Lindemann, *Medicine and Society in Early Modern Europe* (Cambridge, UK: Cambridge University Press, 2010).
17. S. B. Nuland, *Doctors: The Biography of Medicine* (New York: Vintage, 1995).
18. K. S. Makarova, Y. I. Wolf, E. V. Koonin, "Comparative genomics of defense systems in archaea and bacteria," *Nucleic Acids Research* 41(8) (2013) 4360–4377.
19. R. D. Magnuson, "Hypothetical functions of toxin-antitoxin systems," *Journal of Bacteriology* 189(17) (2007) 6089–6092.
20. W. C. Summers, "Bacteriophage research: Early history," in *Bacteriophages: Biology and applications* ed. E. Kutter, A. Sulakvelidze (Boca Raton, FL: CRC Press, 2005), 5–27.
21. M. S. Kinch, *A Prescription For Change: The Looming Crisis in Drug Discovery*, (Chapel Hill, NC: UNC Press, 2016).
22. K. S. Makarova, Y. I. Wolf, E. V. Koonin, "Comparative genomics of defense systems in archaea and bacteria," *Nucleic Acids Research* 41(8) (2013) 4360–4377.
23. F. A. Ran, P. D. Hsu, J. Wright, V. Agarwala, D. A. Scott, F. Zhang, "Genome engineering using the CRISPR-Cas9 system," *Nature Protocols* 8(11) (2013) 2281–2308.
24. A. V. Wright, J. K. Nuñez, J. A. Doudna, "Biology and applications of CRISPR systems: Harnessing nature's toolbox for genome engineering." *Cell.* 164(1–2) (2016): 29–44.
25. L. Margulis, *Symbiosis in Cell Evolution: Life and its Environment on the Early Earth,* (New York: W.H. Freeman and Co., 1981).
26. T. C. Bosch, R. Augustin, F. Anton-Erxleben, S. Fraune, G. Hemmrich, H. Zill, P. Rosenstiel, G. Jacobs, S. Schreiber, M. Leippe, "Uncovering the evolutionary history of innate immunity: the simple metazoan Hydra uses epithelial cells for host defence," *Developmental & Comparative Immunology* 33(4) (2009) 559–569.
27. T. C. Bosch, "Cnidarian-microbe interactions and the origin of innate immunity in metazoans," *Annual Review of Microbiology* 67 (2013) 499–518.
28. A. Isaacs, J. Lindenmann, "Virus interference. I. The interferon," *Proceedings of the Royal Society of London*. 147(927) (1957) 258–67.
29. A. Isaacs, J. Lindenmann, R. C. Valentine, "Virus interference. II. Some properties of interferon," *Proceedings of the Royal Society of London*, 147(927) (1957) 268–273.
30. T. Taniguchi, "Aimez-vous Brahms? A story capriccioso from the discovery of a cytokine family and its regulators," *Nature Immunology* 10(5) (2009) 447.
31. E. De Maeyer, J. F. Enders, "An Interferon Appearing in Cell Cultures Infected with Measles Virus," *Proceedings of the Society for Experimental Biology and Medicine* 107(3) (1961) 573–578.
32. E. Baron, S. Narula, "From cloning to a commercial realization: Human alpha interferon," *Critical Reviews in Biotechnology*, 10 (1990), 179–90.
33. K. Sikora, "Does interferon cure cancer?" *British Medical Journal* 281(6244) (1980) 855.
34. M. F. Flajnik, M. Kasahara, "Origin and evolution of the adaptive immune system: genetic events and selective pressures," *Nature Reviews*. 11(1) (2010) 47–59.
35. W. F. Bynum, *Science and the Practice of Medicine in the Nineteenth Century*, (Cambridge, UK: Cambridge University Press, 1994).
36. M. M. Shoja, R. S. Tubbs, M. Loukas, G. Shokouhi, M. R. Ardalan, "Marie-François Xavier Bichat (1771–1802) and his contributions to the foundations of

pathological anatomy and modern medicine," *Annals of Anatomy-Anatomischer Anzeiger* 190(5) (2008) 413–420.
37. Ibid.
38. G. A. Lindeboom, "François Joseph Victor Broussais; 1772–1838," *Nederlands Tijdschrift Voor Geneeskunde*, 99(13) (1955), 955–63.
39. J. F. Lobstein, *A Treatise on the Structure, Functions and Diseases of the Human Sympathetic Nerve* (Philadelphia, PA: JG Auner, 1831).
40. I. S. Whitaker, J. Rao, D. Izadi, P. Butler, "Historical Article: Hirudo medicinalis: ancient origins of, and trends in the use of medicinal leeches throughout history," *British Journal of Oral and Maxillofacial Surgery* 42(2) (2004) 133–137.
41. S. I. Hajdu, "The discovery of blood cells," *Annals of Clinical & Laboratory Science* 33(2) (2003) 237–238.
42. G. Andral, *Précis d'Anatomie Pathologique*, (Brussels: Societe Typographique Belge, 1837).
43. L. Doyle, "Gabriel Andral (1797–1876) and the first reports of lymphangitis carcinomatosa," *Journal of the Royal Society of Medicine* 82(8) (1989) 491.
44. A. Kay, "The early history of the eosinophil," *Clinical & Experimental Allergy* 45(3) (2015) 575–582.
45. J. J. Beer, *The Emergence of the German Dye Industry* (Urbana, IL: University of Illinois Press, 1959).
46. C. Weigert, "Über die pathologischen Gerinnungsvorgänge," *Archiv für pathologische Anatomie und Physiologie und für klinische Medicin* 79(1) (1880) 87–123.
47. P. Valent P, B. Groner, U. Schumacher, G. Superti-Furga, M. Busslinger, R. Kralovics, C. Zielinski, J. M. Penninger, D. Kerjaschki, G. Stingl, J. S. Smolen, R. Valenta, H. Lassmann, H. Kovar, U. Jäger, G. Kornek, M. Müller, F. Sörgel. "Paul Ehrlich (1854–1915) and his contributions to the foundation and birth of translational medicine." *Journal of Innate Immunity*, 8 (2016), 111–20.
48. F. H. Garrison, "Edwin Klebs (1834-1913)," *Science* 38(991) (1913) 920–921.
49. G. A. Silver, "Virchow, the heroic model in medicine: health policy by accolade," *American Journal of Public Health* 77(1) (1987) 82–88.
50. H. Schramm-Macdonald, *Ein Pereat den Duellen!: Zugleich ein Beitrag zur Geschichte des Duells*, (Leipzig, Denicke, 1869).
51. R. Austrian, "The Gram stain and the etiology of lobal pneumonia, an historical note," *Bacteriological Reviews* 24(3) (1960) 261–265.
52. A. Kay, "The early history of the eosinophil," *Clinical & Experimental Allergy* 45(3) (2015) 575–582.
53. T. D. Brock, *Robert Koch: A Life in Medicine and Bacteriology* (Washington: National Society for Microbiology, 1999).
54. R. Koch, "Investigations into bacteria: V, The etiology of anthrax, based on the ontogenesis of Bacillus anthracis." *Cohns Beitrage zur Biologie der Pflanzen* 2(2) (1876) 277–310.
55. T. D. Brock, *Robert Koch: A Life in Medicine and Bacteriology* (Washington D.C.: American Society for Microbiology, 1999).
56. W. Hesse, D. Gröschel, "Walther and Angelina Hesse-early contributors to bacteriology," *American Society for Microbiology News* 58(8) (1992) 425–428.

57. R. Edwards, "Poison-tip umbrella assassination of Georgi Markov reinvestigated," *The Telegraph (London)*, June 19, 2008. Web. February 14, 2018.
58. V. Kostov, *The Bulgarian Umbrella, The Soviet direction and Operations of the Bulgarian secret service in Europe.* ed. B. Reynolds. (New York: Harvester, 1988).
59. J. R. Tisoncik, M. J. Korth, C. P. Simmons, J. Farrar, T. R. Martin, M. G. Katze, "Into the eye of the cytokine storm," *Microbiology and Molecular Biology Reviews* 76(1) (2012) 16–32.
60. P. J. Bjorkman, M. Saper, B. Samraoui, W. S. Bennett, J. L. Strominger, D. Wiley, "Structure of the human class I histocompatibility antigen, HLA-A2," *Nature* 329(6139) (1987) 506–512.
61. R. H. Schwartz, "T cell anergy," *Annual Review of Immunology* 21(1) (2003) 305–334.
62. E. M. Leroy, P. Rouquet, P. Formenty, S. Souquiere, A. Kilbourne, J.-M. Froment, M. Bermejo, S. Smit, W. Karesh, R. Swanepoel, "Multiple Ebola virus transmission events and rapid decline of central African wildlife," *Science* 303(5656) (2004) 387–390.
63. H. Fausther-Bovendo, S. Mulangu, N. J. Sullivan, "Ebolavirus vaccines for humans and apes," *Current Opinion in Virology* 2(3) (2012) 324–329.

Chapter 4: The Wurst Way to Die

1. M. T. Varro, *Delphi Complete Works of Varro.* ed. H. B. Ash. (East Sussex, UK: Delphi Classics, 2017).
2. G. Rosen, P. J. Imperato, *A History of Public Health* (Baltimore: Johns Hopkins University Press, 2015).
3. V. Nutton, "The reception of Fracastoro's Theory of contagion: the seed that fell among thorns?" *Osiris* 6 (1990) 196–234.
4. L. J. Snyder, *Eye of the Beholder: Johannes Vermeer, Antoni Van Leeuwenhoek, and the Reinvention of Seeing* (New York: WW Norton & Company, 2015).
5. H. Houtzager, "Reinier de Graaf and his contribution to reproductive biology," *European Journal of Obstetrics & Gynecology and Reproductive Biology* 90(2) (2000) 125–127.
6. L. J. Snyder, *Eye of the Beholder: Johannes Vermeer, Antoni Van Leeuwenhoek, and the Reinvention of Seeing* (New York: WW Norton & Company, 2015).
7. H. Gest,"The discovery of microorganisms by Robert Hooke and Antoni van Leeuwenhoek, Fellows of the Royal Society," *The Royal Society Journal of the History of Science,* 58(2) (2004), 187–201.
8. A. M. Bauer, "The Symbolae Physicae and the herpetology of Hemprich and Ehrenberg's expedition to Egypt and the Middle East," *International Society for the History and Bibliography of Herpetology* 2(1) (2000) 8–16.
9. W. Klausewitz, "Frankfurt versus Berlin: The Red Sea explorers Wilhelm Hemprich, Christian Ehrenberg and Eduard Rüppell," *Zoology in the Middle East* 27(1) (2002) 7–12.
10. J. G. Olson, "Epidemic Typhus: a Forgotten but Lingering Threat," in *Emerging Infections 3.* ed W. A. Craign, J. M. Hughes, (Washington, D.C.: American Society for Microbiology, 1999) 67–72.
11. D. Raoult, O. Dutour, L. Houhamdi, R. Jankauskas, P.-E. Fournier, Y. Ardagna, M. Drancourt, M. Signoli, V.D. La, Y. Macia, "Evidence for louse-transmitted

diseases in soldiers of Napoleon's Grand Army in Vilnius," *Journal of Infectious Diseases* 193(1) (2006) 112–120.

12. S. Talty, *The Illustrious Dead: The Terrifying Story of how Typhus Killed Napoleon's Greatest Army* (Portland, OR: Broadway Books, 2009).
13. R. F. Brenner, *Writing as Resistance: Four Women Confronting the Holocaust: Edith Stein, Simone Weil, Anne Frank, and Etty Hillesum* (University Park, PA: Penn State Press, 2010).
14. D. H. Stapleton, "A lost chapter in the early history of DDT: The development of anti-typhus technologies by the Rockefeller Foundation's louse laboratory, 1942–1944," *Technology and Culture* 46(3) (2005) 513–540.
15. A. M. Bauer, "The Symbolae Physicae and the herpetology of Hemprich and Ehrenberg's expedition to Egypt and the Middle East," *International Society for the History and Bibliography of Herpetology* 2(1) (2000) 8–16.
16. W. Klausewitz, "Frankfurt versus Berlin: The Red Sea explorers Wilhelm Hemprich, Christian Ehrenberg and Eduard Rüppell," *Zoology in the Middle East* 27(1) (2002) 7–12.
17. H. Klencke, G. Schlesier, J. Bauer, *Lives of the Brothers Humboldt, Alexander and William* (London: Ingram, Cooke & Co., 1853).
18. S. Rebok, *Humboldt and Jefferson: A Transatlantic Friendship of the Enlightenment* (Charlottesville: University of Virginia Press, 2014).
19. A. von Humboldt, W. MacGillivray, *The Travels and Researches of Alexander Von Humboldt: Being a Condensed Narrative of His Journeys in the Equinoctial Regions of America, and in Asiatic Russia: Together with Analyses of His More Important Investigations* (New York: Harper, 1833).
20. A. Wulf, *The Invention of Nature: Alexander von Humboldt's New World* (New York: Knopf, 2015).
21. W. A. Sarjeant, "Hundredth year memoriam Christian Gottfried Ehrenberg 1795–1876," *Palynology,* 2(1) (1978), 209–11.
22. P. Debré, *Louis Pasteur* (Baltimore: Johns Hopkins University Press, 2000).
23. W. de Blécourt, *The Werewolf, the Witch, and the Warlock: Aspects of Gender in the Early Modern Period, Witchcraft and Masculinities in Early Modern Europe*, (New York: Springer, 2009), 191–213.
24. P. Debré, *Louis Pasteur* (Baltimore: Johns Hopkins University Press, 2000).
25. H. Flack, Louis Pasteur's discovery of molecular chirality and spontaneous resolution in 1848, together with a complete review of his crystallographic and chemical work, *Acta Crystallographica,* 65(5) (2009) 371–389.
26. L. Carroll, *Through the Looking Glass: And what Alice found there*, (Chicago: Rand, McNally, 1917).
27. J. Blish, *Spock Must Die!* (New York: Bantam Books, 1972).
28. P. Debré, *Louis Pasteur* (Baltimore: Johns Hopkins University Press, 2000).
29. L. Fitzharris, *The Butchering Art* (New York: Scientific American, 2017).
30. K. A. Smith, "Louis Pasteur, the Father of Immunology?" *Frontiers in Immunology* 3 (2012) 68.
31. L. Pasteur, R. Chamberland, "Summary report of the experiments conducted at Pouilly-le-Fort, near Melun, on the anthrax vaccination, 1881," *The Yale Journal of Biology and Medicine* 75(1) (2002) 59.

32. M. Best, D. Neuhauser, "Ignaz Semmelweis and the birth of infection control," *Quality and Safety in Health Care* 13(3) (2004) 233–234.
33. K. C. Carter, B. R. Carter, *Childbed Fever: A Scientific Biography of Ignaz Semmelweis* (Santa Barbara, CA: ABC–CLIO, 1994).
34. S. B. Nuland, *Doctors: The Biography of Medicine* (New York: Vintage, 1995).
35. H. Wykticky, M. Skopec, "Ignaz Philipp Semmelweis, the prophet of bacteriology," *Infection Control* 4(05) (1983) 367-370.
36. Ibid.
37. S. B. Nuland, *The Doctors' Plague* (NYC: WW Norton & Co., 2003).
38. Ibid.
39. Ibid.
40. S. Tougher, *The Reign of Leo VI (886-912): Politics and People* (Leiden: Brill, 1997).
41. F. J. Erbguth, "The pretherapeutic history of botulinum neurotoxin," in *Manual of Botulinum Toxin Therapy* (Cambridge, UK: Cambridge University Press, 2014).
42. F. J. Erbguth, M. Naumann, "Historical aspects of botulinum toxin: Justinius Kerner (1786–1862) and the 'sausage poison.'" *Neurology,* 53(8) (1999), 8.
43. F. J. Erbguth, "The pretherapeutic history of botulinum neurotoxin," in *Manual of Botulinum Toxin Therapy* (Cambridge, UK: Cambridge University Press, 2014).
44. F. J. Erbguth, M. Naumann, "Historical aspects of botulinum toxin: Justinius Kerner (1786–1862) and the 'sausage poison.'" *Neurology,* 53(8) (1999), 8.
45. O. Grüsser, "Die ersten systematischen Beschreibungen und tierexperimentellen Untersuchungen des Botulismus: Zum 200. Geburtstag von Justinus Kerner am 18. Sept. 1986," *Sudhoffs Archiv* (1986) 167–187.
46. F. J. Erbguth, "The pretherapeutic history of botulinum neurotoxin," in *Manual of Botulinum Toxin Therapy* (Cambridge, UK: Cambridge University Press, 2014).
47. Ibid.
48. F. J. Erbguth, M. Naumann, "Historical aspects of botulinum toxin: Justinius Kerner (1786–1862) and the 'sausage poison.'" *Neurology,* 53(8) (1999), 8.
49. L. G. W. Christopher, L. T. J. Cieslak, J. A. Pavlin, E. M. Eitzen, "Biological warfare: a historical perspective," *Journal of the American Medical Association,* 278(5) (1997) 412–417.
50. E. Calic, *Reinhard Heydrich,* (New York: William Morrow & Company, 1985).
51. C. A. MacDonald, *The Killing of Obergruppenführer Reinhard Heydrich: 27 May 1942,* (London: Macmillan, 1989).
52. E. Calic, *Reinhard Heydrich,* (New York: William Morrow & Company, 1985).
53. C. A. MacDonald, *The Killing of Obergruppenführer Reinhard Heydrich: 27 May 1942,* (London: Macmillan, 1989).
54. J. Paxman, R. Harris, *A Higher Form of Killing,* (New York: Hill and Wang, 1982).
55. Centers for Disease Control and Prevention, "Diphtheria" in *Epidemiology and Prevention of Vaccine-Preventable Diseases, 15th Edition* (Atlanta, GA: Centers for Disease Control and Prevention, 2015).
56. J. Barry. *The Great Influenza: The Story of the Deadliest Pandemic in History* (London: Penguin, 2005).
57. E. Laval, "The strangling of children (diphtheria) in Spain (16th and 17th centuries)," *Revista Chilena de Infectología: Organo Oficial de la Sociedad Chilena de Infectologia* 23(1) (2006) 78.

58. P. Bretonneau, *Des Inflammations Spéciales du Tissu Muqueux, et en Particulier de la Diphthérite ou Inflammation Pelliculaire* (Paris: Chez Crevot, 1826.)
59. J. M. Packard, *Victoria's Daughters*, (London: Macmillan, 1998).
60. F. H. Garrison, "Edwin Klebs (1834-1913)," *Science* 38(991) (1913) 920-921.
61. P. Fildes, "Richard Friedrich Johannes Pfeiffer. 1858–1945," *Biographical Memoirs of Fellows of the Royal Society* 2 (1956) 237–247.
62. T. Proft, J. D. Fraser, "Bacterial superantigens," *Clinical and Experimental Immunology* 133(3) (2003) 299–306.
63. J. Jui, "Chapter 146: Septic Shock." In *Tintinalli's Emergency Medicine, 7th Edition* Ed. Tintinalli, Judith E.; Stapczynski, J. Stephan; Ma, O. John; Cline, David M.; et al. (New York: McGraw-Hill, 2011). 1003–14.
64. C. Hollabaugh, L. H. Burt, A. P. Walsh, "Carboxymethylcellulose. Uses and applications," *Industrial & Engineering Chemistry* 37(10) (1945) 943–947.
65. A. Fetters, "The Tampon: A History," *The Atlantic, June 1, 2015* (2015) Web. February, 14, 2018.
66. S. L. Vostral, "Rely and Toxic Shock Syndrome: a technological health crisis," *The Yale Journal of Biology and Medicine* 84(4) (2011) 447.
67. Ibid.
68. K. N. Shands, G. P. Schmid, B. B. Dan, D. Blum, R. J. Guidotti, N. T. Hargrett, R. L. Anderson, D. L. Hill, C. V. Broome, J. D. Band, "Toxic-shock syndrome in menstruating women: association with tampon use and Staphylococcus aureus and clinical features in 52 cases," *New England Journal of Medicine* 303(25) (1980) 1436–1442.
69. R. J. Dubos, *Mirage of Health: Utopias, Progress, and Biological change* (New Brunswick, NJ: Rutgers University Press, 1987).
70. C. Reed. "David Brower," *The Guardian (London)*, November 8, 2000. Web. February 15, 2018.
71. R. Buckminster Fuller. *Your Private Sky* (Zurich: Lars Müller Publishers, 1999).
72. W. Stephen, S. Leonard, M. Macdonald, K. Maclean, N. Gupta. *Think Global, Act Local* (Edinburgh, UK: Luath Press Limited, 2017).
73. O. T. Avery, R. Dubos, "The protective action of a specific enzyme against type III pneumococcus infection in mice," *Journal of Experimental Medicine* 54(1) (1931) 73–89.
74. T. Saey, "Body's bacteria don't outnumber human cells so much after all," *Science News* 189(3) (2016) 6.
75. V. D'Argenio, F. Salvatore, "The role of the gut microbiome in the healthy adult status," *Clinica Chimica Acta* 451 (2015) 97–102.
76. M. Blaser, *Missing Microbes: How the Overuse of Antibiotics Is Fueling Our Modern Plagues* (New York: Henry Holt and Co., 2014).
77. The Human Microbiome Project Consortium, "Structure, function and diversity of the healthy human microbiome," *Nature* 486(7402) (2012) 207–214.
78. H. Tilg, A. Kaser, "Gut microbiome, obesity, and metabolic dysfunction," *The Journal of Clinical Investigation* 121(6) (2011) 2126–2132.
79. P. J. Turnbaugh, F. Bäckhed, L. Fulton, J. I. Gordon, "Diet-induced obesity is linked to marked but reversible alterations in the mouse distal gut microbiome," *Cell Host & Microbe* 3(4) (2008) 213–223.

80. M. Kwa, C. S. Plottel, M. J. Blaser, S. Adams, "The Intestinal Microbiome and Estrogen Receptor-Positive Female Breast Cancer," *Journal of the National Cancer Institute* 108(8) (2016).
81. M. Blaser, *Missing Microbes: How the Overuse of Antibiotics Is Fueling Our Modern Plagues* (New York: Henry Holt and Co., 2014).
82. R. Higdon, R. K. Earl, L. Stanberry, C. M. Hudac, E. Montague, E. Stewart, I. Janko, J. Choiniere, W. Broomall, N. Kolker, "The promise of multi-omics and clinical data integration to identify and target personalized healthcare approaches in autism spectrum disorders," *Omics: a Journal of Integrative Biology* 19(4) (2015) 197–208.

Chapter 5: Spreading Like Viruses

1. P. J. Crutzen, The "Anthropocene," in *Earth System Science in the Anthropocene.* Ed. E. Ehlers, T. Krafft. (New York: Springer, 2006).
2. L. P. Villarreal, "Are viruses alive?," *Scientific American* 291 (2004) 100–105.
3. P. J. Livingstone Bell, "Viral eukaryogenesis: was the ancestor of the nucleus a complex DNA virus?" *Journal of Molecular Evolution* 53(3) (2001) 251–256.
4. M. C. Horzinek, "The birth of virology," *Antonie van Leeuwenhoek* 71(1) (1997) 15–20.
5. A. J. Levine, "The origins of virology," *Fields Virology* (Philadelphia, PA: Lippincott-Raven, 1996) 1–14.
6. L. Pauling, J. Sturdivant, "The structure of cyameluric acid, hydromelonic acid and related substances," *Proceedings of the National Academy of Sciences* 23(12) (1937) 615–620.
7. A. Mayer, "Ueber die Mosaikkrankheit des Tabaks," *Die Landwirtschaftlichen Versuchs-Stationen* 32 (1886) 451–467.
8. D. Ivanovsky, "Über die Mosaikkrankheit der Tabakspflanze," *Zentralblatt für Bakteriologie* 5 (1899).
9. A. J. Levine, "The origins of virology," *Fields Virology* (Philadelphia, PA: Lippincott-Raven, 1996) 1–14.
10. W. Stanley, E. G. Valens, *Viruses and the Nature of Life* (Minneapolis, MN: Dutton, 1961).
11. K. M. Wylie, G. M. Weinstock, G. A. Storch, "Emerging view of the human virome," *Translational Research* 160(4) (2012) 283–290.
12. P. Biagini, M. Bendinelli, S. Hino, L. Kakkola, A. Mankertz, C. Niel, H. Okamoto, S. Raidal, C. Teo, D. Todd, Anelloviridae, In *Virus Taxonomy: Classification and Nomenclature of Viruses: Ninth Report of the International Committee on Taxonomy of Viruses. 1st ed.* (San Diego: Elsevier, 2011) 326–341.
13. R. Hewlett, "Dr. E. H. Hankin," *Nature* 143 (1939) 711–712.
14. S. T. Abedon, C. Thomas-Abedon, A. Thomas, H. Mazure, "Bacteriophage prehistory: is or is not Hankin, 1896, a phage reference?" *Bacteriophage* 1(3) (2011) 174–178.
15. J. Venn, *Alumni Cantabrigienses: a Biographical List of All Known Students, Graduates and Holders of Office at the University of Cambridge, from the Earliest Times to 1900* (Cambridge, UK: Cambridge University Press, 2011).
16. Victoria Street Society. "Zoophilist, Notes and Notices," *The Zoophilist* 16(2) (1896) 18–19.
17. E. H. Hankin, "L'action bactericide des eaux de la Jumna et du Gange sur le vibrion du cholera," *Annals of the Institut Pasteur* 10(5) (1896) 2.

18. F. W. Twort, "An investigation on the nature of ultra-microscopic viruses," *The Lancet* 186(4814) (1915) 1241–1243.
19. W. C. Summers, *Felix dHerelle and the Origins of Molecular Biology* (New Haven, CT: Yale University Press, 1999).
20. W. C. Summers, "Bacteriophage research: Early history," in *Bacteriophages: Biology and applications* ed. E. Kutter, A. Sulakvelidze (Boca Raton, FL: CRC Press, 2005), 5–27.
21. R. Atenstaedt, *The Medical Response to the Trench Diseases in World War One* (Cambridge, UK: Cambridge Scholars Publishing, 2011).
22. J. Ellis, *Eye-Deep in Hell* (Oxford, UK: Taylor & Francis, 1976).
23. A. F. Trofa, H. Ueno-Olsen, R. Oiwa, M. Yoshikawa, "Dr. Kiyoshi Shiga: Discoverer of the dysentery bacillus," *Clinical Infectious Diseases* 29(5) (1999) 1303–1306.
24. F. D'Herelle, "On an invisible microbe antagonistic toward dysenteric bacilli: brief note by Mr. F. D'Herelle, presented by Mr. Roux. 1917," *Research in Microbiology* 158(7) (2007) 553–4.
25. D. E. Fruciano, S. Bourne, "Phage as an antimicrobial agent: d'Herelle's heretical theories and their role in the decline of phage prophylaxis in the West," *The Canadian Journal of Infectious Diseases & Medical Microbiology* 18(1) (2007) 19–26.
26. S. T. Abedon, C. Thomas-Abedon, A. Thomas, H. Mazure, "Bacteriophage prehistory: is or is not Hankin, 1896, a phage reference?" *Bacteriophage* 1(3) (2011) 174–178.
27. W. C. Summers, "Bacteriophage research: Early history," in *Bacteriophages: Biology and Applications* ed. E. Kutter, A. Sulakvelidze (Boca Raton, FL: CRC Press, 2005), 5–27.
28. S. Lewis, *Arrowsmith*, (San Diego, CA: Harcourt, Brace, 1945).
29. A. Sulakvelidze, Z. Alavidze, J. G. Morris, "Bacteriophage therapy," *Antimicrobial Agents and Chemotherapy* 45(3) (2001) 649–659.
30. A. Kuchment, *The Forgotten Cure: The Past and Future of Phage Therapy*, (Berlin: Springer Science & Business Media, 2011).
31. Ibid.
32. T. Van Helvoort, "History of virus research in the twentieth century: The problem of conceptual continuity," *History of Science* 32(2) (1994) 185–235.
33. S. T. Abedon, C. Thomas-Abedon, A. Thomas, H. Mazure, "Bacteriophage prehistory: is or is not Hankin, 1896, a phage reference?" *Bacteriophage* 1(3) (2011) 174–178.
34. S. T. Abedon, "The murky origin of Snow White and her T-even dwarfs," *Genetics* 155(2) (2000) 481–486.
35. S. E. Luria, M. Delbrück, T.F. Anderson, "Electron microscope studies of bacterial viruses," *Journal of Bacteriology* 46(1) (1943) 57.
36. D. Matthews, "The world's deadliest and most infectious diseases, in one chart," *Vox*, https://www.vox.com/xpress/2014/10/17/6993851/diseases-deadly-infectious-reproduction-information-beautiful, 2014. Web. February 15, 2018.
37. W. H. Price, "The isolation of a new virus associated with respiratory clinical disease in humans," *Proceedings of the National Academy of Sciences* 42(12) (1956) 892–896.
38. C. Curtis, *Restless Ambition*, (Oxford, UK: Oxford University Press, 2015).
39. Ibid.
40. Editor, "Triggers for Catching Cold," *Science News Letter* 67(2) (1955) 19.
41. C. Curtis, *Restless Ambition*, (Oxford, UK: Oxford University Press, 2015).

42. W. H. Price, "The isolation of a new virus associated with respiratory clinical disease in humans," *Proceedings of the National Academy of Sciences* 42(12) (1956) 892–896.
43. C. Curtis, *Restless Ambition*, (Oxford, UK: Oxford University Press, 2015).
44. D. T. Fleming, G. M. McQuillan, R. E. Johnson, A. J. Nahmias, S. O. Aral, F. K. Lee, M. E. "St. Louis Herpes Simplex Virus Type 2 in the United States, 1976 to 1994," *New England Journal of Medicine* 337(16) (1997) 1105–1111.
45. T. R. A. Thomas, D. P. Kavlekar, P. A. LokaBharathi, "Marine drugs from sponge-microbe association—A review," *Marine Drugs* 8(4) (2010) 1417–1468.
46. W. Bergmann, R. J. Feeney, "The isolation of a new thymine pentoside from sponges," *Journal of the American Chemical Society* 72(6) (1950) 2809–2810.
47. M. S. Kinch, *A Prescription For Change: The Looming Crisis in Drug Discovery*, (Chapel Hill, NC: University of North Carolina Press, 2016).
48. G. B. Elion. *The Nobel Prizes 1988* ed. T. Frangsmyr (Stockholm: Nobel Foundation, 1989).
49. G. B. Elion. "Gertrude B. Elion, M.Sc.," *Academy of Achievement*. Web. Retrieved February 15, 2018 from http://www.achievement.org/achiever/gertrude-elion/.
50. Ibid.
51. Ibid.
52. Ibid.
53. G. H. Hitchings. *The Nobel Prizes 1988* ed. T. Frangsmyr (Stockholm: Nobel Foundation, 1989).
54. F. Barre-Sinoussi, J. Chermann, F. Rey, M. Nugeyre, S. Chamaret, J. Gruest, C. Dauguet, "Isolation of T-lymphotropic retrovirus from a patient at risk for acquired immune defficiency syndrome (AIDS)," *Revista de Investigación Clínica* 56(2) (2004) 126–129.
55. G. Kolata, "FDA approves AZT," *Science* (New York, NY) 235(4796) (1987) 1570.
56. T. D. Meek, G. B. Dreyer, "HIV-1 protease as a potential target for anti-AIDS therapy," *Annals of the New York Academy of Sciences* 616 (1990) 41–53.
57. P. Handover, "The 'Wicked' Bible and the King's Printing House," *Times (London) House Journal* (1958) 215–218.
58. J. D. Roberts, K. Bebenek, T. A. Kunkel, "The accuracy of reverse transcriptase from HIV-1," *Science* 242(4882) (1988) 1171–3.
59. L. Zhang, B. Ramratnam, K. Tenner-Racz, Y. He, M. Vesanen, S. Lewin, A. Talal, P. Racz, A. S. Perelson, B. T. Korber, "Quantifying residual HIV-1 replication in patients receiving combination antiretroviral therapy," *New England Journal of Medicine* 340(21) (1999) 1605-1613.
60. C. Gorman, "Dr. David Ho: The Disease Detective," *Time magazine,* December 30, 1996. Web. February 15, 2018.
61. S. Schmitz, S. Scheding, D. Voliotis, H. Rasokat, V. Diehl, M. Schrappe, "Side effects of AZT prophylaxis after occupational exposure to HIV-infected blood," *Annals of Hematology* 69(3) (1994) 135–138.
62. G. F. Vanhove, J. M. Schapiro, M. A. Winters, T. C. Merigan, T. F. Blaschke, "Patient compliance and drug failure in protease inhibitor monotherapy," *Journal of the American Medical Association,* 276(24) (1996) 1955–1956.
63. J. D. Siliciano, J. Kajdas, D. Finzi, T. C. Quinn, K. Chadwick, J. B. Margolick, C. Kovacs, S. J. Gange, R. F. Siliciano, "Long-term follow-up studies confirm the

stability of the latent reservoir for HIV-1 in resting CD4+ T cells," *Nature Medicine* 9(6) (2003) 727–728.
64. Z. Abdellah, A. Ahmadi, S. Ahmed, M. Aimable, R. Ainscough, J. Almeida, "International human genome sequencing consortium," *Nature* 409 (2004) 860–921.

Chapter 6: A Sense of Humors

1. P. Ehrlich, *Beiträge für Theorie und Praxis der Histologischen Färbung*, Doctoral Dissertation. Leipzig University, 1878.
2. K. Strebhardt, A. Ullrich, "Paul Ehrlich's magic bullet concept: 100 years of progress," *Nature Reviews Cancer*, 8(6) (2008) 473–480.
3. L. Hood, D. W. Talmage, "Mechanism of antibody diversity: germ line basis for variability," *Science*, 168(3929) (1970) 325–334.
4. S. Tonegawa, "Somatic generation of antibody diversity," *Nature*, 302(5909) (1983) 575–581.
5. B. Alberts, J. Lewis *et al. Molecular Biology of the Cell, 4th Edition* (New York: Garland Science, 2002).
6. L. Hood, D. W. Talmage, "Mechanism of antibody diversity: germ line basis for variability," *Science* 168(3929) (1970) 325–334.
7. "Paul Ehrlich—Biographical" *Nobelprize.org*. Nobel Media AB 2014. Web. February 15, 2018.
8. "Emil von Behring—Biographical" *Nobelprize.org*. Nobel Media AB 2014. Web. February 15, 2018.
9. M. Jučas, J. Everatt, *The Battle of Grünwald* (ed. Albina Strunga), (Vilnius: Lithuanian National Museum, 2009).
10. E. Ludendorff, *My War Memories, 1914–1918* (London: Hutchinson & Co., 1919).
11. "Emil von Behring—Biographical" *Nobelprize.org*. Nobel Media AB 2014. Web. February 15, 2018.
12. D. S. Linton, "*Emil von Behring: Infectious Disease, Immunology, Serum Therapy*," (Philadelphia, PA: American Philosophical Society, 2005).
13. J. Toland, *Adolf Hitler: The Definitive Biography* (Norwell, MA: Anchor, 2014).
14. E. Binz, "Memoirs: Protoplasmic Movement and Quinine," *Journal of Cell Science* 2(96) (1884) 682–684.
15. "Emil von Behring—Biographical" *Nobelprize.org*. Nobel Media AB 2014. Web. February 15, 2018.
16. S. H. Kaufmann, "Immunology's foundation: the 100-year anniversary of the Nobel Prize to Paul Ehrlich and Elie Metchnikoff," *Nature Immunology* 9(7) (2008) 705–712.
17. E. von Behring, S. Kitasato, "The mechanism of immunity in animals to diphtheria and tetanus," *Deutsche Med. Wochenschr* 16 (1890) 1113–1114.
18. Ibid.
19. S. S. Kantha, "A Centennial Review; the 1890 Tetanus Antitoxin Paper of von Behring and Kitasato and the Related Developments," *The Keio Journal of Medicine* 40(1) (1991) 35–39.
20. T. N. K. Raju, "Emil Adolf von Behring and serum therapy for diphtheria," *Acta Paediatrica* 95(3) (2006) 258–259.

21. F. J. Grundbacher, "Behring's discovery of diphtheria and tetanus antitoxins," *Immunology Today* 13(5) (1992) 188–90.
22. S. S. Kantha, "A Centennial Review; the 1890 Tetanus Antitoxin Paper of von Behring and Kitasato and the Related Developments," *The Keio Journal of Medicine* 40(1) (1991) 35–39.
23. S. H. E. Kaufmann, "Remembering Emil von Behring: from Tetanus Treatment to Antibody Cooperation with Phagocytes," *American Society for Microbiology,* 8(1) (2017) 1–6.
24. "Emil von Behring—Biographical" *Nobelprize.org.* Nobel Media AB 2014. Web. February 15, 2018.
25. S. H. E. Kaufmann, "Remembering Emil von Behring: from Tetanus Treatment to Antibody Cooperation with Phagocytes," *American Society for Microbiology,* 8(1) (2017) 1–6.
26. W. Slenczka, H. D. Klenk, "Forty years of Marburg virus," *Journal of Infectious Diseases* 196 (Supplement 2) (2007) S131–S135.
27. A. Shelokov, "Viral hemorrhagic fevers," *The Journal of Infectious Diseases* 122(6) (1970) 560–562.
28. W. Slenczka, H. D. Klenk, "Forty years of Marburg virus," *Journal of Infectious Diseases* 196(Supplement 2) (2007) S131–S135.
29. M. B. Oren, *Six Days of War: June 1967 and the Making of the Modern Middle East* (New York: Presidio Press, 2003).
30. W. Slenczka, H. D. Klenk, "Forty years of Marburg virus," *Journal of Infectious Diseases* 196(Supplement 2) (2007) S131–S135.
31. M. T. Osterholm, K. A. Moore, N. S. Kelley, L. M. Brosseau, G. Wong, F. A. Murphy, C. J. Peters, J. W. LeDuc, P. K. Russell, M. Van Herp, "Transmission of Ebola viruses: what we know and what we do not know," *MBio* 6(2) (2015) e00137–15.
32. R. Preston, *The Hot Zone—A Terrifying New Story* (New York: Random House, 1994).
33. E. Johnson, N. Jaax, J. White, P. Jahrling, "Lethal experimental infections of rhesus monkeys by aerosolized Ebola virus," *International Journal of Experimental Pathology* 76(4) (1995) 227–236.
34. F. J. Grundbacher, "Behring's discovery of diphtheria and tetanus antitoxins," *Immunology Today* 13(5) (1992) 188–90.
35. The Lancet Special Commission, "On the relative strengths of diphtheria antitoxic serums." *The Lancet* 148(3803), 182–195.
36. H. Markel, "Long Ago Against Diphtheria, the Heroes Were Horses," *New York Times*, New York, NY, July 10, 2007, p.D6.
37. R. E. DeHovitz, "The 1901 St Louis Incident: The First Modern Medical Disaster," *Pediatrics* 133(6) (2014) 964–965.
38. Anonymous, "St. Louis, the largest stock owner in Missouri owns 2699 horses and mules," *St Louis Post-Dispatch*, St Louis, MO, January 8, 1899, p.1.
39. J. M. Morris, *Pulitzer: a Life in Politics, Print, and Power* (New York: Harper, 2010).
40. P. D. Noguchi, "From Jim to Gene and Beyond: An Odyssey of Biologics Regulation," *Food & Drug Law Journal* 51 (1996) 367–73.
41. M. Liu, K. Davis, *A Clinical Trials Manual from the Duke Clinical Research Institute: Lessons from a Horse Named Jim* (Hoboken, NJ: John Wiley & Sons, 2011).

42. R. E. DeHovitz, "The 1901 St Louis Incident: The First Modern Medical Disaster," *Pediatrics* 133(6) (2014) 964–965.
43. Ibid.
44. Ibid.
45. Editor, "Four Cases for Tetanus Investigation," *St Louis Republic*, St Louis, October 31, 1901, p. 1.
46. R. E. DeHovitz, "The 1901 St Louis Incident: The First Modern Medical Disaster," *Pediatrics* 133(6) (2014) 964–965.
47. Ibid.
48. M. E. Dixon, "Why Nine Camden Children Died from Smallpox Vaccines in 1901," *Main Line Today (Camden, NJ)*, http://www.mainlinetoday.com/Main-Line-Today/September-2016/Why-Nine-Camden-Children-Died-from-Smallpox-Vaccines-in-1901/, Web. February 15, 2018.
49. D. E. Lilienfeld, "The first pharmacoepidemiologic investigations: national drug safety policy in the United States, 1901–1902," *Perspectives in Biology and Medicine* 51(2) (2008) 188–198.
50. M. E. Dixon, "Why Nine Camden Children Died from Smallpox Vaccines in 1901," *Main Line Today (Camden, NJ)*, http://www.mainlinetoday.com/Main-Line-Today/September-2016/Why-Nine-Camden-Children-Died-from-Smallpox-Vaccines-in-1901/, Web. February 15, 2018.
51. T. S. Coleman, "Early Development in the Regulation of Biologics," *Food & Drug Law Journal* 71 (2016) 544.
52. R. A. Kondratas, "Biologics control act of 1902" in *The Early Years of Federal Food and Drug Control* ed. J. H. Young (Madison, WI: American Institute for the History of Pharmacy, 1982) 8–27.
53. J. C. Burnham, *Health Care in America: A History* (Baltimore: Johns Hopkins University Press, 2015).
54. M. S. Kinch, *A Prescription For Change: The Looming Crisis in Drug Discovery* (Chapel Hill, NC: University of North Carolina Press, 2016).
55. L. Owens, "Inventing the NIH: federal biomedical research policy, 1887–1937," *Science,* 236 (1987) 985–987.
56. N. Wade, "Division of Biologics Standards: The boat that never rocked," *Science* 175(4027) (1972) 1225–1230.
57. R. Wagner, *Clemens von Pirquet: His Life and Work* (Baltimore: The Johns Hopkins Press, 1968).
58. S. T. Shulman, "Clemens von Pirquet: A Remarkable Life and Career," *Journal of the Pediatric Infectious Diseases Society,* 6(4) (2016) 376–9.
59. T. Escherich, *Die Darmbacterien des Neugeborenen und Säuglings* (Stuttgart. Germany: Verlag von Ferdinand Enke, 1886).
60. J. Benedict. *Poisoned: The True Story of the Deadly E. Coli Outbreak that Changed the Way Americans Eat* (New York: February Books, 2011).
61. J. Turk, "Von Pirquet, allergy and infectious diseases: a review," *Journal of the Royal Society of Medicine* 80(1) (1987) 31.
62. R. G. Eccles, *A Darwinian Interpretation of Anaphylaxis*, (New York: W. Wood & Company, 1911).
63. J. T. Edsall, "Edwin Joseph Cohn (1892–1953)," *Biographical Memoirs* 35 (1961) 47–83.

64. D. M. Surgenor. *Edwin J. Cohn and the Development of Protein Chemistry.* (Cambridge, MA: Harvard University Press, 2002).
65. J. M. Prutkin, W. B. Fye, "Edward G. Janeway, clinician and pathologist," *Clinical Cardiology* 29(8) (2006) 376–377.
66. P. M. Gayed, "Toward a Modern Synthesis of Immunity: Charles A. Janeway Jr. and the Immunologist's Dirty Little Secret," *The Yale Journal of Biology and Medicine* 84(2) (2011) 131–138.
67. R. S. Geha, C. A. Janeway and F. S. Rosen: "The discovery of gamma globulin therapy and primary immunodeficiency diseases at Boston Children's Hospital," *Journal of Allergy and Clinical Immunology* 116(4) (2005) 937.
68. Editor, "Hieronymus Fabricius," *New England Journal of Medicine* 229(15) (1943) 600–601.
69. D. Ribatti, E. Crivellato, A. Vacca, "The contribution of Bruce Glick to the definition of the role played by the bursa of Fabricius in the development of the B cell lineage," *Clinical and Experimental Immunology* 145(1) (2006) 1–4.
70. Ibid.
71. G. Köhler, C. Milstein, "Continuous cultures of fused cells secreting antibody of predefined specificity," *Nature* 256(5517) (1975) 495–497.
72. J. McCafferty, A. D. Griffiths, G. Winter, D. J. Chiswell, "Phage antibodies: filamentous phage displaying antibody variable domains," *Nature* 348(6301) (1990) 552.
73. N. Lonberg, L. D. Taylor, F. A. Harding, M. Trounstine, K. M. Higgins, S. R. Schramm, C.-C. Kuo, R. Mashayekh, K. Wymore, J. G. McCabe, "Antigen-specific human antibodies from mice comprising four distinct genetic modifications," *Nature* 368(6474) (1994) 856.
74. G. Winter, A. D. Griffiths, R. E. Hawkins, H. R. Hoogenboom, "Making antibodies by phage display technology," *Annual Review of Immunology* 12(1) (1994) 433–455.
75. C. Mantoux, "Intradermo-réaction de la tuberculine," *Comptes Rendus de l'Académie des Sciences*, Paris 147 (1908) 355–357.
76. S. T. Shulman, "Clemens von Pirquet: A Remarkable Life and Career," *Journal of the Pediatric Infectious Diseases Society,* 6(4) (2016) 376–9.
77. R. Wagner, "Clemens von Pirquet, discoverer of the concept of allergy," *Bulletin of the New York Academy of Medicine* 40(3) (1964) 229–235.
78. B. Schick, "Die Diphtherietoxin-hautreaktion des Menschen als Vorprobe der prophylaktischen Diphtherieheilseruminjektion," *Medizinische Wochenschrift (Munich),* November 25, 1913. p.1.
79. C. Kereszturi, W.H. Park, B. Schick, "Parenteral BCG vaccination," *American Journal of Diseases of Children* 43(2) (1932) 273–283.

Chapter 7: Lost in Translation

1. J. Steinberg, *Bismarck: A life*, (Oxford, UK: Oxford University Press, 2012).
2. W. A. Smith, "Napoleon III and the Spanish Revolution of 1868," *The Journal of Modern History* 25(3) (1953) 211–233.
3. M. Howard, *The Franco-Prussian War: The German Invasion of France 1870–1871* (London: Routledge, 2005).

4. F. Jellinek, *The Paris Commune of 1871*, (Redditch, UK: Read Books Ltd., 2013).
5. S. Dronicz, L. Kawalec, "Dictatorship of the 'Proletariat,'" *Dialogue and Universalism* 21(3) (2011) 137–150.
6. F. Jellinek, *The Paris Commune of 1871* (Redditch, UK: Read Books Ltd., 2013).
7. A. Ullmann, "Pasteur-Koch: Distinctive ways of thinking about infectious diseases," *Microbe* 2(8) (2007) 383.
8. S. M. Blevins, M. S. Bronze, "Robert Koch and the 'golden age'of bacteriology," *International Journal of Infectious Diseases* 14(9) (2010) e744–e751.
9. T. D. Brock, *Robert Koch: A Life in Medicine and Bacteriology* (Washington D.C.: American Society for Microbiology, 1988).
10. G. Richet. "From Bright's disease to modern nephrology: Pierre Rayer's innovative method of clinical investigation. *Kidney International* 39(4) (1991), 787–792.
11. P. Debré, *Louis Pasteur* (Baltimore, MD: Johns Hopkins University Press, 2000).
12. A. Ullmann, "Pasteur-Koch: Distinctive ways of thinking about infectious diseases," *Microbe* 2(8) (2007) 383.
13. A. Mathijsen, E. Oldenkamp, "Jean Joseph Henry Toussaint (1847-1890): Predecessors: Veterinarians from Earlier Times, *Tijdschrift voor Diergeneeskunde* 126(4) (2001), 106–7.
14. H. Bazin, *Vaccination: A History* (Paris: John Libbey Eurotext, 2011).
15. M. Bucchi, "The public science of Louis Pasteur: The experiment on anthrax vaccine in the popular press of the time," *History and Philosophy of the Life Sciences* (1997) 181–209.
16. Ibid.
17. N. Chevallier-Jussiau, "Henry Toussaint and Louis Pasteur. Rivalry over a vaccine," *History of Science and Medicine* 44(1) (2010), 55–64.
18. L. Pasteur, R. Chamberland, "Summary report of the experiments conducted at Pouilly-le-Fort, near Melun, on the anthrax vaccination, 1881," *The Yale Journal of Biology and Medicine* 75(1) (2002) 59.
19. P. Debré, *Louis Pasteur* (Baltimore, MD: Johns Hopkins University Press, 2000).
20. A. Loir, *A l'Ombre de Pasteur* (Paris: *Le Mouvement Sanitaire*, 1938).
21. Ibid.
22. A. Ullmann, "Pasteur-Koch: Distinctive ways of thinking about infectious diseases," *Microbe* 2(8) (2007) 383.
23. K. C. Carter, "The Koch-Pasteur dispute on establishing the cause of anthrax," *Bulletin of the History of Medicine* 62(1) (1988) 42.
24. A. Ullmann, "Pasteur-Koch: Distinctive ways of thinking about infectious diseases," *Microbe* 2(8) (2007) 383.
25. Ibid.
26. R. Koch, *Eine Entgegnung auf den von Pasteur in Genf gehaltenen Vortrag* (Kassel, Germany: Kassel, 1882).
27. P. Debré, *Louis Pasteur* (Baltimore, MD: Johns Hopkins University Press, 2000).
28. D. Lippi, E. Gotuzzo, "The greatest steps towards the discovery of *Vibrio cholerae*," *Clinical Microbiology and Infection* 20(3) (2014) 191–195.
29. R. Koch, "An address on cholera and its bacillus," *British Medical Journal* 2(1236) (1884) 453.

30. D. Lippi, E. Gotuzzo, "The greatest steps towards the discovery of *Vibrio cholerae,*" *Clinical Microbiology and Infection* 20(3) (2014) 191–195.
31. S. A. Waksman, *The Brilliant and Tragic life of WMW Haffkine, Bacteriologist,* (New Brunswick, NJ: Rutgers University Press, 1964).
32. B. J. Hawgood, "Waldemar Mordecai Haffkine, CIE (1860–1930): prophylactic vaccination against cholera and bubonic plague in British India," *Journal of Medical Biography* 15(1) (2007) 9–19.
33. W. M. Haffkine, "A Lecture on Vaccination Against Cholera: Delivered in the Examination Hall of the Conjoint Board of the Royal Colleges of Physicians of London and Surgeons of England, December 18th, 1895," *British Medical Journal* 2(1825) (1895) 1541.
34. W. Haffkine, "A Lecture on Anticholeraic Inoculation: Delivered, by Invitation, at the Laboratories of the Royal Colleges, Victoria Embankment," *British Medical Journal* 1(1676) (1893) 278.
35. B. J. Hawgood, "Waldemar Mordecai Haffkine, CIE (1860–1930): prophylactic vaccination against cholera and bubonic plague in British India," *Journal of Medical Biography* 15(1) (2007) 9–19.
36. E. Hankin, "Remarks on Haffkine's method of protective inoculation against cholera," *British Medical Journal* 2(1654) (1892) 569.
37. B. J. Hawgood, "Waldemar Mordecai Haffkine, CIE (1860–1930): prophylactic vaccination against cholera and bubonic plague in British India," *Journal of Medical Biography* 15(1) (2007) 9–19.
38. S. A. Waksman, *The brilliant and tragic life of WMW Haffkine, bacteriologist,* (New Brunswick, NJ: Rutgers University Press, 1964).
39. G. H. Bornside, "Waldemar Haffkine's cholera vaccines and the Ferran-Haffkine priority dispute," *Journal of the History of Medicine and Allied Sciences* 37(4) (1982) 399.
40. B. J. Hawgood, "Waldemar Mordecai Haffkine, CIE (1860–1930): prophylactic vaccination against cholera and bubonic plague in British India," *Journal of Medical Biography* 15(1) (2007) 9–19.
41. R. A. Baker, R. A. Bayliss, "William John Ritchie Simpson (1855–1931): Public health and tropical medicine," *Medical History,* 31(4) (1987), 450–465.
42. W. M. Haffkine, *Protective Inoculation Against Cholera, Protective Inoculation Against Cholera.* (Calcutta: Thacker, Spink & Co., 1913).
43. M. Lombard, P. Pastoret, A. Moulin, "A brief history of vaccines and vaccination," *Revue Scientifique et Technique-Office International des Epizooties* 26(1) (2007) 29–48.
44. J. Théodoridès, "Pasteur and rabies: the British connection," *Journal of the Royal Society of Medicine* 82(8) (1989) 488.
45. C. Mérieux, "1879–1979. It is now one hundred years since Victor Galtier, a professor of Veterinary School in Lyon, presented a paper on the prophylaxis of rabies to the Academy of Sciences," *Bulletin de l'Academie Nationale de Medecine* 163(2) (1979) 125.
46. T. M. Dolan, *The Nature and Treatment of Rabies Or Hydrophobia: Being the Report of the Special Commission Appointed by the Medical Press and Circular, with Valuable Additions,* (Paris: Baillière, Tindall, and Cox, 1878).
47. P. V. Galtier, "Physiologie Pathologique–Les injections de virus rabique dans le torrent cirulatoire ne provoquent pas l'éclosion de la rage et semblent conférer l'immunitée. La rage peut être transmise par l'ingestion de la matière rabique". Note

de Galtier présenté par M. Bouley," *Comptes Rendus de l'Academie des Sciences* 93 (1881) 284–285.

48. L. Pasteur, M. M. Chamberland, E. Roux, *Sur l'etiologie du charbon*, *Comptes Rendus de l'Academie des Sciences,* 91 (1880), 315.
49. P. Debré, *Louis Pasteur* (Baltimore, MD: Johns Hopkins University Press, 2000).
50. G. L. Geison, "Pasteur's work on rabies: reexamining the ethical issues," *Hastings Center Report* 8(2) (1978) 26–33.
51. G. L. Geison, "Pasteur, Roux, and Rabies: Scientific versus clinical mentalities," *Journal of the History of Medicine and Allied Sciences* 45(3) (1990) 341.
52. P. Berche, "Louis Pasteur, from crystals of life to vaccination," *Clinical Microbiology and Infection* 18 1–6.
53. H. D. Dufour, S. B. Carroll, "History: Great myths die hard," *Nature* 502(7469) (2013) 32–33.
54. A. Ullmann, "Pasteur-Koch: Distinctive ways of thinking about infectious diseases," *Microbe* 2(8) (2007) 383.
55. S. M. Blevins, M. S. Bronze, "Robert Koch and the 'golden age' of bacteriology," *International Journal of Infectious Diseases* 14(9) (2010) e744–e751.
56. H. Mollaret, "Contribution to the knowledge of relations between Koch and Pasteur," *NTM Schriftenr. Geschichte Naturwissenschaft,* 20(1) (1983), 57–65.
57. "Lives Saved," *ScienceHeroes.com*, *Lives Saved.* Web. February 15, 2018.
58. Ibid.
59. D. Butler, "Close but no Nobel: the scientists who never won," *Nature*, http://www.nature.com/news/close-but-no-nobel-the-scientists-who-never-won-1.20781, 2016. Web. February 15, 2018.
60. F. B. Rogers, R. J. Maloney, "Gaston Ramon: 1886–1963," *Archives of Environmental Health*, 7(6) (1963), 723–5.
61. "Our History" *Pasteur.fr/en.* Institut Pasteur. Web. February 15, 2018.
62. Ibid.
63. J. D. Bredin, *The Affair: The Case of Alfred Dreyfus* (London: Sidgwick & Jackson, 1986).
64. "Our History" *Pasteur.fr/en.* Institut Pasteur. Web. February 15, 2018.
65. "Biographical Sketch: Gaston Ramon (1886–1963)." *Pasteur.fr/en.* Institut Pasteur. Web. February 15, 2018.
66. P. Bonanni, J. I. Santos, "Vaccine evolution," *Perspectives in Vaccinology* 1(1) (2011) 1–24.
67. "Biographical Sketch: Gaston Ramon (1886–1963)." *Pasteur.fr/en.* Institut Pasteur. Web. February 15, 2018.
68. G. Ramon, "Sur la toxine et sur l'anatoxine diphtheriques," *Annals of the Institut Pasteur* 38(1) (1924) 13.
69. C. Oakley, "Alexander Thomas Glenny. 1882–1965," *Biographical Memoirs of Fellows of the Royal Society* 12 (1966) 163–180.
70. H. J. Parish, *A History of Immunization*, (San Diego, CA: Harcourt Brace, 1965).
71. M. M. Levine, R. Lagos, "Vaccines and vaccination in historical perspective," *New Generation Vaccines,* 2 (1990) 1–11.
72. J. M. Keith, *Bacterial Protein Toxins Used in Vaccines, Vaccine design: Innovative Approaches and Novel Strategies.* (Norfolk, UK: Caister Academic Press, 2011), 109–137.

73. D. Baxby. "The discovery of diphtheria toxoid and the primary and secondary immune response," *Epidemiology and Infection* 133(S1) (2005) S21–2.
74. H. J. Parish, *A History of Immunization*, (San Diego, CA: Harcourt Brace, 1965).
75. S. L. Plotkin, S. A. Plotkin, "A short history of vaccination," *Vaccines* 5 (2004) 1–16.
76. K. A. Ungermann, *The Race to Nome* (New York: Harper & Row, 1963).
77. G. L. Armstrong, L. A. Conn, R. W. Pinner, "Trends in infectious disease mortality in the United States during the 20th century," *Journal of the American Medical Association* 281(1) (1999) 61–66.
78. "Prevention," *Diphtheria, Centers for Disease Control and Prevention,* Web. February 15, 2018.
79. P. Descombey, "L'anatoxine tétanique," *Comptes Rendus de l'Academie des Sciences* 91 (1924) 239–241.
80. S. A. Waksman, S. A. Waksman, *The Brilliant and Tragic life of WMW Haffkine, Bacteriologist*, (New Brunswick, NJ: Rutgers University Press, 1964).
81. W. Rosen, *Justinian's Flea: Plague, Empire, and the Birth of Europe* (New York: Viking, 2007).
82. S. Schama, *A History of Britain*, (London: BBC Worldwide, 2000).
83. C. A. Benedict, *Bubonic Plague in Nineteenth-Century China* (Redwood City, CA: Stanford University Press, 1996).
84. D. G. Atwill, *The Chinese Sultanate: Islam, ethnicity, and the Panthay Rebellion in Southwest China, 1856–1873*, (Redwood City, CA: Stanford University Press, 2005).
85. G.-F. Treille, A. Yersin, "La peste bubonique à Hong Kong," *VIIIe Congrès international d'hygiène et de démographie*, 1894, 310–311.
86. N. Howard-Jones, "Was Shibasaburo Kitasato the co-discoverer of the plague bacillus?" *Perspectives in Biology and Medicine* 16(2) (1973) 292–307.
87. I. J. Catanach, *Plague and the tensions of empire: India, 1896–1918, Imperial Medicine and Indigenous Societies* (Manchester, UK: Manchester University Press, 1988).
88. S. A. Waksman, *The Brilliant and Tragic life of WMW Haffkine, Bacteriologist*, (New Brunswick, NJ: Rutgers University Press, 1964).
89. Ibid.
90. Ibid.
91. B. J. Hawgood, "Waldemar Mordecai Haffkine, CIE (1860–1930): prophylactic vaccination against cholera and bubonic plague in British India," *Journal of Medical Biography* 15(1) (2007) 9–19.
92. E. Hankin, "Remarks on Haffkine's Method of Protective Inoculation Against Cholera," *British Medical Journal* 2(1654) (1892) 569.
93. E. Chernin, "Ross defends Haffkine: the aftermath of the vaccine-associated Mulkowal Disaster of 1902," *Journal of the History of Medicine and Allied Sciences* 46(2) (1991) 201.
94. S. A. Waksman, *The Brilliant and Tragic life of WMW Haffkine, Bacteriologist*, (New Brunswick, NJ: Rutgers University Press, 1964).
95. R. Ross, "The Inoculation Accident at Mulkowal," *Nature* 75 (1907) 486–487.
96. E. Chernin, "Ross defends Haffkine: the aftermath of the vaccine-associated Mulkowal Disaster of 1902," *Journal of the History of Medicine and Allied Sciences* 46(2) (1991) 201.

97. B. J. Hawgood, "Waldemar Mordecai Haffkine, CIE (1860–1930): prophylactic vaccination against cholera and bubonic plague in British India," *Journal of Medical Biography* 15(1) (2007) 9–19.
98. S. A. Waksman, *The Brilliant and Tragic life of WMW Haffkine, Bacteriologist*, (New Brunswick, NJ: Rutgers University Press, 1964).
99. G. H. Bornside, "Waldemar Haffkine's cholera vaccines and the Ferran-Haffkine priority dispute," *Journal of the History of Medicine and Allied Sciences* 37(4) (1982) 399.

Chapter 8: Breathing Easier

1. C. Oakley, "Jules Jean Baptiste Vincent Bordet. 1870–1961," *Biographical Memoirs of Fellows of the Royal Society* 8 (1962) 19–25.
2. N. Guiso, "*Bordetella pertussis* and pertussis vaccines." *Clinical Infectious Diseases*, 49(10) (2009), 1565–9.
3. *The Cambridge World History of Human Disease*, ed. K.F. Kiple (Cambridge, UK: Cambridge University Press, 1993).
4. A. Aslanabadi, K. Ghabili, K. Shad, M. Khalili, M. M. Sajadi, "Emergence of whooping cough: notes from three early epidemics in Persia," *The Lancet Infectious Diseases* 15(12) (2015) 1480–1484.
5. J. D. Cherry, "The present and future control of pertussis," *Clinical Infectious Diseases* 51(6) (2010) 663–667.
6. J. D. Cherry, "The History of Pertussis (Whooping Cough); 1906–2015: Facts, Myths, and Misconceptions," *Current Epidemiology Reports* 2(2) (2015) 120–130.
7. A. Aslanabadi, K. Ghabili, K. Shad, M. Khalili, M. M. Sajadi, "Emergence of whooping cough: notes from three early epidemics in Persia," *The Lancet Infectious Diseases* 15(12) (2015) 1480–1484.
8. N. Guiso, "*Bordetella pertussis* and pertussis vaccines," *Clinical Infectious Diseases* 49(10) (2009) 1565–1569.
9. Ibid.
10. J. Freeman, "Vaccine Therapy: its Treatment, Value, and Limitations," *Proceedings of the Royal Society of Medicine* (1910) 97–101.
11. J. Zahorsky, "Pertussis Vaccine," *Interstate Medical Journal* 19 (1909) 844.
12. L. W. Sauer, "Immunization with bacillus pertussis vaccine," *Journal of the American Medical Association* 101(19) (1933) 1449–1453.
13. J. D. Cherry, "The History of Pertussis (Whooping Cough); 1906–2015: Facts, Myths, and Misconceptions," *Current Epidemiology Reports* 2(2) (2015) 120–130.
14. C. G. Shapiro-Shapin, "'A whole community working together': Pearl Kendrick, Grace Eldering, and the Grand Rapids pertussi trials, 1932–1939," *The Michigan Historical Review* (2007) 59–85.
15. C. G. Shapiro-Shapin, "Pearl Kendrick, Grace Eldering, and the Pertussis Vaccine," *Emerging Infectious Diseases*, 16(8) (2010) 1273–8.
16. Ibid.
17. C. G. Shapiro-Shapin, "'A whole community working together': Pearl Kendrick, Grace Eldering, and the Grand Rapids pertussi trials, 1932–1939," *The Michigan Historical Review* (2007) 59–85.

18. Ibid.
19. C. G. Shapiro-Shapin, "Pearl Kendrick, Grace Eldering, and the Pertussis Vaccine," *Emerging Infectious Diseases,* 16(8) (2010) 1273–8.
20. C. G. Shapiro-Shapin, "'A whole community working together': Pearl Kendrick, Grace Eldering, and the Grand Rapids pertussi trials, 1932-1939," *The Michigan Historical Review* (2007) 59–85.
21. C. L. Oakley, "Alexander Thomas Glenny. 1882–1965," *Biographical Memoirs of Fellows of the Royal Society* 12 (1966) 163–180.
22. P. Marrack, A. S. McKee, M. W. Munks, "Towards an understanding of the adjuvant action of aluminium," *Nature Reviews Immunology* 9(4) (2009) 287–293.
23. P. L. Kendrick, "A field study of alum-precipitated combined pertussis vaccine and diphtheria toxoid for active immunization," *American Journal of Epidemiology* 38(2) (1943) 193–202.
24. M. Kulenkampff, J. Schwartzman, J. Wilson, "Neurological complications of pertussis inoculation," *Archives of Disease in Childhood* 49(1) (1974) 46–49.
25. J. Berg, "Neurological complications of pertussis immunization," *British Medical Journal* 2(5087) (1958) 24.
26. M. Kulenkampff, J. Schwartzman, J. Wilson, "Neurological complications of pertussis inoculation," *Archives of Disease in Childhood* 49(1) (1974) 46–49.
27. G. R. Noble, R. H. Bernier, E. C. Esber, M. C. Hardegree, A. R. Hinman, D. Klein, A. J. Saah, "Acellular and whole-cell pertussis vaccines in Japan: report of a visit by US scientists," *Journal of the American Medical Association,* 257(10) (1987) 1351–1356.
28. J. P. Baker, "The pertussis vaccine controversy in Great Britain, 1974–1986," *Vaccine* 21(25) (2003) 4003–4010.
29. "DPT: Vaccine Roulette", *WRC-TV,* Washington, D.C. ed. L. Thompson April 19, 1982. Television.
30. S. Mnookin, "The whole cell pertussis vaccine, media malpractice, and the long-term effects of avoiding difficult conversations." *The Panic Virus,* September 13, 2012. http://blogs.plos.org/thepanicvirus/2012/09/13/the-whole-cell-pertussis-vaccine-media-malpractice-and-the-long-term-effects-of-avoiding-difficult-conversations/
31. S. Mnookin, *The Panic Virus: A True Story of Medicine, Science, and Fear,* (New York: Simon and Schuster, 2011).
32. S. Mnookin, "The whole cell pertussis vaccine, media malpractice, and the long-term effects of avoiding difficult conversations." *The Panic Virus,* September 13, 2012. http://blogs.plos.org/thepanicvirus/2012/09/13/the-whole-cell-pertussis-vaccine-media-malpractice-and-the-long-term-effects-of-avoiding-difficult-conversations/
33. P. A. Offit, *Deadly Choices: How the Anti-Vaccine Movement Threatens Us All,* (New York: Basic Books, 2015).
34. S. Mnookin, *The Panic Virus: A True Story of Medicine, Science, and Fear,* (New York: Simon and Schuster, 2011).
35. H. L. Coulter, *Divided Legacy: the Conflict Between Homeopathy and the American Medical Association: Science and Ethics in American Medicine 1800–1910* (Berkeley, CA: North Atlantic Books, 1982).

36. S. Mnookin, *The Panic Virus: A True Story of Medicine, Science, and Fear* (New York: Simon and Schuster, 2011).
37. H. L. Coulter, B.L. Fisher, *DPT: A Shot in the Dark* (London: Penguin, 1985).
38. Centers for Disease Control and Prevention, "Control, Prevention, National Childhood Vaccine Injury Act: requirements for permanent vaccination records and for reporting of selected events after vaccination," *Morbidity and Mortality Weekly Report* 37(13) (1988) 197–200.
39. H. V. Fineberg, C. J. Howe, C. P. Howson, *Adverse Effects of Pertussis and Rubella Vaccines*, (Washington, D.C.: National Academies Press, 1991).
40. G. S. Golden, "Pertussis vaccine and injury to the brain," *Journal of Pediatrics* 116(6) (1990) 854–61.
41. J. D. Cherry, "'Pertussis vaccine encephalopathy': it is time to recognize it as the myth that it is," *Journal of the American Medical Association*, 263(12) (1990) 1679–1680.
42. D. L. Miller, R. Alderslade, E. M. Ross, "Whooping cough and whooping cough vaccine: the risks and benefits debate," *Epidemiology Reviews*, 4 (1982), 1–24.
43. H. V. Fineberg, C. J. Howe, C. P. Howson, *Adverse Effects of Pertussis and Rubella Vaccines*, (Washington, D.C.: National Academies Press, 1991).
44. N. Madge, J. Diamond, D. Miller, E. Ross, C. McManus, J. Wadsworth, W. Yule, The National Childhood Encephalopathy Study: A 10-year follow-up (London: Mac Keith Press, 1993).
45. H. V. Fineberg, C. J. Howe, C. P. Howson, *Adverse Effects of Pertussis and Rubella Vaccines*, (Washington, D.C.: National Academies Press, 1991).
46. G. Stewart, "Effect of penicillin on Bacillus proteus," *The Lancet* 246(6379) (1945) 705–707.
47 R. Bud, *Penicillin: Triumph and Tragedy*, (Oxford, UK: Oxford University Press, 2007).
48. G. T. Stewart, "Limitations of the germ theory," *The Lancet* 291(7551) (1968) 1077–1081.
49. E. Papadopulos-Eleopulos, V. F. Turner, J. M. Papadimitriou, G. Stewart, D. Causer, "HIV antibodies: Further questions and a plea for clarification," *Current Medical Research and Opinion* 13(10) (1997) 627–634.
50. G. T. Stewart, "The epidemiology and transmission of AIDS: a hypothesis linking behavioural and biological determinants to time, person and place," *Genetica* 95(1–3) (1995) 173–193.
51. J. Fenton, "Shame on the professional Aids doubters," *The Independent (London)*, April 11, 1993. Web. February 16, 2018.
52. C. L. Decoteau, *Ancestors and antiretrovirals: the biopolitics of HIV/AIDS in post-apartheid South Africa* (Chicago: University of Chicago Press, 2013).
53. G. T. Stewart, "The Durban Declaration is not accepted by all," *Nature* 407(6802) (2000) 286–286.
54. P. Chigwedere, G. R. Seage, 3rd, S. Gruskin, T. H. Lee, M. Essex, "Estimating the lost benefits of antiretroviral drug use in South Africa," *Journal of Acquired Immune Deficiency Syndromes* (1999) 49(4) (2008) 410–5.
55. G. Stewart, "Toxicity of pertussis vaccine: frequency and probability of reactions," *Journal of Epidemiology and Community Health* 33(2) (1979) 150–156.

56. "DPT: Vaccine Roulette", *WRC-TV*, Washington, D.C. Ed. L. Thompson April 19, 1982. Television.
57. P. A. Offit, *Deadly Choices: How the Anti-Vaccine Movement Threatens Us All*, (New York: Basic Books, 2015).
58. Ibid.
59. Ibid.
60. G. T. Stewart, "The law tries to decide whether whooping cough vaccine causes brain damage: Professor Gordon Stewart gives evidence," *British Medical Journal*, 293(6540) (1986) 203.
61. P. A. Offit, *Deadly Choices: How the Anti-Vaccine Movement Threatens Us All*, (New York: Basic Books, 2015).
62. G. S. Golden, "Pertussis vaccine and injury to the brain," *Journal of Pediatrics* 116(6) (1990) 854–61.
63. P. A. Offit, *Deadly Choices: How the Anti-Vaccine Movement Threatens Us All*, (New York: Basic Books, 2015).
64. J. P. Baker, "The pertussis vaccine controversy in Great Britain, 1974–1986," *Vaccine* 21(25) (2003) 4003–4010.
65. P. A. Offit, *Deadly Choices: How the Anti-Vaccine Movement Threatens Us All*, (New York: Basic Books, 2015).
66. C. Dyer, "Judge "not satisfied" that whooping cough vaccine causes permanent brain damage," *British Medical Journal* 296(6630) (1988) 1189.
67. C. Bowie, "Lessons from the pertussis vaccine court trial," *The Lancet* 335(8686) (1990) 397–399.
68. P. A. Offit, *Deadly Choices: How the Anti-Vaccine Movement Threatens Us All*, (New York: Basic Books, 2015).
69. V. E. Schwartz, L. Mahshigian, "National Childhood Vaccine Injury Act of 1986: an ad hoc remedy or a window for the future," *Ohio State Law Journal*, 48 (1987) 387.
70. Y. Sato, K. Izumiya, H. Sato, J. Cowell, C. Manclark, "Role of antibody to leukocytosis-promoting factor hemagglutinin and to filamentous hemagglutinin in immunity to pertussis," *Infection and Immunity* 31(3) (1981) 1223–1231.
71. Y. Sato, M. Kimura, H. Fukumi, "Development of a pertussis component vaccine in Japan," *The Lancet* 323(8369) (1984) 122–126.
72. G. R. Noble, R. H. Bernier, E. C. Esber, M. C. Hardegree, A. R. Hinman, D. Klein, A. J. Saah, "Acellular and whole-cell pertussis vaccines in Japan: report of a visit by US scientists," *Journal of the American Medical Association*, 257(10) (1987) 1351–1356.
73. A. Schaffer, "Why Are Babies Dying of Old-Fashioned Whooping Cough?" *Slate*, September 5, 2012. Web. February 16, 2018.
74. M. Falco, "10 infants dead in California whooping cough outbreak." *CNN, October 20,* 2010. Web. February 16, 2018.
75. M. Chan, L. Ma, D. Sidelinger, L. Bethel, J. Yen, A. Inveiss, M. Sawyer, K. Waters-Montijo, J. Johnson, L. Hicks, "The California pertussis epidemic 2010: a review of 986 pediatric case reports from San Diego county," *Journal of the Pediatric Infectious Diseases Society* 1(1) (2012) 47–54.
76. M. McKenna, "Why Whooping Cough Vaccines Are Wearing Off." *Scientific American*, October 2, 2013. Web. February 16, 2018.

77. M. Park, "Where vaccine doubt persists." *CNN*, October 20, 2010. Web. February 16, 2018.
78. Ibid.

Chapter 9: Three Little Letters

1. M. M. Zarshenas, A. Mehdizadeh, A. Zargaran, A. Mohagheghzadeh, "Rhazes (865–925 A.D.)," *Journal of Neurology* 259(5) (2012) 1001–1002.
2. M. Meyerhof, "Thirty-three clinical observations by Rhazes (circa 900 A.D.)," *Isis* 23(2) (1935) 321–372.
3. M. M. Zarshenas, A. Mehdizadeh, A. Zargaran, A. Mohagheghzadeh, "Rhazes (865–925 A.D.)," *Journal of Neurology* 259(5) (2012) 1001–1002.
4. Y. Furuse, A. Suzuki, H. Oshitani, "Origin of measles virus: divergence from rinderpest virus between the 11th and 12th centuries," *Virology Journal* 7(1) (2010) 52.
5. M. Thrusfield, *Veterinary Epidemiology*, (Hoboken, NJ: Blackwell, 2013).
6. M. Jacoby, "The fifth plague of Egypt," *Journal of the American Medical Association,* 249(20) (1983) 2779–2780.
7. M. Greger, "Their Bugs Are Worse than Their Bite: Emerging Infectious Disease and the Human-Animal Interface," *The State of Animals 2007* Ed. D. J. Salem, A. N. Rowan. (2007), p111–27. (Washington, D.C: Humane Society Press, 2007).
8. F. Fenner, D. Henderson, I. Arita, Z. Jezek, I. Ladnyi, *The history of smallpox and its spread around the world. OMS Suiza* (1988) 209–43.
9. N. D. Cook, *Born to Die: Disease and New World Conquest, 1492–1650* (Cambridge, UK: Cambridge University Press, 1998).
10. P. L. Panum, J. J. Petersen, *Observations Made During the Epidemic of Measles on the Faroe Islands in the Year 1816* (Washington, D.C.: American Public Health Association, 1940).
11. S. T. Shulman, D. L. Shulman, R. H. Sims, "The tragic 1824 journey of the Hawaiian King and queen to London: history of measles in Hawaii," *The Pediatric Infectious Disease Journal* 28(8) (2009) 728–733.
12. A. D. Cliff, P. Haggett, *The Spread of Measles in Fiji and the Pacific* (Canberra, Australia: Australian National University, 1985).
13. O. Steichen, S. Dautheville, "Koplik spots in early measles," *Canadian Medical Association Journal,* 180(5) (2009) 583.
14. H. Koplik, "The diagnosis of the invasion of measles from a study of the exanthema as it appears on the buccal mucous membrane," *Archives of Pediatrics* 79 (1962) 162.
15. F. C. Robbins, "John Franklin Enders: February 10, 1897–September 8, 1985, "*Proceedings of the American Philosophical Society*, 135(3) (1991), 453–7.
16. Ibid.
17. M. Wortman, *The Millionaires' Unit: The Aristocratic Flyboys who Fought the Great War and Invented American Air Power,* (New York: PublicAffairs, 2007).
18. F. C. Robbins, "John Franklin Enders: February 10, 1897–September 8, 1985, "*Proceedings of the American Philosophical Society*, 135(3) (1991), 453–7.
19. J. F. Enders, "John Franklin Enders papers" in *Yale University Manuscripts and Archives,* (New Haven, CT: Yale University Press, 1988).

20. F. Fenner, P. M. De Burgh, *Hugh Kingsley Ward (1887–1972), Bacteriologist* (Carlton, Australia: Australian Dictionary of Biography, 2002).
21. Ibid.
22. J. F. Enders, "John Franklin Enders papers" in *Yale University Manuscripts and Archives*, (New Haven, CT: Yale University Press, 1988).
23. A. E. Feller, J. F. Enders, T. H. Weller, "The prolonged existence of vaccinia virus in high titre and living cells in roller tube cultures of chick embryonic tissues." *The Journal of Experimental Medicine,* 72(4) (1940), 367–88.
24. J. F. Enders, "John Franklin Enders papers" in *Yale University Manuscripts and Archives*, (New Haven, CT: Yale University Press, 1988).
25. C. D. Johnson, E. W. Goodpasture, "An investigation of the etiology of mumps," *Journal of Experimental Medicine,* 59(1) (1934), 1–22.
26. K. Habel, "Cultivation of mumps virus in the developing chick embryo and its application to studies of immunity to mumps in man," *Public Health Reports,* 60(8), (1945), 201–12.
27. M. Wadman, *The Vaccine Race* (New York: Viking, 2017).
28. M. Vogel, *Gene Tierney: A Biography* (Jefferson, NC: McFarland, 2014).
29. A. Davidson, "Wakefield's Vaccine Follies." *The New Yorker,* May 26, 2010. Web. February 16, 2018.
30. A. Christie, R. Leach, *The Mirror Crack'd from Side to Side* (Glasgow, UK: Collins, 1962).
31. M. Wadman, *The Vaccine Race* (New York: Viking, 2017).
32. R. L. King, "This man's infected blood created the world's first measles vaccine." *The Toronto Star,* February 18, 2015. Web. February 16, 2018.
33. J. F. Enders, S. L. Katz, A. Holloway, "Development of Attenuated Measles Virus Vaccines: A Summary of Recent Investigation," *American Journal of Diseases of Children* 103(3) (1962) 335–340.
34. "Lives Saved," *ScienceHeroes.com, Lives Saved*, Web. February 15, 2018
35. "Miles City, Montana," *Wikipedia, last modified January 31, 2018*, https://en.wikipedia.org/wiki/Miles_City,_Montana.
36. M. Sandoz, *The Battle of the Little Bighorn*, (Lincoln, NE: University of Nebraska Press, 1978).
37. N. A. Miles, *Personal Recollections and Observations of General Nelson A. Miles* (New York: Werner Company, 1897).
38. "Anna Uelsmann Hillemann" FindAGrave.com. Web. February 16, 2018.
39. P. A. Offit, *Vaccinated: One Man's Quest to Defeat the World's Deadliest Diseases* (New York: Harper Collins, 2007).
40. Ibid.
41. Anonymous, "Maurice Hilleman," *The Telegraph (London)* April 14, 2005. Web. February 16, 2018.
42. S. Armstrong, *p53: The Gene that Cracked the Cancer Code* (London: Bloomsbury, 2014).
43. Anonymous, "Maurice Hilleman," *The Telegraph (London)* April 14, 2005. Web. February 16, 2018.
44. L. Newman, "Maurice Hilleman," *British Medical Journal*, 330(7498) (2005) 1028.
45. P. A. Offit, *Vaccinated: One Man's Quest to Defeat the World's Deadliest Diseases* (New York: Harper Collins, 2007).

46. R. Coniff, "A Forgotten Pioneer of Vaccines," *New York Times*, May 6, 2013. Web. February 16, 2018.
47. Anonymous, "Jeryl L. Hilleman Wed in California," *New York Times*, March 20, 1988. Web. February 16, 2018.
48. M. Wadman, *The Vaccine Race* (New York: Viking, 2017).
49. P. A. Offit, *Vaccinated: One Man's Quest to Defeat the World's Deadliest Diseases* (New York: Harper Collins, 2007).
50. S. S. Sprigge, *The Life and Times of Thomas Wakley* (Malabar, FL: Krieger Publishing Company, 1974).
51. "Profile: Dr Andrew Wakefield," *BBC News,* January 27, 2010. Web. February 16, 2018.
52. A. J. Wakefield, R. M. Pittilo, R. Sim, S. L. Cosby, J. R. Stephenson, A. P. Dhillon, R. E. Pounder, "Evidence of persistent measles virus infection in Crohn's disease," *Journal of Medical Virology* 39(4) (1993) 345–53.
53. S. Ghosh, E. Armitage, D. Wilson, P. Minor, M. Afzal, "Detection of persistent measles virus infection in Crohn's disease: current status of experimental work," *Gut* 48(6) (2001) 748–752.
54. A. J. Wakefield, S. H. Murch, A. Anthony, J. Linnell, D. M. Casson, M. Malik, M. Berelowitz, A. P. Dhillon, M. A. Thomson, P. Harvey, A. Valentine, S. E. Davies, J. A. Walker-Smith, "RETRACTED: Ileal-lymphoid-nodular hyperplasia, non-specific colitis, and pervasive development disorder in children. *The Lancet*, 351(9103) (1998), 637–41.
55. E. Bleuler. "Autistic thinking." In D. Rapaport, *Organization and Pathology of Thought: Selected Sources*. (New York: Columbia University Press, 1951), 399–437.
56. R. Kuhn, C. H. Cahn, "Eugen Bleuler's Concepts of Psychopathology," *History of Psychiatry* 15(3) (2004) 361–366.
57. "Autism," *Oxford English Dictionary*, Web. February 16, 2018.
58. C. J. Newschaffer, L. A. Croen, J. Daniels, E. Giarelli, J. K. Grether, S. E. Levy, D. S. Mandell, L. A. Miller, J. Pinto-Martin, J. Reaven, "The epidemiology of autism spectrum disorders," *Annual Reviews of Public Health* 28 (2007) 235–258.
59. S. Lundström, A. Reichenberg, H. Anckarsäter, P. Lichtenstein, C. Gillberg, "Autism phenotype versus registered diagnosis in Swedish children: prevalence trends over 10 years in general population samples," *British Medical Journal,* 350 (2015) h1961.
60. T. Hodgkin, *The History of England from the Earliest Times to the Norman Conquest* (London: Longmans, Green and Company, 1906).
61. B. Deer, "Re: Quick Question." Message to Michael S. Kinch. May 9, 2017. E-mail.
62. B. Deer, "Rights for gays in the face of NHS 'homophobia,'" *Health and Social Service Journal* 89(4643) (1979) 618.
63. M Briggs, M. Briggs, "Oral contraceptives and vitamin nutrition," *The Lancet* 303(7868) (1974) 1234–1235.
64. B. Deer, "The pill: professor's safety tests were faked," *The Sunday Times (London)*, September 28, 1986. Web. February 16, 2018.
65. B. Deer, "Top-selling drug may have killed hundreds in Britain," *The Sunday Times (London)*, February 27, 1994. Web. February 16, 2018.
66. B. Deer, "Hard Sell," *The Sunday Times (London)*, March 6, 1994. Web. February 16, 2018.

67. B. Deer, "Re: Quick Question." Message to Michael S. Kinch. May 9, 2017. E-mail.
68. B. Deer, "When needs outweighs blame—After a 'vaccine-damage' court ruling last week, children will not get help. *The Sunday Times (London)*, April 3, 1988. Web. February 16, 2018.
69. B. Deer, "Re: Quick Question." Message to Michael S. Kinch. May 9, 2017. E-mail.
70. B. Deer, "The Vanishing Victims," *The Sunday Times (London)*, November 1, 1998. Web. February 16, 2018.
71. B. Deer, "AidsVax: the long shot," *The Sunday Times (London)*, October 3, 1999. Web. February 16, 2018.
72. B. Deer, "Re: Quick Question." Message to Michael S. Kinch. May 9, 2017. E-mail.
73. Ibid.
74. Ibid.
75. B. Deer, "Fresh doubts cast on MMR study data," *The Sunday Times (London)*, February 22, 2004. Web. February 16, 2018.
76. B. Deer, "MMR: the truth behind the crisis," *The Sunday Times (London)*, February 22, 2004. Web. February 16, 2018.
77. Ibid.
78. "MMR—what they didn't tell you," *Dispatches*. Channel 4 (United Kingdom), London. November 18, 2004. Television.
79. N. Banks-Smith, "Let them eat cake," *The Guardian (London)*, November 21, 2004. Web. February 16, 2018.
80. "Eady Judgement" *BrianDeer.com*, http://briandeer.com/wakefield/eady-judgment.htm. Web. February 16, 2018.
81. B. Deer, "Exposed: Andrew Wakefield and the MMR-autism fraud." *The Sunday Times (London)*, February 8, 2009. Web. February 16, 2018.
82. F. Godlee, J. Smith, H. Marcovitch, "Wakefield's article linking MMR vaccine and autism was fraudulent," *British Medical Journal*, 342, (2011), 7452.
83. Editors of The Lancet, "Retraction—Ileal-lymphoid-nodular hyperplasia, non-specific colitis, and pervasive developmental disorder in children," *The Lancet* 375(9713) (2010) 445.
84. S. Mnookin, "The problems with the BMJ's Wakefield-fraud story." *The Panic Virus*, The problems with the BMJ's Wakefield-fraud story. January 6, 2011. sethmnookin.com/2011/01/06/the-problems-with-the-bmjs-wakefield-fraud-story/
85. S. Mnookin, *The Panic Virus: A True Story of Medicine, Science, and Fear* (New York: Simon and Schuster, 2011).
86. S. Mnookin, "The problems with the BMJ's Wakefield-fraud story." *The Panic Virus*, The problems with the BMJ's Wakefield-fraud story. January 6, 2011. sethmnookin.com/2011/01/06/the-problems-with-the-bmjs-wakefield-fraud-story/
87. E. Kohn, "'Vaxxed: From Cover-Up to Catastrophe' is Designed to Trick You," *IndieWire*, April 1, 2016, Web. February 16, 2018.
88. P. Belluck, M. Ryzik, "Robert De Niro Defends Screening of Anti-Vaccine Film at Tribeca Festival," *New York Times*, March 25, 2016. Web. February 16, 2018.
89. S. Goodman, "Robert De Niro Pulls Anti-Vaccine Documentary From Tribeca Film Festival," *New York Times*, March 26, 2016. Web. February 16, 2018.
90. The Hollywood Reporter Staff, "Hollywood's Biggest Anti-Vaccine Proponents." *The Hollywood Reporter*. September 10, 2014. Web. February 16, 2018.

91. C. Ross, Andrew Wakefield appearance at Trump inaugural ball triggers social media backlash, *STAT News*, January 21, 2017. Web. February 16, 2018.
92. M. McKee, S. L. Greer, D. Stuckler, "What will Donald Trump's presidency mean for health? A scorecard," *The Lancet* 389(10070) (2017) 748–754.
93. L. Garratt, "Donald Trump and the Anti-Vaxxer Conspiracy Theorists," *Foreign Policy*, January 11, 2017. Web. February 16, 2018.
94. M. Lerner, "Anti-vaccine doctor meets with Somalis," *Star Trbune* (*Minneapolis*), March 24, 2011. Web. February 16, 2018.
95. K. Almond, "Somalis finding their place in Minnesota." *CNN*, February 1, 2017. Web. February 16, 2018.
96. E. R. Wolff, D. J. Madlon-Kay, "Childhood vaccine beliefs reported by Somali and non-Somali parents," *The Journal of the American Board of Family Medicine* 27(4) (2014) 458–464.
97. L. H. Sun, "Anti-vaccine activists spark a state's worst measles outbreak in decades," *Washington Post*, May 5, 2017. Web. February 16, 2018.
98. J. Howard, "Anti-vaccine groups blamed in Minnesota measles outbreak," *CNN,* May 8, 2017. Web. February 16, 2018.
99. M. J. Smith, S. S. Ellenberg, L. M. Bell, D. M. Rubin, "Media coverage of the measles -mumps-rubella vaccine and autism controversy and its relationship to MMR immunization rates in the United States," *Pediatrics* 121(4) (2008) e836–e843.
100. Y. T. Yang, P. L. Delamater, T. F. Leslie, M. M. Mello, "Sociodemographic pre dictors of vaccination exemptions on the basis of personal belief in California," *American Journal of Public Health,* 106(1) (2016) 172–177.
101. H. Abbey, "An examination of the Reed-Frost theory of epidemics," *Human Biology* 24(3) (1952) 201.
102. A. W. Hedrich, "Monthly estimates of the child population "susceptible" to measles, 1900–1931, Baltimore, MD." *American Journal of Epidemiology,* 17(3) (1933) 613–36.
103. A. Hedrich, "The 'normal' for epidemic diseases," *American Journal of Public Health* 17(7) (1927) 691–698
104. P. E. Fine, "Herd immunity: history, theory, practice," *Epidemiologic Reviews* 15(2) (1993) 265–302.
105. D. Adams, R. Jajosky, U. Ajani, J. Kriseman, P. Sharp, D. Onwen, A. Schley, W. Anderson, A. Grigoryan, A. Aranas, Summary of notifiable diseases—United States, 2012, *Morbidity and mortality weekly Report,* 61(53) (2014) 1–121.
106. J. Bixler, G. Botelho, "361 cases of mumps in central Ohio," *CNN*, May 16, 2014. Web. February 16, 2018.
107. A. Stapleton, D. Goldschmidt, "60 reported cases of mumps at University of Illinois since April," *CNN*, July 31, 2015. Web. February 16, 2018.
108. The Associated Press, "University of Missouri mumps outbreak passes 200-cases mark." *St Lois Post-Dispatch*, December 16, 2016. Web. February 16, 2018.

Chapter 10: When Future Shocks Become Current Affairs

1. S. David, *Operation Thunderbolt: Flight 139 and the Raid on Entebbe Airport, the Most Audacious Hostage Rescue Mission in History* (New York: Little, Brown and Company, 2015).

2. A. Rice, *The Teeth May Smile but the Heart Does Not Forget: Murder and Memory in Uganda* (London: Macmillan, 2009).
3. S. David, *Operation Thunderbolt: Flight 139 and the Raid on Entebbe Airport, the Most Audacious Hostage Rescue Mission in History* (New York: Little, Brown and Company, 2015).
4. Anonymous, "Climate: Entebbe." *Climate-Data.Org,* en.climate-data.org /location/765748/.
5. M. R. Holbrook, "Historical Perspectives on Flavivirus Research," *Viruses* 9(5) (2017).
6. G. Dick, S. Kitchen, A. Haddow, "Zika virus (I). Isolations and serological specificity," *Transactions of the Royal Society of Tropical Medicine and Hygiene* 46(5) (1952) 509–520.
7. D. Gatherer, A. Kohl, "Zika virus: a previously slow pandemic spreads rapidly through the Americas," *Journal of General Virology* 97(2) (2016) 269–273.
8. H. F. Dobyns, "Disease transfer at contact," *Annual Review of Anthropology* 22(1) (1993) 273–291.
9. M. A. Smith, "Andrew Brown's 'Earnest Endeavor': The Federal Gazette's Role in Philadelphia's Yellow Fever Epidemic of 1793," *The Pennsylvania Magazine of History and Biography* 120(4) (1996) 321–342.
10. G. Symcox, "Louis XIV and the Outbreak of the Nine Years War" in *Louis XIV and Europe* (New York: Springer, 1976), 179–212.
11. M. S. Pernick, "Politics, parties and pestilence: epidemic yellow fever in Philadelphia and the rise of the first party system. *The William and Mary Quarterly,* 29(4) (1972), 559–86.
12. K. R. Foster, M. F. Jenkins, A. C. Toogood, "The Philadelphia yellow fever epidemic of 1793," *Scientific American* 279(2) (1998) 88.
13. J. H. Powell, *Bring out Your Dead: The Great Plague of Yellow Fever in Philadelphia in 1793* (Philadelphia: University of Pennsylvania Press, 1993).
14. J. A. Del Regato, "Carlos Juan Finlay (1833–1915)," *Journal of Public Health Policy* 22(1) (2001) 98–104.
15. E. Chaves-Carballo, "Carlos Finlay and yellow fever: triumph over adversity," *Military Medicine* 170(10) (2005).
16. D. P. Pentón, "Celebridades médicas y acontecimientos políticos, una mirada desde la historia de Francia," *Panorama Cuba y Salud* 9(1) (2014) 20–28.
17. A. Reyes-Santos, *Our Caribbean Kin: Race and Nation in the Neoliberal Antilles* (New Brunswick, NJ: Rutgers University Press, 2015).
18. C. Finlay, "Carlos Finlay and Yellow Fever," *The Journal of Parasitology* 28(2) (1942) 172–174.
19. W. B. Bean, *Walter Reed: A Biography,* (Charlottesville, VA: University Press of Virginia, 1982).
20. H. A. Kelly, *Walter Reed and Yellow Fever* (New York: McClure, Phillips, 1907).
21. W. B. Bean, *Walter Reed: A Biography,* (Charlottesville, VA: University Press of Virginia, 1982).
22. W. Reed, J. Carroll, A. Agramonte, J. W. Lazear, "The etiology of yellow fever—a preliminary note," *Public Health Papers and Reports* 26 (1900) 37.
23. W. Reed, *Propagation of Yellow Fever: Observations Based on Recent Researches* (New York: William Wood & Company, 1901).

24. C. Finlay, "Carlos Finlay and Yellow Fever," *The Journal of Parasitology* 28(2) (1942) 172–174.
25. J. G. Frierson, "The yellow fever vaccine: a history," *Yale Journal of Biology and Medicine* 83(2) (2010) 77–85.
26. H. Noguchi, "Yellow fever research, 1918–1924: A summary," *Journal of Tropical Medicine and Hygiene* 28(10) (1925).
27. T. P. Monath, "Yellow fever vaccine," *Expert Review of Vaccines* 4(4) (2005) 553–574.
28. M. Theiler, H. H. Smith, "The use of yellow fever virus modified by in vitro cultivation for human immunization," *The Journal of Experimental Medicine* 65(6) (1937) 787.
29. E. Norrby, "Yellow fever and Max Theiler: the only Nobel Prize for a virus vaccine," *Journal of Experimental Medicine* 204(12) (2007) 2779–2784.
30. S. B. Halstead, "Dengue virus–mosquito interactions," *Annu. Rev. Entomol.* 53 (2008) 273–291.
31. J. G. Rigau-Pérez, "The early use of break-bone fever (Quebranta huesos, 1771) and dengue (1801)," *The American Journal of Tropical Medicine and Hygiene* 59(2) (1998) 272–274.
32. B. Rush, *An Account of the Bilious Remitting Fever as it Appeared in Philadelphia, Medical Inquiries and Observations* (Philadelphia: Prichard and Hall, 1789), 104–107.
33. W. R. Smart, "On dengue or dandy fever," *British Medical Journal* 1(848) (1877) 382.
34. R. Preston, *The Hot Zone—A Terrifying New Story* (New York: Random House, 1994).
35. M. K. Bhattacharya, S. Maitra, A. Ganguly, A. Bhattacharya, A. Sinha, "Dengue: a growing menace—a snapshot of recent facts, figures & remedies," *International Journal of Biomedical Science,* 9(2) (2013) 61–7.
36. B. Rush, *An Account of the Bilious Remitting Fever as it Appeared in Philadelphia, Medical Inquiries and Observations* (Philadelphia: Prichard and Hall, 1789), 104–107.
37. J. G. Rigau-Pérez, "The early use of break-bone fever (Quebranta huesos, 1771) and dengue (1801)," *The American Journal of Tropical Medicine and Hygiene* 59(2) (1998) 272–274.
38. D. J. Gubler, "Dengue and dengue hemorrhagic fever," *Clinical Microbiology Reviews* 11(3) (1998) 480–496.
39. P. K. Lumsden, "William Hepburn Russell Lumsden," *British Medical Journal* 324(7352) (2002) 1527.
40. W. Lumsden, "An epidemic of virus disease in Southern Province, Tanganyika territory, in 1952–1953 II. General description and epidemiology," *Transactions of the Royal Society of Tropical Medicine and Hygiene* 49(1) (1955) 33–57.
41. M. C. Robinson, "An epidemic of virus disease in Southern Province, Tanganyika territory, in 1952–1953," *Transactions of the Royal Society of Tropical Medicine and Hygiene* 49(1) (1955) 28–32.
42. F. Ludwig, *Church and State in Tanzania: Aspects of a Changing Relationship, 1961–1994*, (Leiden: Brill, 1999).
43. Ibid.
44. G. Kuno, "A re-examination of the history of etiologic confusion between dengue and chikungunya," *PLoS Neglected Tropical Diseases* 9(11) (2015).
45. M. S. Kinch, *A Prescription For Change: The Looming Crisis in Drug Discovery* (Chapel Hill, NC: University of North Carolina Press, 2016).

46. K. F. Smith, M. Goldberg, S. Rosenthal, L. Carlson, J. Chen, C. Chen, S. Ramachandran, "Global rise in human infectious disease outbreaks," *Journal of The Royal Society Interface* 11(101) (2014).
47. A. Mack, E. R. Choffnes, M. A. Hamburg, D. A. Relman, *Microbial Evolution and Co-Adaptation: A Tribute to the Life and Scientific Legacies of Joshua Lederberg: Workshop Summary*, (Washington D.C.: National Academies Press, 2009).
48. M. E. Woolhouse, R. Howey, E. Gaunt, L. Reilly, M. Chase-Topping, N. Savill, "Temporal trends in the discovery of human viruses," *Proceedings of the Royal Society of London,* 275(1647) (2008) 2111–2115.
49. T. Daniel, "Leon Charles Albert Calmette and BCG vaccine," *The International Journal of Tuberculosis and Lung Disease* 9(9) (2005) 944–945.
50. A. Calmette, C. Guérin, A. Boquet, L. Nègre, "La vaccination préventive contre la tuberculose par le BCG," *American Journal of Public Health,* 18(8) (1928), 1075.
51. B. Lange, "Die Calmettesche Schutzimpfung und die Säuglingserkrankungen in Lübeck," *Deutsche Medizinische Wochenschrift* 56(22) (1930) 927–929.
52. M. Kwa, C. S. Plottel, M. J. Blaser, S. Adams, "The Intestinal Microbiome and Estrogen Receptor-Positive Female Breast Cancer," *Journal of the National Cancer Institute* 108(8) (2016).
53. M. Blaser, *Missing Microbes: How the Overuse of Antibiotics Is Fueling Our Modern Plagues*, (New York: Henry Holt and Co., 2014).
54. Ibid.
55. J. O'Neill, *Tackling Drug-Resistant Infections Globally: Final Report and Recommendations* (London: Review on Antimicrobial Resistance, 2016).
56. Anonymous, "HHS Secretary Tom Price Says States Should Decide on Vaccines." *The Daily Beast*, March 7, 2017. Web. February 17, 2018.
57. M. S. Kinch, J. Merkel, S. Umlauf, "Trends in pharmaceutical targeting of clinical indications: 1930–2013," *Drug Discovery Today* 19(11) (2014) 1682–5.
58. R. Horesh, " Technical aspects of trade negotiations." *New Zealand Branch, Australian Agricultural Economics Society Conference*, Lincoln College, Canterbury, New Zealand, July 1988.
59. A. Travis, "Will social impact bonds solve society's most intractable problems?" *The Guardian (London)*, October 6, 2010. Web. February 17, 2018.
60. Anonymous, Private backers fund scheme to cut prisoner reoffending, *BBC News*, September 10, 2010. Web. February 17, 2018.
61. K. Alibeck, S. Handelman, *Biohazard: The Chilling True Story of the Largest Covert Biological Weapons Program in the World*, (New York: Dell Publishing, 1999).
62. T. P. Monath, J. R. Caldwell, W. Mundt, J. Fusco, C. S. Johnson, M. Buller, J. Liu, B. Gardner, G. Downing, P. S. Blum, "ACAM2000 clonal Vero cell culture vaccinia virus (New York City Board of Health strain)–a second-generation smallpox vaccine for biological defense," *International Journal of Infectious Diseases* 8 (2004) 31–44.

Acknowledgments

There are many people, who deserve my deep gratitude. First and foremost is my family, who endured endless nights and missed family time, so I could selfishly complete this book. My wife, Dr. Kelly Carles-Kinch, deserves much credit for her understanding given the countless meals, movies and conversations, when I was distracted in thinking about infected cows, the sources of pus or obscure historical figures. Likewise, my daughter Sarah and son Grant provided the inspiration for this book as both (mostly) listened politely while being subjected to constant waves of medical and historical trivia. I would also like to thank my extended work family, including my boss and professional inspiration, Dr. Holden Thorp, who encouraged and helped expand my horizons and for supporting our work at the Center for Research Innovation in Biotechnology at Washington University in St Louis. Our team, past and present, at Washington University includes fantastic investigators, including Dr. Rebekah Griesenauer, Constantino Schillebeeckx, David Maness, Meredith

Herd, Ryan Moore, and Thomas Krenning. Indeed, this book arose from a project conducted at the Center, where our team was the first to catalog all innovative vaccines ever developed.

This work was partially inspired by multiple conversations with the late Bill Rosen while writing *A Prescription for Change.* In wonderful discussions that often lasted for hours at a time, Bill encouraged me to expand beyond my immediate academic focus and expand into the personalities behind the research. Bill's outstanding books, including *Justinian's Flea* and *Miracle Cure*, set a bar that this work could never hope to match. Another Bill, this one with the last name of Bryson, also set the standard for how to convey technical information to a general audience. In my humble opinion, the most brilliantly-enjoyable book of the past century is Bill Bryson's *A Short History of Nearly Everything*, a work matched only by *Cosmos* by the late Carl Sagan. I would also like to thank my agent, Don Fehr at Trident Media Group and importantly, Jessica Case at Pegasus. This is my first trade book and Jessica has been exceedingly patient in introducing a not-terribly bright academic to the real world of publishing.

Finally, I would like to thank another inspiration, Brian Deer of *The Sunday Times* of London. In a few, short exchanges, Brian conveyed his story of how a self-admitted vaccine skeptic deployed his deep investigational skills to reveal the truth behind the anti-vaccinator movement, not just once (for the DPT scandals), but again years later with the emergence of the MMR charlatans, led by Andrew Wakefield, who preyed upon the fears of well-meaning parents. Brian's investigative reporting has already saved thousands of lives and despite this service, he has had to endure the vocal and often violent disdain of a misguided minority of ardent anti-vaccinators. In a troubling time, where expertise is fundamentally doubted and fundamental facts are disputed by a fringe minority, we increasingly owe our safety and liberty to a small number of heroic investigative reporters such as Brian.

Index